CROP BREEDING

FOUNDATIONS FOR MODERN CROP SCIENCE SERIES

Crops and Man 1975
J. R. Harlan
Introduction to Crop Protection 1979
W. B. Ennis, Jr.
Crop Quality, Storage, and Utilization 1980
C. S. Hoveland
Crop Breeding 1983
D. R. Wood
Physiological Bases for Crop Growth and Development 1983
M. B. Tesar
Ecological Bases for Crop Growth and Development
In preparation
Propagation of Crops In preparation

Cover: Buffalograss, a dioecious grass of the high plains of the USA. The staminate flowers are on elongated stalks of the male plants. The pistillate flowers are within the broad leaf sheaths of the female plants. Improved cultivars have been developed for pasture and range use as well as for lawns.

CROP BREEDING

D. R. Wood
Editor

K. M. Rawal and M. N. Wood
Associate Editors

American Society of Agronomy
Crop Science Society of America
Madison, Wisconsin
1983

Domenic Fuccillo, *Managing Editor*
Ann M. Diliberti, *Assistant Editor*
Buffalograss print for cover designed by Dorothy Udall,
Maridadi West, Fort Collins, Colo.
Cover graphics prepared by Patricia Jeffson

Library of Congress Cataloging in Publication Data

Crop breeding.
Includes bibliographies and index.
1. Plant-breeding. 2. Food crops—Breeding.
I. Wood, D. R. (Donald R.) II. Rawal, K. M. (Kanti M.)
III. Wood, M. N. (Marcile N.)
SB123.C798 1983 631.5′3 83-11733
ISBN 0-89118-036-2

The American Society of Agronomy, Inc. and
the Crop Science Society of America, Inc.
677 S. Segoe Road, Madison, Wisconsin, USA 53711

Printed in the United States of America

CONTENTS

PART II. REMODELING THE DNA

PART III. THE BREEDING PROCESS

FOREWORD

The accelerated pace of research, augmented by sophisticated instrumentation and techniques, and new opinions, imparts to crop science a rapidly changing character as new discoveries replace and/or add to former concepts. New findings force us to reevaluate and often reconstruct the foundations of modern crop science.

The Teaching Improvement Committee of the Crop Science Society of America identified the urgent need for developing contemporary reading materials aimed at upper level undergraduate college students. A current presentation of the dynamic state of modern crop science is a formidable challenge worthy of the best talents of eminent research and teaching personnel in the field. This task necessitates assembling the most capable representatives of the various disciplines within crop science and bringing them together in teams of writers to prepare a series of publications based on contemporary research. The Crop Science Society of America and the American Society of Agronomy have undertaken this large assignment by selecting more than 100 specialists who will contribute to making the Foundations of Modern Crop Science books a reality.

The authors and editors of this series believe that the new approach taken in organizing subject matter and relating it to current discoveries and new principles will stimulate the interest of students. A single book cannot fulfill the different and changing requirements that must be met in various programs and curricula within our junior and senior colleges. Conversely, the needs of the students and the prerogatives of teachers can be satisfied by well-written, well-illustrated, and relatively inexpensive books planned to encompass those areas that are vital and central to understanding the content, state, and direction of modern crop science. The Foundations for Modern Crop Science books represent the translation of this central theme into volumes that form an integrated series but can be used alone or in any combination desired in support of specific courses.

The most important thing about any book is its authorship. Each book and/or chapter in this series on Foundations for Modern Crop Science is written by a recognized specialist in the discipline. The Crop Science Society of America and the American Society of Agronomy join the Foundations for Modern Crop Science Book Writing Project Committee in extending special acknowledgment and gratitude to the many writers of these books. The series is a tribute to the devotion of many important contributors who, recognizing the need, approach this major project with enthusiasm.

A. W. Burger, chairperson
D. R. Buxton
C. O. Qualset
A. A. Hanson
L. H. Smith

PREFACE

From the test tube to the harvest, the science of crop breeding is alive and well. People have always expected crop breeding to enhance agriculture by providing improved crops for all kinds of situations. Recent breakthroughs in recombinant DNA research together with more conventional breeding techniques promise continued progress in the future.

As the science of crop breeding has grown, practitioners have developed a comprehensive literature; their breakthroughs have resulted in major contributions to society; and new disciplines have emerged.

The theoretical foundations of the science are unique and challenging. The landmark discoveries in genetics and evolution by Mendel and Darwin, partly motivated by their interest in the improvement of domestic plants, are still the foundation of crop breeding. As sophisticated molecular genetic techniques, such as gene splicing, come closer to application in plant breeding the synergistic relationship between crop breeding and genetics becomes more important.

Statistical techniques have become increasingly important as tools for plant breeders. Since the early work of "Student," Fisher, Snedecor, and many others in evaluating new strains of crops, statistics has developed as a respected science important to many disciplines beyond agriculture.

Crop breeding must be described in broader terms than applied genetics or directed evolution. It certainly has human and social relevance for our world's survival and well-being. Early plant selections were made to meet basic human needs for food, fiber, and shelter. New uses will continue to be discovered as crop breeders focus on the uses and productivity of our plant resources. Changes in agriculture, crop production, and in standards of food quality will require crop breeders to continue to respond to the needs of agriculture and society.

The centerpiece of crop breeding, and the focus of this book, is the cultivar. The word, a contraction of cultivated variety, has been adopted by the Crop Science Society of America following the *International Code of Nomenclature of Cultivated Plants* as the appropriate word for scientific use. In papers intended for the lay public or non-scientific community, the term variety is the most desirable synonym. (For a discussion of this point, see Weiss, M. G. 1972. Cultivar vs. variety. Crop Sci. 12:551.)

This book was designed to complement textbooks used for crop breeding courses, primarily those for junior and senior undergraduates in agriculture. Outstanding crop breeders were asked to write about specific areas related to their work, including examples and anecdotal material. As you share in the experiences of these breeders you will gain new insights about this important science and we trust you will be rewarded with exciting glimpses of new frontiers in crop breeding.

You will find some of the more technical words defined in a glossary of terms along with a list of scientific names in the back of the book. A mastery of this language will put you on speaking terms with others in our profession.

Donald R. Wood
Marcile N. Wood
— Fort Collins, Colo.

Kanti M. Rawal
— San Leandro, Calif.

CONTRIBUTORS

Burton, G. W., Research Geneticist, USDA, ARS, Coastal Plain Station, Tifton, GA 31793

Busey, Philip, Assistant Professor, Agricultural Research Center, University of Florida, Ft. Lauderdale, FL 33314

Frey, K. J., C. F. Curtiss Distinguished Professor in Agriculture, Department of Agronomy, Iowa State University, Ames, IA 50011

Gengenbach, B. G., Professor, Department of Agronomy and Plant Genetics, University of Minnesota, St. Paul, MN 55108

Harpstead, D. D., Chairman, Department of Crop and Soil Sciences, Michigan State University, East Lansing, MI 48824

Hooker, A. L., Bioscience Director, Dekalb-Pfizer Genetics Inc., St. Louis, MO 63141

Jain, S. K., Professor, Department of Agronomy and Range Science, University of California-Davis, Davis, CA 95616

Jensen, N. F., Emeritus Professor, Department of Plant Breeding and Biometry, Cornell University, Ithaca, NY 14853

Ladd, S. L., Professor, Department of Agronomy, Colorado State University, Ft. Collins, CO 80523

Morris, Rosalind, Professor, Department of Agronomy, University of Nebraska, Lincoln, NE 68583

Paule, M. R., Professor of Biochemistry, Colorado State University, Ft. Collins, CO 80523

Poehlman, J. M., Emeritus Professor, Department of Agronomy, University of Missouri, Columbia, MO 65211

Quick, J. S., Professor, Department of Agronomy, Colorado State University, Ft. Collins, CO 80523

Rasmusson, D. C., Professor, Department of Agronomy, University of Minnesota, St. Paul, MN 55108

Rawal, K. M., Plant Breeder, Del Monte Corporation, San Leandro, CA 94577

Sigurbjörnsson, B., Director, Agricultural Research Institute, Reykjavik, Iceland

Wood, M. N., Vocational Education Consultant, Ft. Collins, CO 80526

PART I.

Design and Preparation

Photo furnished by Philip Busey, University of Florida, Fort Lauderdale.

Chapter 1

Crop Breeding in a Hungry World

J. M. POEHLMAN

University of Missouri
Columbia, Missouri

J. S. QUICK

Colorado State University
Fort Collins, Colorado

It is a disturbing fact that more than a billion of the world's people are hungry or malnourished. Equally disturbing is the prospect that the world's food production, scarcely adequate to feed 4 billion people in 1975, will need to be doubled to feed 8 billion or more people in the year 2015 (Borlaug, 1981) if the present increase in world population continues unabated. Most of this population growth is taking place in the less developed or poor countries. Those countries have neither the resources nor the technology to increase their food production at a pace corresponding to their growth in population. Short-term deficiencies can be made up through importation, assuming that adequate foreign exchange and surplus production in other countries exist. Several reasons make importation a non-solution to food deficit problems among hungry nations. It is in this context that production of food to alleviate world hunger is paramount among the challenging problems confronting the world today (Brown, 1963; Revelle, 1974; Abelson, 1975; Sneep and Hendrikson, 1979, Presidential Commission on World Hunger, 1980).

Hunger is not new. The danger of population overtaking food supplies was predicted by Malthus nearly 200 years ago. The dire predictions of Malthus were forestalled, at least temporarily, by extensive cultivation of new lands and by the development of a modern agricultural science that enables food crops to be produced with far higher yields than Malthus could have ever anticipated. Even so, for the poor and the impoverished, hunger has not been eliminated.

Several critical questions regarding food production are pertinent for tomorrow's agronomists. Will total food production continue to increase at a sufficiently rapid rate to feed the growing world population? Is the scientific knowledge and technology available, or can it be generated, to keep food production in the less developed countries apace with their population growth? Can new technology to increase food production in the hungry nations be implemented fast enough? The objective of this chapter and this book is to describe the key role and techniques of crop breeding in the development of this technology.

Increased food production involves the interrelationships of diverse cultural and economic factors. The factors differ with specific geographic areas and political systems. Certainly, no single technological advance in agriculture can assure adequate food production and its distribution to all parts of the globe, to a geographic or cultural area of the world, or even to a specific country. Yet, among the inputs that will be needed to provide an increasing food supply, none is more critical than the genetic improvement of food plants. Simply stated, this technology involves the breeding of improved crop varieties or cultivars.

BREEDING TO INCREASE CROP PRODUCTIVITY

Increasing the global food supply requires greater crop productivity. The four requisites for greater productivity are: (1) an improved farming system, (2) instruction of farmers, (3) supply of inputs, and (4) availability of markets (Wortman and Cummings, 1978). The central requirement for an improved farming system is improved cultivars, those that will respond to improved water management and optimum utilization of fertilizer and pest control practices. The application of the appropriate technology on a global basis is very complex. An improved cultivar must improve the stability of production, meet local quality standards, fit the local cropping system, and be acceptable for local consumption.

The work to improve cultivars should be considered in a different perspective than techniques for irrigation, fertilization, or pest control. The latter techniques modify the environment, and optimize the microclimate for plant growth. By contrast, a plant with improved genetic potential will thrive and produce a superior yield, or a product of superior quality, in the particular environment to which the plant is adapted. Whereas irrigation, fertilizer, and pesticides must be applied anew with the production of each crop, the benefits from genetic improvements are realized in each crop cycle as long as genetic purity can be maintained.

Genetic improvement of the crop plant alone, as with any other production input, will not solve the world's food problems. For crop productivity to be increased, the planting of high-yielding cultivars must be combined with improved practices of irrigation, fertilization, and pest control. This statement may be illustrated by comparing yields of new high-yielding semidwarf wheats with yields of native wheats grown with different levels of nitrogen fertilizer (Fig. 1.1). Comparisons between improved and native cultivars of rice and maize show similar results (Russel et al., 1970; Athwal, 1971).

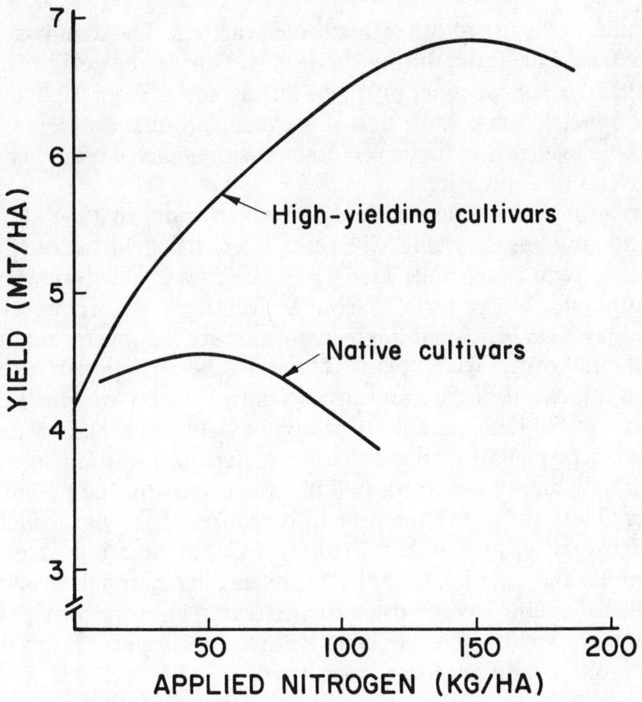

Fig. 1.1. Comparative yield response of native and improved high-yielding cultivars of wheat to applications of nitrogen fertilizer. (Redrawn from Russel, Henshaw, Schauble, and Diamond, 1970).

Why do the yields of new semidwarf wheats exceed those of the native wheats following heavy applications of nitrogen fertilizers? Two reasons may be cited:

1. With heavy fertilization, vegetative growth increases. Because the native wheats are tall with weak culms, they do not support the increased growth, so they fall over or lodge more than the semidwarf wheats. As a result, the number of grains is reduced or they are poorly developed and the lodged plants create a microclimate favorable to the development of certain diseases.

2. The high-yielding semidwarf wheats are more efficient physiologically in their use of available nitrogen and produce more grain than the native unimproved wheats. The native wheats evolved over many years growing in soil of low fertility, so even with more fertilizer, they still have fewer grains per plant and per unit area of land, even when little or no lodging occurs (Ephrat, 1974).

Maximum crop yield will be obtained only if the improved crop cultivar receives and responds to the optimum combination of water, fertilizer, and cultural practices. While a favorable combination of all inputs is important to achieve high crop productivity, it has been demonstrated in India, and elsewhere, that the improved crop cultivar can be the catalyst which stimulates the use of other favorable practices. The semidwarf wheats were introduced into India during the 1960's. During the period from 1965 through 1972, wheat production in India increased from 12.3 million to 26.4 million metric tons. This, and subsequent production increases, were made possible by growing the new cultivars with an accompanying package of improved cultural practices.

Observations of the wheat fields of North India in 1964 showed tall wheat plants and sparse yields. Six years later, the fields were filled with high-yielding, semidwarf cultivars. The change was revolutionary, more than evolutionary, so the term "Green Revolution" was coined (Borlaug, 1972; Randhawa, 1974). What happened was more than just a change in the genetic potential of the wheat plant. There was better use of fertilizer, irrigation, and improved disease and insect control. But it was the new semidwarf wheats, with their potential for higher yield, that made it profitable for the Indian farmer to increase the use of fertilizer and nitrogen. When lack of soil moisture or soil fertility limits plant growth, the potential yield response bred into the crop cannot be fully realized. Likewise, beneficial effects from irrigation and added fertilizer will not be attained unless the wheat plant has the genetic capability to convert the materials available into larger yields of grain. Recent dramatic increases in energy prices and the need to improve yields with reduced resources must be reflected in future breeding objectives and selection procedures.

NATURE OF CROP BREEDING

Improvement of crop plants through breeding procedures is dependent upon the presence of genetic differences among plants within the species. The genetic variations resulted from evolutionary changes that have taken place since plant life first appeared on the earth. Significantly for agriculture, a few plant species have evolved that store substantial amounts of carbohydrates, proteins, and fats in a form that can be used for food. Cultivated food crops now supply around 90% of the world's caloric intake. The plant species and crop groups most important for supplying food are:

cereal grains—wheat, rice, maize, barley, sorghum, millet;
root crops—potato, sweet potato, yam, cassava;
oil seed crops—soybean, cottonseed, oil palm, peanut, sunflower, rape;
protein-rich pulse crops—bean, pea, chickpea, cowpea, lentil, mungbean;
sugar-storing plants—sugarcane, sugarbeet;
fruit crops—banana, plantain, citrus, apple, berries;
vegetable crops—tomato, cabbage, onion, carrot, cucumber.

More than 50% of the calories that people obtain are directly from cereal grains (Fig. 1.2); in some of the developing countries this may be as high as 70%. On the average in the developing countries, rice is more than twice as important as wheat; millets and sorghum are twice as important as barley;

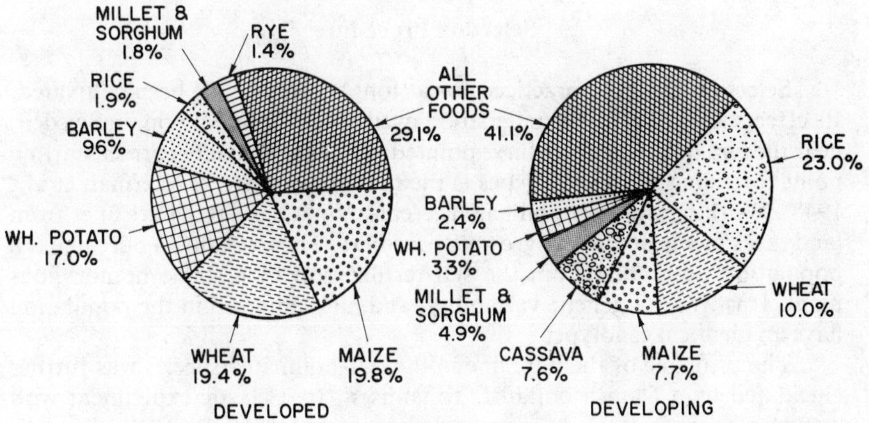

Fig. 1.2. Food crops supplying major portions of the human diet in developing and developed countries (Food and Agriculture Organization Production Yearbook, 1979).

and cassava is more than twice as important as the white potato (Food and Agricultural Organization Production Yearbook, 1979). Several relatively minor crops in the developed nations are major food sources in developing countries, e.g. the non-cereal starchy food plants—white and sweet potato, cassava, other roots and tubers, and banana/plantain provide major basic sustenance (primarily energy) for nearly 25% of the world's people. Additionally, forages, cereals, and food processing byproducts are fed to livestock that are consumed as food.

The method of crop breeding is determined largely by the reproductive system of the crop plant, whether self-pollinated or cross-pollinated (Poehlman, 1979). Self-pollinators are inbreeders and inbreeding leads to homozygosity and genetic stability. Essentially, self-pollinators are maintained as populations of homozygous plants and the progenies are like the parents, or very nearly so. By contrast, cross-pollinated crops are outbreeders; they foster heterozygosity and heterogeneity. Every plant may be inherently different. Progeny plants differ from the parent plants. The genetic limitations of a cultivar maintained by cross-pollination are less rigid and less specific than a cultivar maintained by self-pollination. Within the limits of its genetic variability, a cultivar maintained for several generations by cross-pollination will be changed far more rapidly by environmental stresses than a cultivar maintained by self-pollination.

Many important crop species reproduce vegetatively by asexual propagation. Tuber crops, sugarcane, and some tree fruits are propagated asexually by cuttings. Apomixis, the production of seed without the union of male and female gametes, and nucellar embryony, when seeds result from the nucellus rather than the zygote, are two additional types of asexual reproduction important in some food crops. In asexual reproduction, unless mutations occur the genotypes remain constant. Any desirable genetic combination can be fixed immediately and propagated indefinitely.

Selection Procedure

Selection has been practiced for as long as crops have been cultivated. Its effects were not clearly understood until N. Hjalmar Nilsson, in Sweden, near the end of the 19th century, pointed out that "the only correct starting point for fixing discovered types is the individual plant" (Åkerman et al., 1948). Working with self-pollinated cereals, Nilsson isolated pure lines from land races of wheat and oat grown by farmers. Theoretically, a pure line is a population descended from the self-fertilization of a single homozygous plant. It is without genetic variability, and all plants within the population have an identical genotype.

The principle of the pure line in the self-pollinated species was further elucidated by a Danish botanist, Johannsen. In a classic experiment with common beans in 1903, Johannsen demonstrated that once a pure line was isolated, further selection within the pure line did not result in additional genetic change. As pure lines were isolated from the genetically mixed landraces and their economically useful characteristics were identified, hybridization among the lines was used to obtain new combinations of the desirable characters. Hybridization followed by selection subsequently became the predominant method for development of new cultivars of the self-pollinated crops. These methods have been used to breed thousands of improved cultivars of many self-pollinating species, including the high-yielding cultivars of wheat and rice that have made substantial contributions to food production in many of the food-deficit countries of the world.

The principles utilized in the breeding of cross-pollinated crops were succinctly expressed by G. H. Shull in 1908. Describing his experiments in inbreeding and outbreeding maize, Shull concluded that the individuals (plants) in an ordinary field of maize are very complex hybrids, that deterioration takes place as a result of self-fertilization, and that there is a gradual reduction of the strain to homozygous conditions. He believed that the object of the maize breeder should not be to find the best pure line but to find and maintain the best hybrid combination (Shull, 1908). Dr. Shull's experiments led to the present methods for breeding hybrid maize. The procedure of breeding hybrids has since been extended to many other crops. The principle of utilizing the most productive hybrid combinations is also employed in producing modern synthetic cultivars (Poehlman, 1979), a method extensively used in the breeding of forage crops.

Cultivar Development

The economic end product of genetic improvement is improved cultivars. A cultivar is a population of cultivated plants, related by descent, that have genetic traits in common. Identity and reproducibility are inherent features of cultivars. A particular cultivar may be identified from all other plants of the same species either by morphological features or by performance. Morphological features commonly used for identification vary with the species, but may include height, seed color, flower color, size and nature of specific floral structures, and many others. These characteristics should be genetically stable within the cultivar so they will reappear when the cultivar is grown again. They may or may not affect performance as measured by the yield and quality of the product harvested. Semidwarfness was an identifying characteristic of the new wheat cultivars used in the "Green Revolution" and contributed to higher yields. By contrast, flower color could be useful in identification of a crop cultivar, yet may have no measurable effect on the yield or quality of the harvest.

Performance differences among cultivars are usually more difficult to identify than morphological features because they are more likely to be affected by the environment and are more complex in inheritance. Overall, yield potential is the most important characteristic by which cultivar performance is measured. Yield potential is evaluated by growing the cultivar in the field, and measuring its production in comparison with yields of standard cultivars under similar environmental conditions. Performance is affected by physiological factors such as response to fertilizers, as well as resistance to environmental stress (heat, cold, drought), disease-inciting pathogens, insects, and lodging. The inherent characteristics influencing performance of the improved cultivar need to be genetically stable and reproducible, so that the superior performance will be repeated when the cultivar is grown again.

Special Procedures

Some special tools available to the plant breeder are techniques to produce mutations (changes in the hereditary units) and to alter chromosome numbers. Mutation breeding is based on the knowledge, first reported by L. J. Stadler in 1928, that the natural mutation rates in plants may be increased by exposing seeds to ionizing radiations (Stadler, 1928). More recently, potent chemical mutagens have been identified and used to produce viable mutations. The use of mutation breeding has extended the range of genetic variability available to the breeder. Induced mutations are discussed in detail in Chapter 8.

Polyploidy (duplication of entire sets or complements of chromosomes) became a breeding tool following the development of a procedure for doubling the chromosome number in plant cells (Blakeslee and Avery, 1937). The plant with an increased chromosome number will have characteristics which differ from a typical plant with the normal chromosome number. This procedure also, provides a means to increase the range of genetic variability available to the crop breeder principally by increasing the ease of hybridizing crop plants to their closely related cultivated and wild species. Chromosome sets from two or more species can be combined as, for example, in triticale, the wheat × rye hybrid. The ease of doubling the chromosome complement with colchicine has also led to the utilization of haploid plants in breeding programs.

Before leaving the subject of breeding methods, we should examine the consequences of specific procedures. In general, inbreeding and selection greatly reduce the range of genetic variability within a population. Nilsson practiced selection within native landraces that had a broad spectrum of genetic characteristics adapted to local areas. Landraces have long since been replaced by pure lines in the developed countries. Those landraces that remain will be found in remote areas in the less developed regions of the world. It is inevitable that they too will be replaced, as new cultivars outyield the traditional, and improved cultural practices reduce the wild weedy relatives. The net result is a reduction in genetic variability for the crop.

Hybridization is less restrictive. Using existing gene pools to obtain genetic recombination, crop cultivars with greater yield potential or wider adaptation to climatic differences, may be constructed by combining genes for diverse forms of characteristics from two or more parent cultivars. Mutation breeding and polyploidy are more creative in their approach but difficult to direct toward specific goals. Thus far, mutation breeding and polyploidy have not been as widely used by breeders as the conventional procedures of selection and hybridization.

Recurrent selection breeding methods are applied to a wide range of plant species, including self-pollinated crops. These population improvement methods can be more practically used in developing than in developed countries since the cost of making large numbers of hand crosses is less. Other techniques to implement high rates of crossing in self-pollinated crops involve the introduction of genetic male-sterility to eliminate the time-consuming emasculation procedure.

Finally, conscious effort toward greater genetic diversification within the cultivar has been made. Examples include blends, composites, and multilines in self-pollinated crops, and synthetics in cross-pollinated crops. In these, the breeder strives to maintain a broad genetic base that will provide adaptation over a wide range of environmental conditions.

THE STRATEGY OF CROP BREEDING

The strategy of the crop breeder is to build into the cultivar a superior genetic potential for yield, protection against production hazards, and improved quality. The specific characteristics desired will, of course, vary with the particular crop species, the climate in which it will be grown, the cultural practices to be used in its production, the utility of the product, traditional preparation methods, and many other factors. In breeding the new high-yielding rice and wheat cultivars, the emphasis was on characteristics which gave wide geographic and climatic adaptation with stability in yield in contrast to the breeding of cultivars for a local area (Dalrymple, 1978; Jennings, 1974).

Breeding for high and stable yield potential is generally the major goal in any breeding project, although yield potential may, in some instances, be sacrified for improvement in quality or for characteristics which adapt the cultivar for a specific use. Breeding for high yield potential is inextricably linked with breeding for other characteristics. While semidwarfness, greater floret fecundity, and tillering capacity have contributed to higher yields of wheat and rice, so have early maturity, reduced sensitivity to photoperiod, and resistance to environmental stresses, diseases, and insects. For instance, in tropical regions, climatic conditions are often favorable for growing two or three crops per year. Early maturity shortens the duration of growth and permits cycling the wheat or rice plant into multiple-cropping systems. Reduced sensitivity to the photoperiod permits a rice cultivar to be grown in more than one season of the year and in a wider range of latitudes, thus broadening adaptation. However, photoperiod sensitivity may be desirable in complex cropping systems which fit monsoonal rainfall patterns.

Breeding for Resistance to Production Hazards

Stability of yield is augmented by breeding for resistance to (1) a wider spectrum of disease organisms and insect pests, (2) unfavorable soil problems, and (3) season-to-season fluctuation. One example is the progress made at the International Rice Research Institute (IRRI) in breeding rice cultivars for resistance to production hazards. The 'IR-8' cultivar of rice, released from IRRI in 1966, was resistant to one major insect pest and moderately resistant to one disease and two soil-deficiency conditions. Released in 1975, IR-28 is resistant to four diseases, three insect pests, and four soil-deficiencies (Fig. 1.3). The increased resistance was designed to stabilize production of IR-28 over a wider range of seasonal and climatic variations than of IR-8 (International Rice Research Institute, 1975).

BREEDING FOR QUALITY

In a crop breeding program, improvement of quality of the product is also important. The emphasis may be on physical or nutritional quality, or a combination of the two. In the different rice-improvement programs of the tropical countries, cultivars are being developed with the eating and cooking qualities desired by people of different cultures. In addition, emphasis is being placed on improvement of the nutritional qualities, such as higher protein and better nutritional balance in the amino acids. Genes to increase the content of lysine, an essential amino acid normally deficient in the cereal grains, are being incorporated into maize, sorghum, barley, and wheat. Higher lysine content would reduce the dependency on supplemental sources of protein to provide nutritional balance needed by the people who consume the cereal grain as a major portion of their diet. It is possible, also, that utilization of high lysine grains may lead to adverse nutritional effects. For example, greater dependence on cereals as the major source of protein, just because they are better balanced nutritionally, may not achieve the goal of better nutrition. If smaller quantities of protein supplements are eaten,

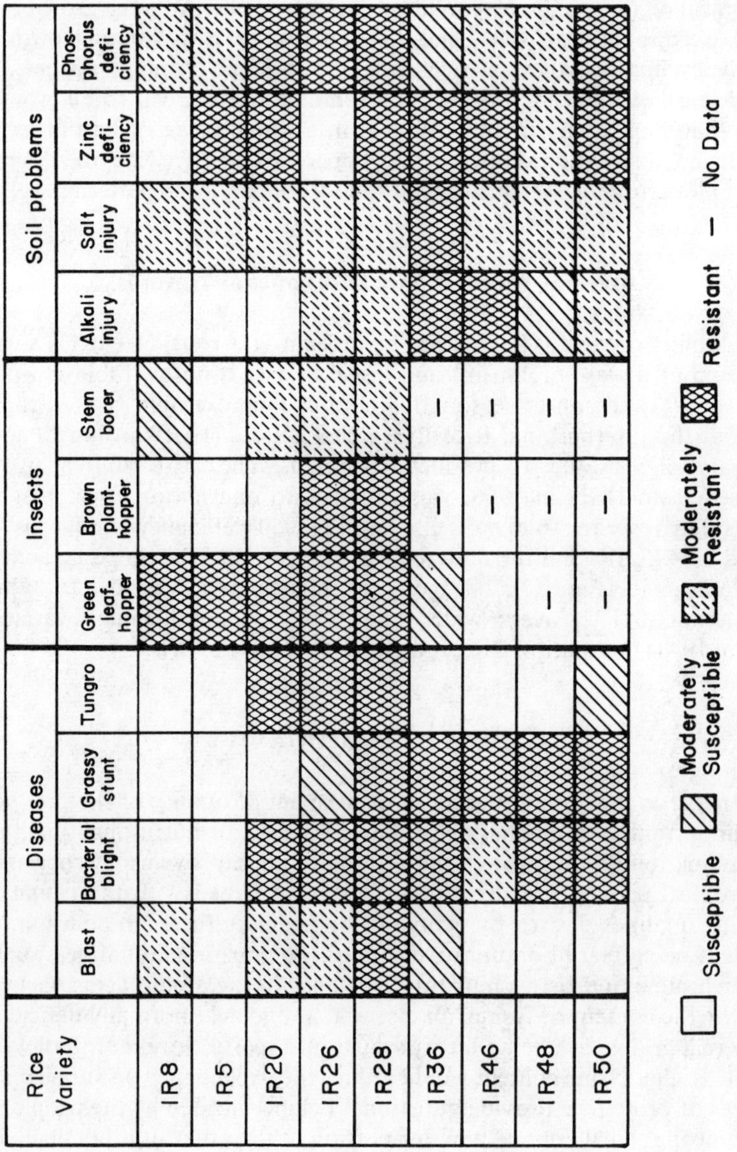

Fig. 1.3. Rice cultivars released from the International Rice Research Institute have become successively more resistant to diseases, insects, and soil problems. (From the IRRI Reporter, 1975, and IRRI Annual Report, 1979).

the total protein intake is reduced. Furthermore, reducing the diversity of foods may reduce the intake of minerals and vitamins. Breeders should determine if protein content can be increased in the starchy tuber crops such as cassava. The possibility of reducing toxic constituents also needs further investigation. The breeder needs to give careful attention to the social and economic constraints on diets in any attempt to improve the nutritional value and diversity of plant products as a means of improving the nutrition of people. For a more complete discussion of breeding for nutritional value see Chapter 12.

THE CROP BREEDING HERITAGE

Present evidence suggests that crop plants have been subjected to cultivation for perhaps more than 8,000 years (Hutchinson et al., 1976). The time period, no doubt, differs with the species and the geographic area. During most of this period, changes in the cultivated species evolved slowly. Today plant breeders are accelerating the evolutionary process to increase the usefulness of plants by exploiting genetic differences within a species. This rapid development occurred once the science of genetics was rediscovered and a scientific basis for developing crop breeding procedures was established.

The science of genetics is young. The foundation was laid by Gregor Mendel, the Austrian monk, who in 1865 reported elegant experiments with garden peas in which he demonstrated that inheritance of plant characters followed definite and specific laws. Mendel's research went unnoticed for 35 years. Upon its rediscovery in 1900, the science of genetics was born. The crop breeder largely depends upon the discipline of genetics for an understanding of ways to create new cultivars. But the disciplines of agronomy, botany, plant pathology, plant physiology, entomology, and statistics all contribute to an understanding of the plants that the crop breeder strives to improve.

Germplasm Resources

It is axiomatic that crop breeding is dependent upon genetic variability within the crop species. The breeder cannot improve the rust resistance of a cultivar of wheat, for example, unless there is an available germplasm resource of genes which, when incorporated into an adapted cultivar, will enable the wheat plant to resist, escape, or endure the invasion of the wheat-rust pathogen. Before initiating a breeding program with any crop species, the breeder, after defining the objectives of the program, needs to survey the range of genetic variability available to meet those objectives.

In a breeding program, breeders continually look for genetic variability in those traits that affect yield and quality of the species. Such traits may be morphological, such as height, straw or stalk strength, presence or absence of awns in cereals, husk covering in maize, or seed size and shape. The traits may be physiological and related to the ability of the plant to grow and flourish under adverse environmental conditions, such as heat, cold, drought, wind, and soil salinity, acidity, or aluminum toxicity. Certain characteristics may permit the plant to be productive in different climatic areas, or to extend or change the harvest season, as through reduced photoperiod sensitivity. They may enable the plant to resist disease pathogens or insect pests. They may improve the nutritional quality and increase the lysine content, as in maize and sorghum, or reduce the presence of toxic substances in the plant or seed.

An enormous reservoir of useful genes has been accumulated through normal evolutionary processes in all of the major crop species. A challenge for present-day crop scientists is to collect, identify, and preserve this vast pool of genes for use in the future (Frankel, 1977). The local cultivars or landraces that evolved over centuries of cultivation in different climatic areas provided the major source of the germplasm used earlier by crop breeders. From the landraces of wheat, Nilsson at Svalof, Sweden, selected the early improved Swedish wheat cultivars and in England, the 'Squarehead' cultivar of wheat was developed. The landraces of wheat brought to the USA by early immigrants formed the basic germplasm used in the development of the improved wheat cultivars grown today, In the USA, native strains of maize obtained from the Indians were the basis for development of the open-pollinated maize cultivars grown in the early part of this century. From this germplasm resource, inbreds were extracted and used in breeding modern maize hybrids.

Today, the local or landraces of wheat, maize, and other crops are no longer grown in the developed world (Harlan, 1972). They have been swept aside by development and utilization of new and improved cultivars. In the process, the loss of genetic variability has been great. Except for local cultivars and landraces systematically maintained by public or private agencies, much of the genetic variability available earlier no longer exists. As already pointed out, landraces can be found only in the less developed areas of the world. These are rapidly disappearing, and a serious effort needs to be made to preserve them.

Other sources of genetic variability are the wild relatives of the major crop plants. The wild relatives are generally found most abundantly in or near the geographic area where the crop originated (Harlan, 1975). Representative plants collected of the wild species may have useful genes which the crop breeder can transfer to the cultivated species by the appropriate procedures discussed in Chapter 9. Recent collections of wild species of oat in Israel have been the source of useful genes for rust resistance and higher protein in the cultivated species. But extinction of the wild relatives, too, is threatened by the invasion of improved cultivars. Many of the wild relatives grow as weeds in waste areas, and these may be eliminated as the cultivation of crops is extended or as cultural practices are improved.

The question may be asked, "What is being done to preserve the vast genetic resources which are needed by the crop breeder?" The answers, contradictory as they seem, are "Much" and "Not nearly enough". The

USDA established a Division of Plant Introduction to collect and evaluate crop germplasm as early as 1898 (Klose, 1950; Creech and Reitz, 1971; USDA, 1979). Although much effort was expanded to collect and test germplasm from other countries, early collections were not always adequately maintained, and many of the introduced strains were eventually lost. About 1946, the resources in the USA to test and maintain germplasm of economically valuable crop species were expanded, with new plant introduction stations established in different climatic areas. Germplasm collections of seeds and plant materials may now be evaluated and maintained within the climatic conditions most suited for that particular species. These facilities have been augmented with a National Seed Storage Laboratory established in 1958 at Fort Collins, Colo. for long-term storage of seed resources. Later, a system of germplasm repositories for clonally reproduced fruit and nut crops was established. These units are now beginning to function. Extensive germplasm collections are maintained by the USDA in these facilities and the number of accessions is increased yearly. Improved coordination and development activities of the USDA germplasm system was accomplished by the establishment of the National Plant Germplasm Committee in 1974 (USDA, 1979). An international overview of the worldwide state of genetic resource conservation was obtained by the establishment in 1974 of an International Board for Plant Genetic Resources (IBPGR) with headquarters in the Food and Agricultural Organization of the United Nations in Rome, Italy.

Germplasm Centers

Outside the USA, extensive germplasm collections are also assembled and maintained. These collections normally contain accessions of the local crops for use in the national breeding programs. In addition, germplasm collections of specific crops are maintained at several international research institutes: maize at the International Wheat and Maize Improvement Center in Mexico, rice at the International Rice Research Institute in the Philippines, sorghum and millet at the International Center for Research in the Semi-arid Tropics in India, potato at the International Potato Center in Peru, starchy food plants at the International Institute for Tropical Agriculture in Nigeria, and various grain legumes at the international centers in India, Colombia, Nigeria, and Taiwan.

While these collections of germplasm may seem extensive, they include but a small fraction of the variability that has evolved over centuries in nature and which is gradually being eroded away and lost or destroyed. Furthermore, the numbers of items in storage may be misleading, for in each collection there is duplication of strains, or large numbers of strains with quite similar genetic characteristics. None of the collections was assembled by systematically sampling a species and its wild relatives (Harlan, 1972). Since each item in a germplasm collection must be grown at regular intervals to renew its viability, the maintenance of viable seeds of a large collection is indeed a huge task. It is especially difficult and expensive with species that propagate vegetatively and must be maintained as living plants.

Human Resources

Crop breeding began with the cultivation of wild plants. Over long periods of time there was a transition from the collection of wild plants for food to the selection of those to be cultivated. Through this selection process, people, consciously or unconsciously, began to guide the evolutionary processes. Now plant breeders accelerate the evolution of the major crop species through skillful manipulation of breeding procedures. The history of plant breeding is filled with spectacular accomplishments. Here, only a few will be cited.

In 1801, Franz Carl Achard, a French scientist, perfected a procedure for extracting sugar, a prized and expensive product, from the wild fodder beet. Achard started selecting fodder beets with uniform root conformation and high sucrose content. He and his successors, over a period of 175 years, increased the sugar content from about 7% in the wild fodder beet to around 12 to 18% in modern cultivars of sugarbeets.

In 1890, N. Hjalmar Nilsson became the cereal breeder and director of the Swedish Seed Association (Åkerman et al., 1948). Nilsson soon discovered the principle of individual plant selection in pure line selection experiments. Nilsson's son, H. Nilsson-Ehle, developed the concept of combining characteristics of diverse parents through hybridization. From their research and the research of many others, the procedures for present-day breeding of self-pollinated crops evolved.

In 1904, George H. Shull began selfing and crossing maize at the Station for Experiment Evolution, Cold Spring Harbor, N.Y. He had neither a preconception of the results, nor any thought of the economic benefits that might occur. Yet, 5 years later, as a result of his research, he proposed a hybrid breeding procedure that revolutionized the breeding of maize. The procedure has since been extended to the breeding of hybrids in sorghum, wheat, barley, pearl millet, sunflower, sugarbeet, onion, tomato, and many other crops.

In 1956, E. R. Sears, working at the University of Missouri, described an intricate procedure by which he had transferred a single gene for resistance to the wheat leaf rust pathogen from a wild relative of the common bread wheat species (Sears, 1956). This procedure involved (1) crossing a tetraploid wheat species ($2n = 4x = 28$ chromosomes) to a wild diploid species ($2n = 2x = 14$ chromosomes), (2) doubling the chromosome number of the triploid hybrid ($3x = 21$ chromosomes), (3) backcrossing to substitute wheat chromosomes for those of the wild wheat relative except for the one chromosome which carried a gene for rust resistance, then (4) irradiating pollen to induce a substitution of a portion of the alien chromosome carrying the resistance gene for a portion of a wheat chromosome. By this sophisticated technique in chromosome engineering, the rust resistance gene from the wild species was transferred to cultivated wheat ready for use in the farmer's fields and in breeding programs.

Most successful research does not result from the work of a single individual, but is the culmination of a sequence of events involving the accumulated research of many individuals. The development of the high-yielding semidwarf wheats is an excellent example (Reitz and Salmon, 1968). In 1946, S. C. Salmon distributed strains of a semidwarf wheat bred in Japan to USA wheat research workers. Orville A. Vogel, a USDA scientist working at Washington State University, Pullman, crossed one of these semidwarf wheats, 'Norin 10', with Washington wheat cultivars. Strains derived from these crosses had short straw and increased number of seeds per plant. The semidwarf winter wheat cultivars, 'Gaines' and 'Nugaines', developed by Vogel from the Norin 10 crosses, set world yield records in the wheat-growing areas of Washington and adjacent states. Seeds from the Norin 10 crosses were sent to Mexico where Norman E. Borlaug utilized them in an advanced spring wheat breeding program (Borlaug, 1968). Semidwarfed spring wheat cultivars from the Mexican program have since set new yield records in many regions of the world.

These are only a few of the significant crop breeding achievements made possible by people who were intelligent, innovative, knowledgeable, and dedicated to the improvement of agriculture. They are examples of the human resources that have contributed to the crop breeding heritage. Who will take their place in the future? The challenge is great for those who choose crop breeding as a way to contribute to the feeding of a hungry world.

Research workers qualified to staff the crop breeding programs of local, regional, and international crop breeding programs are being sought. They must have a (1) comprehension of the technical aspects of crop breeding, (2) an interest in the practical applications of new breeding discoveries (3) a desire to help reduce the food deficit, and (4) an understanding of cultural differences among people. A difficult task, but one that is rewarding.

Institutional Resources

In the beginning, crop breeding developed largely through individual efforts, but as breeding procedures evolved, the need for interdisciplinary approaches increased and crop breeding research became more institutionalized. Crop breeding began very early in Europe, often under the auspices of commercial seed production enterprises. The Swedish Seed Association at Svalof, Sweden, was mentioned earlier. The seed firms of de Vilmorin in France, Weibullsholm in Sweden, Rimpau in Germany, and others initiated crop breeding programs in the 19th century. In Great Britain, research at the Plant Breeding Institute at Cambridge and the Plant Breeding Station at Aberystwyth contributed to early knowledge in the breeding of cereals and forages. An extensive collection of wheat and other crops made by Vavilov and maintained in Russia brought worldwide acclaim and provided an institutional basis for collection, conservation, and utilization of germplasm. Many other examples could be cited.

In the USA, crop breeding, except for cotton, began largely as a tax-supported endeavor with breeding programs in most State Agricultural Experiment Stations and in the USDA. This pattern changed with the advent of hybrid maize where inbred lines were initially developed by public institutions and utilized to produce hybrids by private companies. Competition to produce superior hybrids led to extensive maize-breeding research by private seed companies. Increases in the breeding of sorghum and wheat by private companies occurred as methods for breeding hybrids of those crops were developed. With the implementation of a Plant Variety Protection Act in the USA in 1974, private breeding was expanded in forages, cereals, soybean, and other crops. The activities of the private companies strengthen the total crop breeding effort and offer a large number of opportunities for prospective crop breeders.

International Research Centers

In recent years, a group of international research centers have developed (Wade, 1975). By providing improved crop cultivars or germplasm materials to developing countries and by assisting those countries in strengthening their research programs, the international centers have become effective institutional crop breeding resources. Two of the best known, partially because they were the earliest in the field and hence have a longer history in supplementing national programs, are the International Maize and Wheat Improvement Center (CIMMYT) located in Mexico and the International Rice Research Institute (IRRI) located in the Philippines. CIMMYT, an outgrowth of a cooperative program between the Mexican government and the Rockefeller Foundation established in 1943, was expanded to an international institute in 1966; IRRI was started in 1962. Comparable research institutions have been developed and located in Colombia, Nigeria, India, Peru, Taiwan, Ethiopia, Kenya, Syria, and other countries. These international oriented research institutes focus on specific crops or regions, with crop breeding an important component of their research programs. In addition, they conduct training programs for agricultural scientists from developing countries. These international research programs offer additional opportunities for young and innovative crop breeders.

CAN A HUNGRY WORLD BE FED?

Can a hungry world be fed? To quote from Toynbee (1963): "Today mankind's future is at stake in a formidable race between population growth and famine." Which will win? Both population control and increased food production will be needed. The nature, strategy, and heritage of plant breeding have been outlined and their roles in increasing food production emphasized.

Recent events and developments will influence the future outlook. The energy crunch is causing many nations to examine the possibility of developing biomass-derived sources of fuel. Once the more desirable sources are identified, there will be a rapidly expanding need for plant breeders, explorers, and other scientists to help improve plants and microorganisms as sources of liquid and solid fuels.

In many developing regions, new high yielding cultivars are available, but the inputs, especially fertilizers and plant protectants necessary to allow superior performance, are unavailable. Genetic improvement and management should stress reliability or stability of production by maximizing resistances and tolerances, improved adaptation, and ability to thrive under less than optimal soil and water conditions.

Additional efforts must be made to improve genotypes for complex multiple-cropping systems. Very little effort has been made to evaluate genotypes of different species for their ability to perform in association with each other. The implications and potential of breeding multiple crop genotypes has been reviewed by Francis (1981). Multiple cropping systems make efficient use of land and labor, two limiting resources through much of the cropping cycle. Relevant research in these systems requires: (1) knowledge of the cropping systems, (2) understanding of the nature and variation of climatic and soil conditions, growth potentials, and stress effects, (3) practical experience with the limiting production constraints, and (4) a multidisciplinary approach to decisions on priorities in a breeding program.

New biological advances will have an impact on breeding. An overview of the role of plant genetic engineering in crop improvement was outlined by a Rockefeller Foundation Conference in 1980 (Rachie and Lyman, 1981) and reviewed in detail in Chapter 7. Plant breeders have a great need for more effective selection methods. Selection in tissue culture or at the cell level would be an important tool to increase efficiency of selection and to reduce the time required to produce new cultivars.

The opportunity to incorporate nitrogen-fixing ability into cereal crops is very attractive. The symbiotic relationships are very specific and some success has been achieved. Unless new facts come to light, however, the potential contribution to the total nitrogen economy is at best a minor one (Sneep and Hendricksen, 1979). Established principles and new discoveries will both be critical to continued improvement. Burton (1981) summarized a plant breeder's activities with six words. They are: variate, isolate, evaluate, intermate, multiplicate, and disseminate. These are action words—requirements for reducing world hunger. Success will depend, in large measure, upon the fundamental scientific knowledge and financial resources available to the crop breeder, and upon the breeder's originality, dedication, and persistence. To quote Wortman and Cummings (1978): "While the food-poverty-population problem is massive and complex and will be extremely difficult and time-consuming to resolve, the existence of new capabilities provides a magnificent opportunity, perhaps a fleeting one, to deal with it effectively—if governments have the wisdom and the will to act". The future of a hungry world is at stake.

LITERATURE CITED

Abelson, P. H. (ed.). 1975. Food: politics, economics, nutrition and research. Amer. Assoc. Adv. Soc., Washington, D.C.

Åkerman, A., O. Tedin, and K. Froier, (ed.). 1948. Svalof, 1886–1946. K. Bloms Boktryckeri A. B., Lund, Sweden.

Athwal, D. S. 1971. Semidwarf rice and wheat in global food needs. Q. Rev. Biol. 46:1–34.

Blakeslee, A. F., and A. G. Avery. 1937. Methods of inducing doubling of chromosomes in plants. J. Hered. 28:393–411.

Borlaug, N. E. 1968. Wheat breeding and its impact on world food supply. p. 1–36. *In* K. W. Finlay and K. W. Shepherd (ed.) 3rd Int. Wheat Genet Symp., Canberra, Australia. 5–9 Aug. 1968. Butterworth and Co., Sydney, Australia.

———. 1972. The green revolution, peace, and humanity. CIMMYT Reprint and Translation Series No. 3, Int. Maize and Wheat Imp. Center, Mexico D.F.

———. 1981. Increasing and stabilizing food production. p. 467–492. *In* K. J. Frey (ed.) Plant Breeding II. Iowa State University Press, Ames.

Brown, L. R. 1963. Man, land, and food. Foreign Agr. Econ. Rep., No. 11, Economic Research Service, USDA, Washington, D.C.

Burton, G. W. 1981. Meeting human needs through plant breeding: past progress and prospects for the future. p. 433–466. *In* K. J. Frey (ed.) Plant Breeding II. Iowa State University Press, Ames.

Creech, J. L., and L. P. Reitz. 1971. Germplasm now and for tomorrow. Adv. Agron. 23:1–49.

Dalrymple, D. G. 1978. Development and spread of high-yielding varieties of wheat and rice in the less developed nations. Foreign Agric. Econ. Rep. No. 95, 6th ed., Econ. Res. Serv., USDA, Washington, D.C.

Ephrat, J. 1974. Some ideas about wheat breeding for higher yields after the "Green Revolution" breakthrough. Z. Pflanzenzuecht. 72:39–45.

Food and Agricultural Organization Production Yearbook. 1979. Food and Agricultural Organization of the United Nations. Vol. 33. Rome, Italy.

Francis, C. A. 1981. Developing plant genotypes for multiple cropping systems. p. 179–231. *In* K. J. Frey (ed.) Plant Breeding II. Iowa State University Press, Ames.

Frankel, O. H. 1977. Natural variation and its conservation. p. 21–44. *In* Amir Muhammed, Rustem, Aksel, and R. C. von Borstel (ed.). Genetic diversity in plants. Plenum Press, New York.

Harlan, J. R. 1972. Genetics of disaster. J. Environ. Qual. 1:212–215.

———. 1975. Crops and man. Am. Soc. Agron., Madison, Wis.

Hutchinson, J., G. Clark, E. M. Jope, and R. Riley (ed.). 1976. The early history of agriculture. Phil. Trans. R. Soc. Long. B. 275–213.

International Rice Research Institute. 1975. Annual report for 1974. Los Banos, Philippines.

Jennings, P. R. 1974. Rice breeding and world food production. Science 186:1085–1088.

Klose, Nelson. 1950. America's crop heritage. Iowa State University Press, Ames.

Poehlman, J. M. 1979. Breeding field crops. 2nd ed. AVI Publishing Co., Inc., Westport, Conn.

Presidential Commission on World Hunger. 1980. Overcoming world hunger—the challenge ahead. U.S. Government Printing Office, Washington, D.C.

Rachie, K. O., and J. M. Lyman (ed.). 1981. Genetic engineering for crop improvement. Working papers. The Rockefeller Foundation, New York, N.Y.

Randhhawa, M. S. 1974. Green revolution. John Wiley and Sons., New York.

Reitz, L. P., and S. C. Salmon. 1968. Origin, history, and use of Norin 10 wheat. Crop Sci. 8:686–689.

Revelle, Roger. 1974. Food and population. Sci. Am. 231:160–170.

Russel, D. A., D. M. Henshaw, C. E. Schauble, and R. B. Diamond. 1970. High-yielding cereals and fertilizer demand. Bull. Y-4. Natl. Fertilizer Development Center, Muscle Shoals, Alabama.

Sears, E. R. 1956. The transfer of leaf rust resistance from *Aegilops umbellulata* to wheat. p. 1–21. *In* Genetics in plant breeding, Brookhaven Symp. Biol. No. 9, Brookhaven Natl. Lab., Upton, New York.

Shull, G. H. 1908. The composition of a field of maize. Am. Breeders Assoc. Annu. Rep. 4:296–301.

Sneep, J., and A. J. T. Hendriksen (ed.). 1979. Plant breeding perspectives. Center for Agricultural Publishing and Documentation, Wageningen, Netherlands.

Stadler, L. J. 1928. Mutations in barley induced by X-rays and radium. Science 68: 186–187.

Toynbee, A. J. 1963. Man and hunger—The perspective of history. World Food Congress, Washington, D.C.

U.S. Department of Agriculture. 1979. Plant genetic resources: conservation and use. Prepared by Natl. Plant Genetics Resources Board. USDA, Washington, D.C.

Wade, N. 1975. International agricultural research. p. 91–95. *In* P. H. Abelson (ed.) Food: politics, economics, nutrition, and research. Am. Assoc. Adv. Sci., Washington, D.C.

Wortman, Sterling, and R. W. Cummings, Jr. 1978. To feed this world. Johns Hopkins Univ. Press, Baltimore, Md.

SUGGESTED READINGS

Frey, K. J. (ed.). 1981. Plant Breeding II. Iowa State University Press, Ames.

Simmonds, N. W. 1979. Principles of crop improvement. Longman Inc., New York.

Welsh, J. R. 1981. Fundamentals of plant genetics and breeding. John Wiley and Sons, New York.

Chapter 2

Crop Breeding as a Design Science

N. F. JENSEN

Cornell University
Ithaca, New York

Crop breeding is one of several scientific disciplines that contribute to crop improvement and to the high quantity and quality of food and fiber production in the world. Because crop breeding is an important discipline for developing improved cultivars, crop breeders are in a position to influence changes in crop improvement. Crop breeders cooperate with and integrate their work into other scientific disciplines (e.g., plant pathology and plant physiology), and their efforts can be called a design science (Jensen, 1967).

The visible evidence of crop breeding surrounds us everywhere with fields of maize, potato, cotton, barley, flowers, sorghum, tomato, wheat—in fact, a total listing would read like a litany of the world's food, fiber, fruit, and flower crops. Crop breeding as an organized occupation is only about a century old. The desire of pioneer agriculturists to improve the landrace received the necessary scientific foundation with the rediscovery of the principles of Mendelian genetics at the beginning of the 20th century. Today, genetics remains the solid base underlying crop breeding but our knowledge, innovation, and technology have become increasingly sophisticated. Moreover, crop breeding has grown as a gainful career occupation. Once found largely in the domain of the U.S. Land Grant universities with their agricultural schools and experiment stations, crop breeding today is also a vigorous part of the corporate and free enterprise system. Research organizations specialize in genetic engineering and tissue culture research, which offer both the hope of changes at the micro (gene) level and macro changes in hybridization possibilities between previously isolated plant entities. These exciting new developments, in most cases, must be submitted to traditional plant breeding procedures, often involving further hybridization, to adapt the new discoveries to acceptable agricultural form and performance levels. Crop breeding today is an exciting, challenging, and remunerative occupation with working environments as diverse as universities and corporations.

In addition to these new directions in genetics and crop breeding, all of the factors that affect crop performances such as weather, fertilizer, diseases and insects, nutritional quality, and industrial use, are potential elements to be considered in the design of a new cultivar. Crop breeding is the management of genetic variability to meet crop production goals; it creates and channels new genetic combinations to meet the design criteria of the desired cultivar. Genetic variability is an essential resource for the development of new cultivars. Genetic variability exists in collections of breeding lines, composites of lines, and interbreeding populations maintained by plant breeders all over the world. Plants found in primitive agriculture and in the wild are important sources of variability which are constantly in danger of being lost. New genes, which arise by mutation, and new gene combinations, which arise by genetic recombination, are also important sources of genetic variability.

Although the genetic principles are universal, the breeding problems and techniques may be quite different from crop to crop. For example, a breeder of a self-pollinating small grain works hard to get the initial genetic variability by making hybrids, but thereafter self-pollination in successive generations provides the breeder with the opportunity to select homozygous lines or "inbreds". On the other hand, a breeder of maize finds the variability ready-made in cross-pollinating populations and works hard to produce true-breeding inbreds. There is a tendency for breeders of different crops to associate themselves most closely to colleagues working and publishing research results on the same crop. This practice tends to create zones of relative intellectual isolation, for example, meetings and newsletters are organized along crop lines. A sizable gulf may exist between breeders of self-pollinated crops such as soybean and wheat; and even between breeders of the same crop having different climatic and geographic requirements, such as spring and winter grains. These barriers constitute impediments to crop design. Breeders who disregard these barriers reap the added bonus of an expanded horizon in planning. To illustrate, barley breeding historically lies within the orbit of self-pollinated crops, yet an important breeding technique, recurrent selection using male-sterile lines, is more closely allied with the experience and literature of the cross-pollinated crops. Such barriers do not limit breeders with an active curiosity who follow an omnivorous reading program.

In a crop breeding project, new genetically variable plant material is fed into the program each season. The overall project may be segmented by years or generations. After 10 to 15 years, the finished product may appear in the form of a cultivar to be released. Of course, much effort and attention has been given to the material at each stage of development and most has been discarded. Thus, a crop breeding project that has been in operation for a decade or more will have the dynamic characteristic of showing each year the current products of the plant breeding cycle that originated 10 or more years earlier, plus all the annual segments (cohorts) in between. It is difficult to overemphasize the close linkage between success at the end of the program and the planning decisions at the initial planning stage. The annually renewing characteristics permit new design changes.

Crop breeding deals with population dynamics. Much time is devoted to the synthesis of genetically variable populations and selecting desired phenotypes from a population. Breeders must also decide how to handle populations in the field; for example, whether to use the pedigree or composite method in self-pollinated crops, or, in cross-pollinated crops, what form of recurrent selection should be used.

What is there about crop breeding that qualifies it as a design science? Surely the clearest example is the creation of a crop cultivar. Crop breeders usually visualize the need for a new cultivar. Sometimes the need is brought to the breeder's attention by producers or users. The breeder designs the future cultivar through a consideration of possible contributing genotypes, which might be brought together in a series of hybridizations to form segregating plant populations, from which true breeding lines later can be selected. The breeder may increase the chances of success by adopting more than one approach. For example, in wheat breeding, where a wheat resistant to a dangerous race of stem rust is urgently needed, the breeder may simultaneously begin (1) a backcross program, (2) a hybridization program involving several parents and recurrent selection, and (3) a mutation program to reduce plant height of an already resistant genotype.

Guidelines for design depend upon a knowledge of production practices important to the farmer, marketing requirements and consumer needs related to the crop. Successful breeders must be aware of all facets that affect goals and timing, including climatic influences, disease problems, and such disparate elements as regional, national, and international influences, industry and world trends in markets, governmental policies, foreign aid programs, and population growth trends. If breeders are aware of these elements, they may find their goals changing. For example, changes in the requirements of grain quality for overseas markets may become a design element. Design science becomes important after these influences become translated into a breeding objective at the project level.

This sense of awareness, so important to a crop breeder, most particularly embraces goals and timing. Time, and how it is used, is one of the most crucial elements in plant breeding. The nature of the process requires time—frequently 15 years from design to realization of a cultivar. Therefore, breeders must plan ahead and anticipate trends in needs so that breeding goals will be appropriate.

Design considerations interact at the operational level of crop breeding. Because we know that breeding objectives can be accomplished using different methodology, we can further partition crop breeding into operational categories: (1) Planning and hybridization, (2) Selection and stabilization, (3) Line testing and evaluation, and (4) Cultivar release.

PLANNING AND HYBRIDIZATION

Planning the choice of parents and the hybrid combinations to be made represents an important part of plant breeding and is crucial to success. Choice of parents and hybrid combinations will ultimately determine the success or failure of the breeding program. Careful planning is necessary to enhance the opportunities for developing superior cultivars. If the breeder has a desired plant improvement goal, or the crop faces a hazard requiring change in the cultivar, good planning implies that the breeder examines possible solutions; e.g., introducing an existing adapted cultivar or making a change in cultural practices. If the solution requires a new cultivar formation, however, the breeder will use hybridization. Creating variability by mutation breeding is an approach to use when the genes we want are not known to exist in germplasm collections. In the early days of crop breeding it was often possible to find a desirable plant by selection out of a farmer's field, but this possibility is less likely today as the landraces of traditional agriculture have been replaced by "pure" cultivars.

Planning for the operation of a breeding program begins with a background of knowledge and assumptions. The general goals of wheat improvement, for example, should be clear to the wheat breeder, as well as the problems involved in wheat growing and the performances and weaknesses of the available commercial cultivars. Is it surprising to say that cultivars are far from perfect? They often are. In practice the crop breeder has a continuously changing array of goals. No matter what immediate problem-solving is involved, the breeder wants higher performance and loathes making a step backward in any cultivar characteristic while improving a specific trait. However, one of the hazards of developing populations by hybridization is that the desired recombinant genotype may not occur or cannot be found.

Planning can be subdivided into simpler, more manageable packets by dealing with specific objectives. For example, a program to raise protein levels of grain includes all the objectives needed for a high performing cultivar, but it focuses on high protein quantity. The design considerations in broad concept, therefore, include searching out the world's high protein germplasm of the grain involved and hybridizing them with high performing and adapted cultivars. A breeder may plan to do this by use of single crosses, backcrosses, three-way crosses, double crosses, a series of sequential crosses, composite methods, or other variations of possible breeding methods. As a contribution and example of ways in which design science is used in crop improvement, I have proposed a diallel selective mating system. In this system, designated single crosses which incorporate the objective (e.g., high protein) are crossed among themselves in all possible combinations (Jensen, 1970, 1978). This design expands the potential genetic combinations, all directed towards the chosen objective. It would be difficult to carry a large number of crosses as separate populations, so the design calls for compositing the hybrid seed. Appropriate sampling techniques can accommodate the large numbers inherent in such composites. In this system, pure lines are available on exactly the same time schedule as in other breeding methods. The selective mating provisions, a form of recurrent selection, permit early screening for the high protein objective and immediate hybridization of high protein individuals to begin a new cycle. The length of the selective mating cycle for the high protein example is 2 years.

The planning stage is the place to look ahead. If a crop, or a particular use of its product, is on the decline, what does the trend line tell us about the future need for any new cultivar which, in any case may not be ready for 12 or 15 years? Future events cast their shadows before them. For instance, barley in New York, was grown on 143,600 ha (1882), declined to 120,000, 80,000, and 40,000 ha in 1890, 1928, and 1946, respectively. Today, barley is grown on about 4,000 to 6,000 ha. The New York breeder reacted to this decline by closing out the spring barley breeding program in the 1950's. The winter barley program persisted longer because of the potential for specialty use in malting. The plant breeder must try to anticipate trends in food or product needs so that breeding goals will be on target.

The distant future also intrudes into design planning. For example, if global mean temperatures should decrease, as some scientists believe, this would have a profound effect upon food production. Some breeders are emphasizing winter hardiness in the design of future wheat cultivars. If, when the new cultivar is ready, the climate trend has eased, the effort need not be considered lost because, even now, more hardy winter wheats are needed in northern production areas.

Let me emphasize again the close relationship between the planning and hybridizing phase of crop breeding and the results that will come from the program many years later. The release of new and improved cultivars is an infrequent occurrence. For example, in the entire history of the New York (Cornell University) winter wheat breeding project, begun in 1907, 11 new cultivars were released, the first in 1920. This means that in more than 60 of the years of this period, no cultivar was released. Of course, the number of cultivars released may not be a good measure of the success of a crop breeding program—success is related more to what the cultivars do in performance and how well they meet the continuing needs of society.

The planning and hybridization phase lasts until the hybrid seeds have been obtained. For convenience, the growing of the hybrid plants (F_1) is included in this phase so that the period usually covers 2 years.

SELECTION AND STABILIZATION

The selection and line stabilization phase in self-pollinated crops begins with the second generation (F_2), whose seeds came from the hybrid plants, and includes the F_2 to F_4 or F_5 generations, covering a period of 3 or 4 years. In the F_2, heterozygosity is at its greatest; thereafter, it decreases with each generation. Theoretically, hybrid populations need to be selfed for nine or ten generations to reach practically complete homozygosity; but plant breeders, for practical reasons, will make their final individual plant selections in the F_4 or F_5, at which time, most selections will exhibit a fair degree of uniformity, or "working homozygosity". The selections are then grown in plant rows in the F_5 or F_6.

In crops such as the potato, the selected genotype can be fixed immediately, since reproduction is asexual through the potato seed piece. A breeder can evaluate selections as soon as sufficient progeny are obtained. Breeders produce homozygous inbreds of maize, a cross-pollinated crop, by selfing successive generations before the inbreds are selected to be used as parents to produce hybrids. As the breeder selects many true-breeding individuals for evaluation, there is always the hope that there will be a superior line that merits release as a cultivar. A very important product is the feedback of information into the planning and hybridization stages as the process begins anew each year.

LINE TESTING AND EVALUATION

This is the period during which the plant breeder tests the products of the previous planning stages. In maize it would be the period when inbreds used to form hybrids are subjected to competition testing in top-cross tests. In wheat, the line evaluation phase begins when the breeder makes selections for harvest from the head or plant row nursery, which may number initially 8,000 to 15,000 rows. In the New York wheat breeding nursery of 10,000 progeny rows, the practice was to select approximately 15%. The 1,500 or so lines selected are planted next season and further field selection of about 33% was made. The 500 lines saved are planted in the field and in this way the discarding process continues with the selections becoming smaller in number. Discarding of inferior lines is important.

Design considerations in this phase require the use of biometrics, important for reasons of precision, accuracy, and efficiency. We know that 99% of all individuals in a breeding program will eventually be discarded (Remember, only 11 cultivars were developed by the Cornell wheat program in 75 years, from more than 100 million plants!). It is good strategy to keep design as simple as possible so as not to waste effort on non-payoff areas. Why plant five replications when three will do?

In the early stages of line evaluation many breeders rely on experience and the "breeder's eye". Applied to the case of the 10,000-line wheat cohort, this meant harvesting only 1,500 head rows. The alternative would have been to harvest the entire nursery of 10,000 and proceed with analysis of yield, 1000-kernel weight, and so forth.

It is easy to complicate decision making in plant breeding. In reality, it may be reduced to two questions: (1) Assuming all unknowns to be favorable, does the line have the potential to be a new cultivar? (2) Is it wanted for any other reason, e.g., as a parent for further crossing? If the answer to both is "no", there is no reason to keep the line. If the answer to either is "yes", or "I don't know", the line should be kept through another season. Each additional year a line is retained results in the accumulation of more data. Thus, in time, decision making can become easier because of the added experience and research data.

CULTIVAR RELEASE

Crop breeding as a design science ends with a cultivar release. The breeder's role does not stop abruptly at this point. In the human family, good parents do not abandon their children at birth; instead they shelter, feed, love, protect, and foster tham until they can make their own way. A new cultivar of a crop also needs a lot of fostering. Seed supplies must be built up from the few grams available when the selection is identified as outstanding, to thousands of kilograms upon release. Usually, plant breeders make the first increases on carefully isolated small fields. For subsequent increases, breeders in the public sector will turn to an agency expressly organized to service the seed industry and the farmer. In the USA, each state has a seed certifying agency working with the Land Grant University and the seed industry to make further increases with careful checks to maintain purity. All phases of planting, growing, and harvesting are closely monitored by trained personnel. Breeders serve as consultants in this process and keep close contact through field inspections, meetings, and so forth. They will continue to be involved for the useful life of the cultivar.

Good design results in productive cultivars. Next time you drive by beautiful fields of hybrid maize, dwarf sorghum, wheat, or soybean, remember that all of this beauty did not happen by accident but rather by design.

DESIGN LIMITS TO CROP IMPROVEMENT

I have never forgotten the strong impression left with me in the 1930's, by a college professor who held the view that the potential gains from crop improvement through breeding had been almost enhausted. Forty years later, we know he was wrong because we have experienced gains through crop breeding that verge on the dramatic. In New York the state average yield of wheat rose more in the decade ending in 1945 than the total gain from the previous 70 years! These gains resulted from genetic and production improvements. Between 1935 and 1975, rising productivity was essentially a straight-line phenomenon; however, the rate since 1975 has slowed. Whether this is a temporary slowing while new technological forces are being marshalled is not known. Even so, we might agree that there must be a finite limit to the productivity of a crop (yield per area). If we accept this assumption, then it follows that continuing improvements of a cumulative nature are using up this reserve of potential. We also recognize a general truism in the assumption that attaining the final increments of productivity may become disproportionately difficult and expensive.

The thrust of crop improvement over the past several decades has been the removal of inherent obstacles to higher yields. For example, reducing plant height increased yield and quality simply by increasing the plant's resistance to lodging. As lodging was eliminated, the remaining factors influencing maximum productivity came into focus more clearly. It is important to remember that a yield ceiling is not fixed, except at its ultimate point. It will rise as we continue to improve our technological efforts. Even more weather features can be circumvented. For example, it is possible to provide water through irrigation in dry climates. In the future we may confidently expect crop breeding and design to be significantly affected by emerging new technology, not the least important of which is the research dealing with the biological and genetical materials themselves.

IS CROP BREEDING AN ART OR A SCIENCE?

If a breeder is successful, is it due to an understanding of genetics, biometrics, and related scientific disciplines? Or is it more related to personal traits such as ways of thinking, prior experiences, intuition, different insights, and inferences—things that don't fit under the neat compartments of scientific disciplines? There is an art as well as a science to crop breeding, and sometimes the art can be as important as the science.

Everyone has heard of Luther Burbank. He popularized crop breeding before it had much standing as a scientific discipline. Largely untrained, Burbank had a keen eye and a sense of commerce—and an understanding of the link between the two. Burbank himself related the story of the "Burbank" potato, calling it his most important discovery. He observed a seed ball maturing on a vine of the 'Early Rose' cultivar in a garden plot and marked it in his mind for harvest. But one morning when he visited the patch the seed ball was gone! Search as he might he could not find the missing seed ball. He did not give up the search, however, and several days later his patience was rewarded when he spotted the seed ball lodged in the soil several feet away. The ball contained 23 seeds. It is remarkable, considering that he planted them directly in the field, that each germinated the next spring and produced a plant. At harvest time in the fall, he dug each seedling hill and examined the tubers. Twenty-one were very ordinary, ranging from poor to average, but 2 vines were different. These were very large, smooth, white potatoes, excelling in all respects any vegetables of their kind that he had ever seen. One was judged better than the other and was sold to a Mr. Gregory from Marblehead, Mass., who gave the name, Burbank, to the cultivar. Today the 'Russet Burbank'' still accounts for about 50% of the potato acreage in the USA.

Burbank characterized the discovery of the Burbank potato as a "simple development," and certainly luck played an important part throughout. It is said, however, that successful people often make their own good luck and Burbank did all the things necessary to expose the greatness of the Russet potato; he was curious and picked the seed ball, preserved it, dried and planted the enclosed seeds, took care of the plants and finally, he had the eye for the one superior plant and its tubers. Burbank epitomized the art in plant breeding. Because the case of Burbank is a rare exception, crop breeders today are trained in the science of the discipline. They work with scientists, who transmit some of the learned art of the craft, and encourage development of innate resources and creative impulses.

Throughout this chapter we have used the term "design science" to apply to cultivar creation by breeding methods. The concept, in fact, is broad in scope and ranges from the international classification and use of germplasm to a search for the "why" answers in biology. Research in plant physiology, for example, may tell us that leaves borne erect are more efficient in light interception than are drooping leaves. This knowledge affects the design for breeding. It also affects crop production practices, since with less shading it is possible to have populations of higher density for higher productivity. The world class rice and wheat cultivars that were a part of the Green Revolution have already been redesigned to conform to long-established eating habits. For example, many people in India prefer white rather than red kernels of wheat for their home use, so the original red-kernelled cultivars were redesigned through mutation and hybridization to meet this need. Crop breeding is influenced by research that makes different formulations of a cultivar possible. In wheat, oat, and barley, the typical cultivar has been a pure line. There are also multi-line and hybrid cultivars. The designs of some hybrid cultivars are based on the development and use of critical cytogenetic stocks. Mutation breeding, in the hands of experts, can produce desired genotypic conformations with a high success ratio. No matter how deeply we probe into fundamental research, we seldom lose sight of the binding link between design and crop breeding.

LITERATURE CITED

Jensen, N. F. 1967. Agrobiology: Specialization or systems analysis? Science 157: 1405–1409.

———. 1970. A diallel selective mating system for cereal breeding. Crop Sci. 10: 629–635.

———. 1978. Composite breeding methods and the DSM system in cereals. Crop Sci. 18:622–626.

Chapter 3

Management of Crop Breeding

PHILIP BUSEY

University of Florida
Fort Lauderdale, Florida

How do you actually do crop breeding? As with any other profession, you must have technical knowledge. But the good crop breeder must also be a manager, must be able to make decisions, to make plans, and to carry them out, and to learn from successes and failures. These abilities do not come from the technical knowledge of breeding, nor do they come easily from practical experience. This chapter provides tools for planning and breeding new crop cultivars. It is written from the viewpoint that crop breeding is an application of management.

Imagine that today is your first day on the job. You, as a new crop breeder, have learned the mechanics of genetics. Through years of study you have trained yourself in agriculture. Let us assume that you know how to grow plants, and have acquired other technical skills—you can make artificial hybrids and can design and analyze a field experiment. You are familiar with the assigned crop and have made a commitment to know it thoroughly. You are competent to select germplasm and develop new hybrids. Your challenge is to plan your breeding program and carry it out, based on appropriate management procedures.

One of the best ways to learn about crop breeding is to study the work of successful breeders. You will find many interesting ideas in their technical articles. Check the "Materials and Methods" section and you will learn some of the "tricks of the trade", and in some cases see how an overall program is organized. Unfortunately, crop breeders do not spell out all of the details of what they do and why they do it. So you should also associate personally with successful crop breeders, as your instructors, supervisors, or colleagues. By combining technical knowledge and practical experience you should develop a strong crop breeding expertise. Although technical limitations of breeding will determine many specific decisions and operations, the problem-solving process is the universal activity of all technical professions. Think of yourself now as a manager. Your success as a crop breeder depends as much upon your ability to organize, to direct, to follow up, and thereby to control, as it does upon your technical proficiency.

WHAT IS MANAGEMENT?

Management is the process of conducting a complex series of tasks requiring the action of several elements. Crop breeding requires management of many interdependent elements. These include the design of the potential cultivars, assembly and manipulation of germplasm, best use of physical resources, and people. Because these elements must work in harmony to develop a new cultivar, it is appropriate to apply modern principles of management.

Objectives and Needs

The crop breeder is confronted with many choices. An overall plan is needed. Decisions must be integrated to form a continuous sequence from cultivar design to its ultimate release. Situations may change during the course of the program, and the breeder must adapt the plan when predetermined objectives no longer appear worthwhile or obtainable. Also, results not previously imagined may be realized in the process of accidental discovery.

When starting to work on a crop plant, write down the genetic improvements that can be made, how involved and expensive such improvements would be, the procedures to use, and anything else that seems pertinent. Such an analysis will outline the context and the possible directions for the breeding program, and will serve as an aid in decision making. The first item should be specific program objective(s). If several objectives are needed, they should be ranked by priority so that the greatest attention can be assigned to the most important areas. It is important to be comprehensive in your vision but specific in your actions. Critical choices must be made to maximize the effectiveness of the program. Low priority objectives may be dropped if time and resources become limiting.

Long-term objectives should be included even though they are difficult to imagine. In addition to improved yield and stability of performance, consideration should be given to adaptability to new technology. Dwarf sorghum and impact-resistant tomatoes are examples of adaptations to machine harvesting technology. Society's needs continually change and tomorrow's technology will place new obligations on the crop breeder.

Factors outside of the breeding program will also affect success. A full understanding of the history and economics of the crop is needed, including the defects of presently available cultivars. Scientists working on cultural practices should be brought into the breeding program. Consulting with entomologists, plant pathologists, taxonomists, and other experts can help you determine major objectives. The more people who review the objectives, the more comprehensive the analysis will be.

Table 3.1. Four dimensions of management activities.†

Activity	Variables that influence activity	Example(s)
Organization design	Number of positions available	Create three levels of supervision
	Classification of work areas (field, greenhouse, seed processing, inventory, etc.)	Assign tasks for all operations
	Technological complexity	Create a cytology laboratory
Planning and control	Breeding system of the crop	Establish a polycross breeding scheme
	Level of documentation needed for cultivar release	Include additional evaluations for insect resistance
Behavioral processes	Capabilities of individuals	Agree on a cooperative research project with a plant pathologist
	Need for others to know results	Write a journal article; organize demonstration plots for field day
Decision making	Problem: program isn't working	Select alternative procedures
	Problem: New cultivar isn't acceptable to farmers	Choose new objectives based on farmer's comments and criticisms

† Adapted from Gannon (1977) with examples from crop breeding.

Outlining the total strategy and possible options will provide a working model for the breeding program. This model will help communicate what you are doing and why you are doing it with your employer, sales people, extension agents, and colleagues. Managers want to know how your work relates to the objectives of the company or the experiment station, in order to determine the kinds of support or involvement needed for the success of the program. If conditions change and the program must change directions, the original analysis will be a good starting point in considering possible alternatives. Be prepared to revise plans occasionally as new information becomes available. Finally, if you have established clear ground rules, it will be possible to measure progress and review costs and benefits at any time.

The Four Dimensions of Management

Now that you have outlined the breeding program, you can begin to manage it. Gannon (1977) described four dimensions of management: (1) organization design, (2) planning and control, (3) behavioral processes, and (4) decision making. These are presented with examples from crop breeding in Table 3.1.

Organization design is the physical structure of the breeding program, including the relationship of personnel, their responsibilities, equipment, field areas, and other facilities.

Planning and control methods must relate to the objectives. Planning is done by strategies (long-term methods) and tactics (short-term methods). In a breeding program, strategies might include methods of creating germplasm diversity while tactics include the specific methods for producing hybrids, setting up field plots, and organizing data. Control in crop breeding is the verification of plans and actions. It is accomplished by setting up efficient systems of data management, including data storage and processing. Control in crop breeding also requires "feedback mechanisms"—double-checks on the accuracy of records, the performance of progenies, and ways to compare advanced breeding lines with present cultivars. Repeated observations are also a form of control.

Behavioral processes involve human relations. This is an important area and includes the management of efficient communications, minimization of interference and interruptions, and ways to use the crop breeder's time and that of other workers as efficiently as possible.

Finally, decision making is the central activity of the breeder, because it affects and determines the other three dimensions. Decision making is choosing a course of action to solve a problem. Examples include selection of genotypes, objectives, and among alternative plans.

DECISION MAKING: THE CENTRAL ACTIVITY OF THE CROP BREEDER

Decision making is problem solving, especially related to the choice of a course of action. The basic steps in decision making, once a problem is recognized, are to: (1) analyze the problem, identify underlying causes and effects; (2) develop alternative solutions; (3) develop and apply evaluation criteria; (4) choose among alternatives; and (5) implement the chosen solutions. These activities may involve personnel assignment, development of work plans, communication methods, and review of performance.

The kinds of problems that you may confront will encompass the other dimensions of management. Examples of problems are:

1. A particular breeding line is not performing adequately.
2. Not enough land is available to test all breeding lines.
3. No one breeding line is best in all respects, but a cultivar must be released.
4. An established cultivar is susceptible to an important disease.
5. A new cultivar specification has been identified by market surveys and by sales personnel.

Several valuable decision aids exist for analyzing and solving these problems.

In a smoothly running breeding program, procedures for selection are planned and followed to simplify the operational tasks. But at certain critical junctures, decisions must be made. The amount of decision making varies with the level, span of control, and the time required for the activity. In a typical breeding situation, the long-range objectives are determined partly from outside of the breeding program. A board of directors or a research director will have a major role in long-range decisions. It is important that someone with authority exercise this role, to ensure continuity of purpose and programs. In the public sector, much of the responsibility for long-range policy and decision making is left up to committees and work groups that answer the major needs for a commodity nationwide. You as a crop breeder will be responsible for making decisions in most of the remaining areas of medium, short, and current range. If you are developing germplasm for a tree crop, remember that medium-range goals could still involve several decades of work. Obviously, such time considerations increase the challenge for management and decision makers. The breeding program could be passed on through many administrators before a single cycle of hybridization and selection is accomplished.

Before considering specific applications of decision making in crop breeding, there are other universal problems in making sound management decisions. Two that will be covered here are bias and uncertainty.

Bias in Decision Making

You as a manager may have biases, or prior conceptions, that will affect your decision making ability. Years of formal schooling may have permitted you to develop a coherent outlook and approach to problem solving. Valuable as that is, such an outlook may carry with it narrowness of thinking or a tendency to ignore key facts. Awareness of personal bias can be an important step in reviewing solutions to problems. Sometimes the discovery of a hidden preconception permits the crop breeder to make a major breakthrough. Set up prior standards for observation, analysis, and decision making. Stand back occasionally and review your assumptions. A favorite genotype to be released as a new cultivar must be given a fair test if later enbarrassment is to be avoided. If you are making a personnel decision, set up evaluation criteria ahead of time. Common sense and fairness dictate a need for openness with others, as well as with yourself.

Uncertainty

Most decisions must be made with some uncertainty about the outcome. As a scientist you can generate qualitative information upon which to base your decisions, thereby reducing uncertainty. Perform a trial run, a preliminary experiment, and consult the literature available on the subject. Do not ignore the commonplace uncertainties that may impede the breeding program, such as delays in planting due to weather, sudden changes in personnel, and breakdown of equipment. Essential functions, such as irrigation or greenhouse operation, must be performed dependably and timely.

A crop breeder must develop a degree of confidence based on a thorough knowledge of how plant materials perform. Learn how to maintain the crop and how to evaluate it under varying conditions. Look at it from many angles and at many times of the day. Feel it, taste it, look at it under the microscope occasionally, and pull it up to look at the roots. With due regard to the possibility of bias, a close involvement with your crop can lead to important discoveries and it can give you an idea of feasible breeding goals. Thorough understanding of your crop may be necessary before you can deal with the second area of uncertainty in decision making, which is the quantitative.

No matter how certain you are of your information, there might still remain a certain randomness in the outcome of your actions. For example, the presence or absence of a rare transgressive segregant in a particular population provides an outcome over which you have little control, yet you may have convincing data on the probabilities of such occurrences. Variability in outcome is classically called "risk". There are statistical methods for recognizing and measuring risk. Quantitative data are relatively easy to analyze statistically and probability statements can be assigned to facilitate decisions. Other advantages of quantitative information that are useful for decision making include: (1) systematic data processing, including easy machine analysis, storage, and retrieval; (2) reduction in management error, that might be due to gaps or duplications of information; and (3) reduction in management error due to bias. If it helps the evaluator to be better organized, quantitative data can reduce uncertainty by permitting the use of a broad information base and of repeatable evaluation procedures. Quantitative models are thus an excellent basis for making better decisions in plant hybridization and selection. Careful analysis also permits better assignment of duties (the organizational dimension of management) and better planning and control (including accounting for progress). Let us look at some problems and techniques in decision making, with emphasis on quantitative methods.

The Decision Tree

Crop breeding decisions are often made in conjunction with other decisions. A well-managed program should outline the kinds of future decisions that must be made, and the basis on which they should be made, before a problem arises. A diagram of decision making (Fig. 3.1) outlines an efficient course of action. A simple situation is diagrammed in this model. Several choices will be made by the plant breeder regarding the kind of breeding program that should be implemented, and what new information is needed to minimize uncertainty. Some of this information may be expensive and time-consuming to obtain. Therefore, the steps have been arranged to minimize branch points.

It would be wasteful and unnecessary to initiate a breeding program unless two key questions can be answered affirmatively (Fig. 3.1):

1. Do genetic differences in performance exist?
2. Is genotype performance a factor in the marketplace? In addition, it would be expedient to know whether a cultivar should be developed for the marketplace as a whole, or for a particular segment of the marketplace. If the latter were possible, then additional questions should be asked:
3. Does specific geographic adaptation exist?
4. Is the market for a new cultivar sufficiently specialized to accept specifically adapted cultivars?

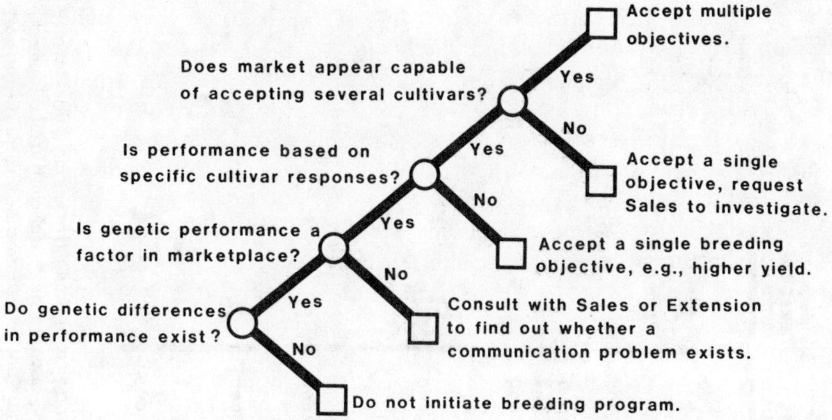

Fig. 3.1. A simplified decision tree applicable to most crop breeding situations. This diagram illustrates a sequence of questions and decisions.

The profitability of developing a new cultivar would depend upon other questions not covered in Fig. 3.1:

5. Would the new cultivar be an improvement over existing ones with greater profit and less risk to the farmer?
6. Would the development provide a net return to the developer and the distributor?

The last point is more complex, and requires the use of a probability model. Let us say that all lines of information indicate that there is a place for a new cultivar, based on an affirmative answer to questions 1, 2, and 5, above. Thus there is sufficient basis to design a new cultivar, but the potential success and profitability require more detailed analysis.

Probability Model

Because the questions of profitability and development cost are interrelated, you analyze these matters in detail. It would be particularly appropriate to use probability in your model of profit and uncertainty. Imagine a breeding plan with several steps and alternate pathways, each with a chance of success or failure. Each pathway opens up new contingencies and new uncertainties. Decisions can be made that will minimize uncertainty or risk, and maximize profit.

Consider the following example which illustrates the use of probability in crop breeding decisions (Fig. 3.2). After careful analysis of the need for and marketability of a new cultivar, you conclude that a potential does exist, and that if it were accomplished it could bring the company $200,000 in gross returns. Based on comments by your experienced colleagues, you conclude that there is only a 50% chance that this new cultivar could be developed successfully. Furthermore, you predict that development costs of

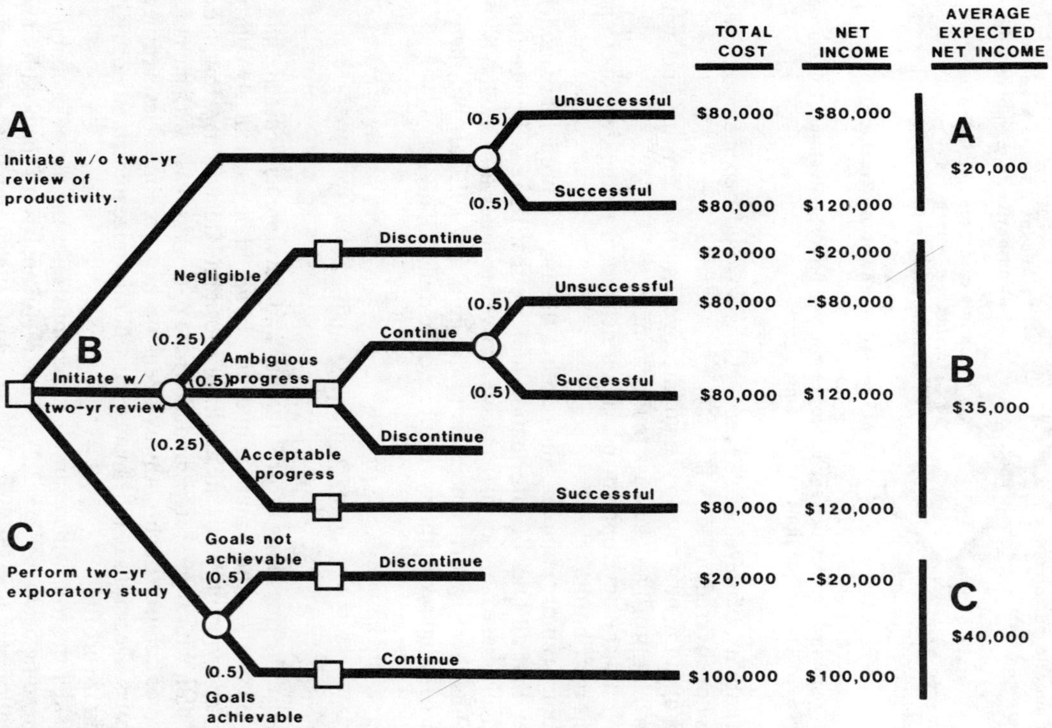

Fig. 3.2. Decision tree for a crop breeding program, with the application of probability. Plan A involves no 2-year review of productivity. Plan B does involve a 2-year review. Plan C starts with a 2-year exploratory study. Net income equals gross return ($200,000 in the case of a successful cultivar development) minus total cost. The discontinuation of the program following ambiguous progress is an alternative which would not be recommended, therefore, no cost figures are presented.

$80,000 would be incurred over an 8-year period. There is also a 50% chance that the cultivar could be a total failure, resulting in $80,000 worth of development costs down the drain. The net expected income to the firm, after development costs are subtracted, is based on the expected chances of success or failure. Thus the net expected income would be:

$$\text{Probability of success} \times \left(\text{Expected gross profit}\right) - \text{Estimated cost} = \text{Net expected income}$$

or

$$0.5 \times (\$200{,}000) - \$80{,}000 = \$20{,}000$$

You recommend to your supervisor that the company initiate this breeding program (Fig. 3.2, plan A).

After several weeks your supervisor comes back with this disappointing news:

> "Unfortunately, the Board of Directors cannot accept your proposal. They feel that the breeding program is far too risky, and the estimated income too low. But the motion was tabled until next month, when they hope you can come up with a more satisfactory proposal. I suggest that you either redesign it to improve the financial return, or change your outlook on the degree of risk the Company can assume."

You realize that you have essentially been challenged to do a better job of management. After thorough consideration of the original proposal, you notice some biases that you had not previously stated. You had assumed that you could get out of the breeding program in a few years if it did not seem to be working out. After 2 years it would be apparent whether the program would succeed or fail. The Board apparently didn't realize this outcome; rather they understood the 50% risk of failure to be based on an 8-year program. How do you explain this to the Board? You develop plan B, which involves a 2-year review of progress, and present that in a form which the Board will understand (Fig. 3.2). This plan predicts a 50% chance of knowing, after 2 years, whether the whole thing is going to be a success or fa failure. You predict a 50% chance of inconclusive progress in 2 years. In that case the Board will again have to decide whether to continue or not. The net expected income from implementing plan B would be $35,000. Plan C might also be proposed including a 2-year exploratory study, costing $20,000, which would let you know with certainty whether the development of the new cultivar would be successful or not. Because the exploratory study would be basic research, it would be added to the 8-year development plan, making a total of 10 years spent on the project. Taking into account the chances of success or failure, the net expected income from plan C would be $40,000.

The Board of Directors can now choose more intelligently from the alternatives. They can measure profitability and uncertainty, plus the possible spin-off benefits of an exploratory study. If your estimates of uncertainty are well documented, then you will have done a good job in presenting the possible outcomes, and have left a part of the final responsibility to your employer.

Optimization

Many decisions are more complex than just selecting the alternative with the greatest expected profits. In the previous example the Board might select option C, which should provide the greatest average profits and the smallest possible loss. Alternatively, the Board might select option B which should provide an earlier return on investment. In either case the decision would be based on choosing the alternative which provides the optimal combination of two or more factors. Decisions which involve multiple factors require careful analysis. Complex models are available to solve these difficult problems of evaluation and probability. Two simplified examples will be presented in order to illustrate possible aids in analyzing complex decisions.

After years of breeding a particular crop, you have recognized a number of improved genotypes which could be introduced to farmers. The seed division of your company wants to produce the "best" one because the marketplace for the proposed new cultivar is capable of handling only one. However, the advanced field testing resulted in a dilemma, with several promising breeding lines (Table 3.2). The proposed cultivars differ remarkably from one another in maximum yield potential as well as tolerance to flooding, drought, and disease. One of these perils strikes the growing region about every second year. Rather than classifying the four lines according to simple resistance and susceptibility, without regard to yield, it would be appropriate to state their actual economic yield under the circumstances of various perils. The potential cultivar capable of the highest yields, line C, tends to yield the worst in unfavorable conditions. Its standard deviation over different circumstances is 658, which indicates high uncertainty of production, high risk to farmers; Line D, on the other hand, has an equally high yield, but has a value of 316 for its standard deviation. Line D is better in both yield and standard deviation than line B. What line D loses in regard to flood tolerance, compared to line B, it more than makes up for in disease

Table 3.2. Yield of potential cultivars of a crop under different production conditions. The probability for each condition is enclosed in parenthesis.

Potential cultivar	"Normal" (0.50)	Floods (0.10)	Disease (0.10)	Drought (0.30)	Mean	SD†
	kg•ha^{-1}					
A	800	800	800	800	800	0
B	1,000	500	100‡	1,000	850	337
C	1,500	100	100	500	900	658
D	1,000	100	1,000	1,000	900	316
Mean	825	325	450	825	785	429
Blend A + D	900	450	900	900	855	142

† Standard deviation, a measure of variability among environments.
‡ A yield of only 100 could not be economically harvested, and is treated as zero.

tolerance. Line A has the poorest yield of the four, but is perfectly stable across all environments. Thus for many farmers line A would be the one to use, but for those more indifferent to risk, line D would be the one to use. Another possibility would be to blend two lines, A and D, to make a medium-yield, medium-dependable cultivar. These possibilities are hypothetical, but they should give you ideas for choices among complex options.

In many crop breeding situations there is a trade-off among opposing objectives. No farmer would accept a higher level of risk at the same level of yield or a lower level of yield at the same level of risk. For that reason, it was easy to eliminate breeding lines B and C, even though in some situations they did better than A or D.

When a large number of selections is available (Fig. 3.3), a majority can often be easily eliminated. Potential cultivars with less than maximum yield at a particular level of risk are eliminated (shaded region). This presumes a precise description of cultivar performance, and for this you must do careful data analysis and interpretation. Many outstanding selections still remain and must be narrowed down to a single cultivar. Once a rationale has been developed it may be presented graphically to demonstrate a solution. Find the appropriate economic relationship between the two variables, yield and instability, and draw a line or function representing that relationship (Fig. 3.3). The slope of this line tells the decision maker that there is a trade-off between y units of yield and x units of risk. The point of tangency of this line to the distribution of cultivars will be the point at which to select.

ORGANIZATION AND CONTROL

Once the objectives are established for a breeding program, a framework is needed to carry it out. A greater amount of information must be processed in crop breeding than in most management tasks. Therefore ac-

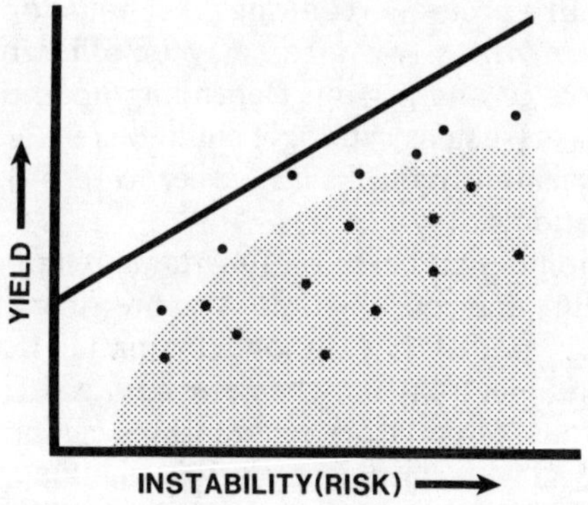

Fig. 3.3. A diagram and function for relating two genetic traits, in this case yield and instability. The shaded region covers the genotypes that are lowest for one trait (yield) at a particular level of the other trait (risk), and are automatically eliminated from further consideration. A function is drawn (sloped line) in order to equate the economic trade-off between the two traits. This function will be drawn tangent to the distribution of genotypes, and the best selection will be made at the point of tangency.

tivities and resources must be appropriately organized in time and space. Organization and control permit timely decision making by reducing information overload. The crop breeder who can quickly classify a problem in terms of subject matter is more likely to solve that problem quickly, reserve it until more data is available, or delegate it to a colleague with appropriate experience. The breeder who can readily visualize the flow of genetic materials toward the breeding objectives can summarize progress, anticipate problems, and make improvements in the methods.

Organization Design

The tasks in a breeding program can be illustrated as in Fig. 3.4. Such a chart should show the major functional areas and their relationships. It will serve as a guide in the delegation of responsibility, in the assignment of work space, and in the filing of information. Details of the organization chart will vary among breeding programs, but some general statements can be made.

The task of reporting and summarizing important progress is crucial to good management. For example, financial records are kept near the hub of most business organizations. In the same sense, critical record-keeping (accession records, hybridization notes, and field books) should be carefully maintained by the crop breeder or by someone under his or her direct supervision. Therefore record-keeping is illustrated in close proximity to program management (Fig. 3.4). The standards for recording key information must initially be decided by the crop breeder. But as the amount of information increases, this task will either become more systematic, evolving into data management, or else will be apportioned to subunits with internal record-keeping for their part of the total operation.

Other areas of work can be diagrammed according to their relationship with one another. In a program requiring greenhouse operations for the hybridization program, these two areas may be allied on the organization chart, and delegated to one person. Depending upon the ease of self-containment, some tasks (such as cytology) could be relatively isolated. By isolating or compartmentalizing tasks it is easier to record progress and other important information.

An organization design based on functional relationships by goals can provide for flexibility and coordination. As a breeding program increases in size, more control is needed, but without eliminating flexibility. With more personnel, rules and hierarchy may be necessary, especially in dealing with exceptional situations. Supervisory relationships can be emphasized (heavy lines, Fig. 3.4) to deal with conflict resolution, planning, and budgeting, but direct contact would be the normal mode of information exchange between and within functional work groups. Some tasks would be coordinated through lateral relationships, e.g., seed storage. The functional organization differs from the organization chart of a large company, since it is decentralized with less emphasis on hierarchy. It should allow individuals to support one another, to agree on actions, and to communicate results, without overloading the crop breeder with information. Some kind of organization design is needed even for a small breeding program because of the complexity of tasks and the large amount of information that must be processed. The manager of the program will use such a chart in personnel and other resource assignments, as well as in filing information.

Communication and Assignment of Responsibility

Management errors often occur in the areas of responsibility and communication. Much has been written about these areas of human relations and personnel management. Unlike the traditional management approach

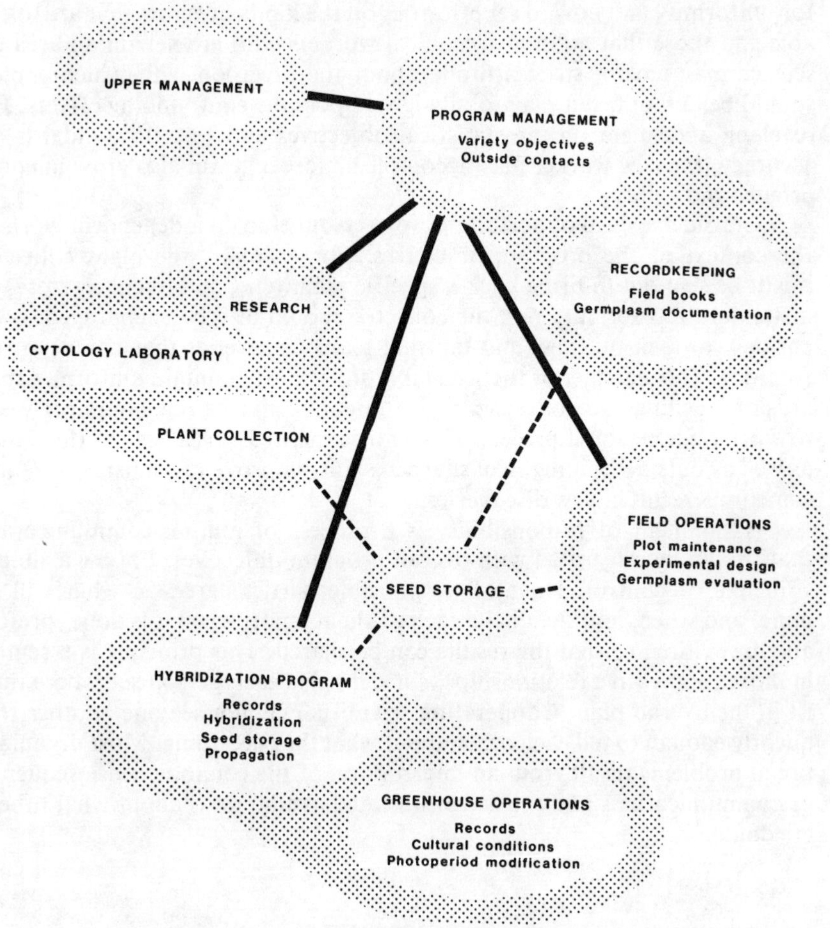

Fig. 3.4. Functional relationships in a crop breeding organization. Functional clusters of activity (shaded regions) are moderately self-contained areas of work and communication. Control relationships (solid lines) are formal supervisory ties. Lateral relationships (dashed lines) involve coordination by a team from several work units of a mutual activity, seed storage in this example.

which considers personnel matters primarily in terms of behavioral processes, I will consider these in the overall context of control and organization. This approach is suggested despite the realization that people have important needs that cannot be treated the same as the other elements of a breeding program.

Leadership styles are observed to vary among individual situations. Some supervisors delegate responsibility in a coercive manner. While this is essential in some situations, it is a form of leadership with little opportunity for organizational growth. In a more open situation, subordinates usually will take an active role in improving the quality of research as they accept responsibility for the achievement of worthwhile goals. For example, to achieve the "local control" needed for good experiments, the person(s) laying out the plots and making the measurements must know the procedure for uniformity and how to set priorities on the kinds of errors that are tolerable and those that are not. Technical workers who are new in an area of science may have a strict attitude about the precision with which a plot should be laid out, but may totally ignore possible confounding effects. By reaching agreement on the statistical objectives and specific standards of accuracy, the new worker may accomplish more and will also grow in competence and ability.

Consider special opportunities for personnel to do independent work in the context of the program objectives. For example, the plant collector might be sent out to bring back a specific plant type, and to collect notes in a standardized format. Yet, the collector should also be encouraged to be curious, to seek out new and unusual plant discoveries that could not be foreseen. The manager of the seed laboratory must maintain uniform standards of labelling and cold storage, yet might be given a few hours per week to work on a personal project of scientific merit. Experiences of this kind, as well as outside training, can sharpen skills, improve job satisfaction, and sometimes result in new discoveries.

Assignment of responsibility is a process of mutual communication that should be integrated with overall program objectives. Follow a simple principle of control: Have all responsible parties agree on what will be done, and when, and then have each of them report when it is done, preferably in writing, so that the results can be shared. This principle is essential in most cooperative relationships, since it helps each person see where they fit in the overall plan. Cooperating scientists may not see one another frequently enough to tell each other about what they are doing. Most organizational problems result from an interruption of the common-sense sequence of communications and the use of unstated assumptions about what others are doing.

Data Management

Plants in nature evolved in a non-directed process. Random mutations were acted upon by selection forces that were variable in magnitude and frequency. The crop breeder can provide direction and accelerate these processes using crop breeding procedures. An integral part of crop breeding is recording information in a format structured for future analysis, summarization, and decision making. Any of a number of good quality record systems can be employed. The principal attributes of good data management are: (1) consistency, (2) accuracy, and (3) immediacy. Computers offer excellent means for achieving these objectives, in part because they force the crop breeder to systematize the parts of the program.

A computerized data system, GENOS, was developed for use on a perennial crop such as turfgrass (Busey, 1978). Perennial crops require evaluations from the same field areas over periods of 2 years or more. These various data sets must be summarized every time a new experiment is planned or a new cycle of hybridization is begun. Because of the clerical problems of getting timely, accurate updates on performance. the minicomputer was chosen to store and assemble information. GENOS is an integrated set of programs, written in BASIC, designed to satisfy various options and needs of the user. Data sets and experiments were titled and indexed. Access to any data set, plot plan, and any of several possible analyses was possible through a simple interactive questionnaire. Special-purpose programs were written to compile and tabulate large data sets for use in numerical taxonomy. Another integrated set of programs was written to store and analyze hybridization records. It was possible to find out which crosses had been made and which were successful, which facilitated the making of as many unique hybrid combinations as possible. The computer system was designed to search all information, and to print out a list of hybrids attempted, hybrids successful, and so forth. It was possible to calculate percent seed set from each of several hundred combinations and to obtain the mean frequency of seeds set from multiple attempts at the same cross.

Well-managed data are especially important in cooperative programs involving scientists from various disciplines. As a breeding program becomes large it is critical that the team members know what other members are doing. A central data base of germplasm records and performance information is an excellent way to unify ongoing efforts. Correlations between resistance data, plant morphology, and origin can be brought together and studied in a systematic fashion. Much more will be accomplished on a planned basis than if analysis is done strictly "after the fact".

Flow Diagrams and Network Analysis

A crop breeding program requires that a sequence of activities be performed according to a schedule. We have already seen how decision points can be diagrammed (Fig. 3.1). Operations should be similarly organized in time so that the breeder does not become confused by the diversity of activities that must be planned and conducted. The flow diagram is ideally suited for visual interpretation of a complex task. Examples of such aids are becoming more commonplace in scientific literature. They range in span of application from a procedure for screening rice for drought tolerance

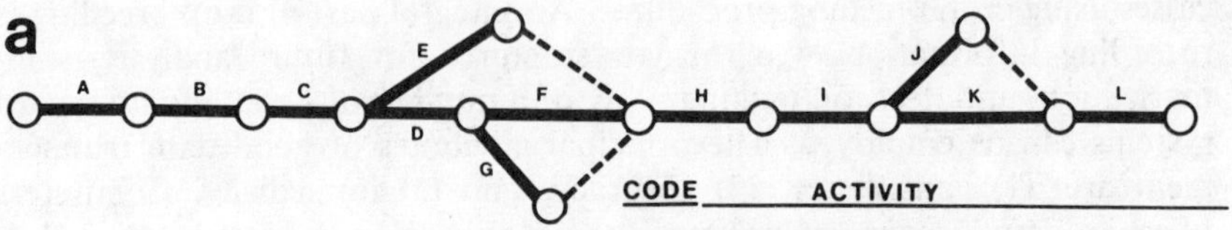

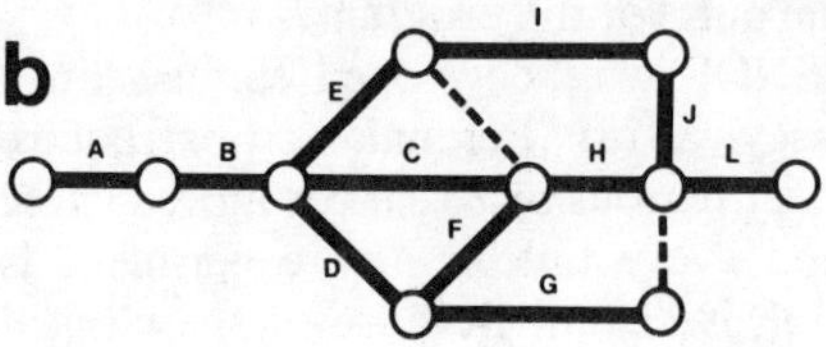

CODE	ACTIVITY
A	Preliminary objectives (1 mo.)
B	Collect germplasm (2 mos.)
C	Preliminary observations (3 mos.)
D	Performance evaluation (1 year)
E	Study cytology, crossability (1 year)
F	Select better parents (1week)
G	Collect more germplasm (6 mos.)
H	Trial hybridization (6 mos.)
I	Progeny trial (1 year)
J	Inheritance evaluation (1 mo.)
K	Collect more germplasm (6 mos.)
L	Design full breeding program (3 mos.)

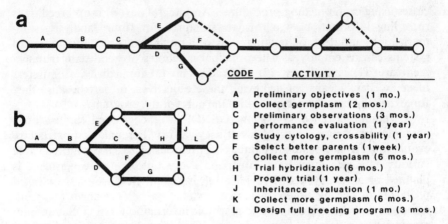

Fig. 3.5. Network diagram of a preliminary breeding program in a crop about which little is known. Plan **a** represents an orderly sequence of activities over an extended time. Plan **b** is compressed into a shorter period, requiring more concurrent operations as well as certain shortcuts.

(O'Toole and Maguling, 1981) to the 6-year pathway by which resistance to two different aphids was increased in red clover (Gorz et al., 1979). The book *Principles of Crop Improvement* (Simmonds, 1979) is extensively illustrated by flow diagrams and is recommended for this and other effective presentations of crop breeding.

In addition to providing ease of visual interpretation, the flow diagram can be analyzed in order to improve an operation. The complexity of breeding activities sometimes involves bottlenecks that delay further progress or restrict the amount of material that can be processed. A crop breeder hybridizing a biennial crop may be concerned with a nearly 2-year delay from seed planted to seed produced. Another breeder may find that spring planting is so delayed that only a small number of progenies can be evaluated each year. Operations such as hybridization and field evaluation can becomc compounded with other restrictions such as recurrent selection which further complicate and slow down a breeding program.

A method to analyze these problems is network analysis, an analysis of interlinked activities. Network analysis begins as a diagram (Fig. 3.5). Interlinked activities form chains, concurrent paths, and branch points. Estimates of time required for activities are used to estimate the total time required for the job, and to make further refinements in the operation. Some activities have "slack", that is, they can be started late without affecting the time required for the total operation. Other activities are "critical" because their delay would delay the entire project. In the diagram (Fig. 3.5a), there is a "critical path" consisting of activities A-B-C-D-G-H-I-K-L, any one of which could delay the project if started late. Network analysis would identify the critical path in order to study whether any activity on it could be speeded up. By redesigning critical paths or using more resources for a critical activity the crop breeder may reduce the total development time for a new cultivar. Time is a crucial problem in the efficiency of breeding programs because of the need to accelerate return on investment and to reduce the total overhead for program maintenance during the development period.

Network analysis was used to propose an acceleration of the preliminary breeding program represented in Fig. 3.5a. A review of activities suggested a possible compression of the program from 51 months to only 31 months (Fig. 3.5b). The critical path is A-B-E-I-J-L. Certain key areas such as the design of the full breeding program (L) depend upon information from cytology, crossability, and small-scale progeny evaluation, all of which require considerable time. By trying to compress those activities into concurrent operations, time may be saved, but some sacrifices may also be necessary. In the compressed version, it was necessary to eliminate step K (the third cycle of germplasm collection). A rigid, compressed schedule for a major program could incur cost overruns and uncertainties, and the risk of discarding germplasm before evaluation is complete. But the timeliness factor may justify the more intensive approach.

Charts of various kinds will help to organize activities in time, improve planning and point out efficient allocation of resources. Annotate your flow diagram with all pertinent information. List the resources and personnel needed and designate job responsibilities. This will be useful for supervision as well. Network analysis can be used to design workable solutions to time problems, to contribute to efficiency by permitting control, and to provide an account of progress. If a nursery was not planted on time, the network analysis should explain why things fell behind. Network analysis gives the crop breeder a feedback mechanism, a way to verify activities and identify problems, as well as an overall conceptual framework in which to consider changes.

EFFICIENCY IN CROP BREEDING

Most of the discussion in this chapter has dealt with fundamental problems of management applied to crop breeding, with examples of aids used by managers in planning and decision making. Crop breeders have searched for solutions to the specialized management problems of field testing, selection, and mating (Burton, 1979; Koch and Rigney, 1951; Soplin et al., 1975). A summary of some of the problems posed by limited resources will be presented here. The objective in each instance will be to encourage the crop breeder to develop the most useful information at the least cost.

Field trials are expensive in person-hours and supplies. Land areas may also be critical. Fortunately, many studies have evaluated efficiency of experimental designs. An excellent review of experimental design in crop breeding is presented by LeClerg (1966). Most of his discussion is devoted to field testing, often the largest single cost in a breeding program. More recent models have integrated probability considerations with cost so that optimum testing plans can be developed.

Unfortunately, many of the options available for reducing error and thereby reducing the costs of quality data have not been developed into a complete systems analysis. Geng and Hills (1978) evaluated the effects of subsampling and number of replications on type I and type II errors. Other approaches might be used to produce quality information at the lowest cost. One could use covariance to reduce the effect of soil heterogeneity (Mak et al., 1978). Burton (1974) showed that phenotypic selection could be based on small, 25-plant groups within a larger population. Although genotypes were not replicated, local control was exercised.

In many cases replication at a single location does not add sufficient information to justify costs in preliminary screenings. In the early stages of a breeding program, differences are expected to be large and may be identified by a few replications. Furthermore, after the majority of entries are eliminated by the first evaluation, a later evaluation can be used to document differences, using more replications. Lyrene (1978) studied selection in sugarcane, a vegetatively propagated crop. In that work it was shown that single-plant selection was effective in isolating superior populations. Even for the most difficult characters, from a selection standpoint, stringent selection based on only one replication resulted in 91% elimination of below average clones. The larger number of plants in seedling populations that could be screened would make rapid genetic improvements possible.

A thorough systems analysis of field testing methods is needed in order to test these concepts on different crops. In general, one should try to use the simplest approach and the fewest replications in field testing that will satisfy research objectives.

Environmental Representation

A crop breeder must determine the geographical area for a proposed new cultivar. A central testing location representative of the production region is necessary. Other locations will be needed for appropriate evaluation tests. Choice of locations can influence the cost of the program and the rate of advance of your populations toward the intended goals. Because differences between genotypes will not be the same at different locations (genotype × environment interaction), multiple locations protect against inappropriate product development and against bad recommendations to growers. In the early stages of a testing program, it has been suggested there be only one replicate per location, with as many locations as possible (LeClerg, 1966). But testing at different locations can be costly, and so an efficient balance of size and number of locations should be designed for the circumstances of the crop and its production region.

Lack of consistency in genotype performance across locations or years provides additional information for the breeder. In addition to justifying the need for additional broad-based testing in different environments, the degree of inconsistency can help predict the variability expected among different farms. If genotype × environment interaction is great, it may even be necessary to partition the cultivar development program with different objectives for different regions and for different weather conditions. Because of the applications of genotype × environment interaction in decision making, the plant breeder should be aware of the techniques for analyzing such relationships.

Stability indices provide a method for comparing the responses of cultivars across a range of environments. A common method is to calculate, for each genotype, the regression coefficient of the yields at various locations, against the mean yield of all genotypes at those locations (Eberhart and Russell, 1966). Analysis of stability has been extended to experiments involving unequal numbers of entries (Pederson et al., 1978) and has been incorporated with multivariate techniques (Tai, 1979).

Despite the many years elapsed since Yates and Cochran (1938) considered regression and other tools for the analysis of groups of experiments, these techniques require careful review as to their value in selection. Ultimately the practical test of genotype by environment interaction is indicated by changes in the rankings (the order of genotypes from best to worst) across locations. If appreciable differences in rankings do not exist, then there is little basis for specific development, testing, or recommendation for specific locations. It has been shown that selection for high stability or tolerance to stress environments "will generally decrease both mean productivity and yield in non-stress environments" (Rosielle and Hamblin, 1981). In upland cotton, selection in a non-stress environment was more effective than selection in a stress environment for improving yield in both environments (Quisenberry et al., 1980). The typical exception in which direct selection for mean productivity should be augmented by selection for tolerance is when a stress factor (disease, insects, or soil, for example) is the primary cause of variability in yield and where genetic variability is greater in the high stress environments.

Evaluation Criteria and Genotype Comparison

There are two problems in selecting within a plant population: (1) to develop and apply appropriate criteria for comparison and (2) to use those comparisons in making decisions. An efficient basis for comparing genotypes should be geared to the specific objectives for a new cultivar. A number of different beneficial traits may be evaluated, but they should later be integrated and simplified. A complicated series of evaluation criteria may cause problems in interpretation, and redundancy wastes effort. Study the correlations among traits, graph them, and try to apply a selection index to the distribution of genotypes. Normally, one particular characteristic, such as economic yield (Table 3.2) will be the most important objective. When this is not appropriate, as in an ornamental crop, a simple economic weighting scheme is needed. Develop an index based on the combined value of various traits to producers and consumers with attention to the expected variabilities of these estimates. If a particular trait shows low heritability, deemphasize its selection, so faster gains can be made for other criteria.

Pest and environmental stress resistance are potentially important aspects of most breeding programs. As the manager of a program you will decide on the emphasis to give to this area. If possible, the degree of emphasis should be based on a realistic budget for a pest resistance screening program, balanced with the monetary benefits of such an effort. In many cases, a better understanding of the relationship between pest resistance in the laboratory and in the field, as well as the relationship of field-level resistance to yield, quality, and risk abatement for the farmer is needed. A systematic approach to incorporate resistance goals should include an assessment of the permanence of resistance (Gould, 1978) and a comparison of the cost and benefits of direct vs. indirect selection.

In the early 1900's, cultivars were selected for yield and quality, with no knowledge of specific environmental stress response. Years later, crop physiologists showed that those cultivars had been unwittingly selected for resistance to an environmental stress. How much more efficient would those breeding programs have been if a crop physiologist had been hired on the team initially? Unfortunately, the crop breeder is frequently not aware of environmental stress factors at the time a program is initiated. Yield evaluations under representative conditions of disease and insect infestation and environmental stresses that affect the production region may be necessary. If these factors are chronic problems, yield improvement may also improve the resistance of the crop. If, however, pest and environmental stresses are infrequent but acute and damaging, an evaluation of risk is appropriate before decisions are made on breeding for resistance.

Genotype selection methods vary with the breeding system of a crop, the time span of the breeding program, and the power of a particular test. The crop breeder simply saves the best plants and throws away the worst. The number to discard at each stage will vary. Discarding too few materials will impede the progress of the program; saving too few materials will risk reducing genetic variation and impede long-range progress. In the earlier stages of population improvement, an optimal selection intensity can be chosen based on its interrelationship with the breeding method.

Breeding Method

Breeding methods have been studied from a quantitative viewpoint, in order to allocate resources efficiently. Models of genetic variation are somewhat predictive but generally entail uncertainties regarding the effects of epistasis, crossing over, and spontaneous mutation. Published reports are available on the improvements realized from various breeding methods, but reports of failures may not be published regularly. Controlled comparisons are the best source of information on rates of genetic improvement obtainable from several breeding methods.

In an excellent review of breeding methods, Hallauer (1981) showed that additive genetic effects predominate in crop breeding populations and that all cyclical breeding methods select for them. The choice of method often had no effect on rate of improvement. The appropriate method would depend on how well the method complements other aspects of the breeding program. Early failure in ear-to-row selection for maize yield can be attributed to the design of field plot procedures rather than inadequate additive genetic variation. Substantial improvements in the efficiency of the ear-to-row method were made by modifying the testing procedures (Webel and Lonnquist, 1967). In an analysis of allocation of resources, Hill and Byers (1979) showed that modified plant-to-row procedures would be most efficient in increasing resistance in alfalfa to the leafminer. Polycross selection methods, although long in vogue for forage crops are often based on too few entries or evaluation in too few environments to permit effective selection. For many crops, composite breeding methods appear to offer more promise than pedigree selection. The composite breeding method makes it possible to widen the germplasm base and to improve economy of operations although individual genotype identities may be lost (Jensen, 1978).

APPROPRIATE TECHNOLOGY

The application of heredity and statistics is the foundation for crop breeding methods, but this scarcely describes what crop breeders actually do. Knowledge of heritabilities, genetic variances, and the breeding system for a particular crop are tremendous assets in determining population sizes and numbers of replications, and in optimizing operations. Given the right information and the right program, a computer can tell you how many individuals are needed per progeny and how many progenies to select. Taken to the extreme, crop breeding may be viewed as a numbers game. But in reality numbers may be the easiest part of the job. Despite what the crisp equations may predict, it is essential to use whatever breeding method is tactically feasible for a particular crop. Quantitative methods are ideal for solving quantitative problems, but the crop breeder must also consider the quality of information, the representativeness of the test sites, the appropriateness of selection goals, and other options for which no scientific test is available. It is not a question of whether or not to use well-organized technology in crop breeding. Rather, we must consider in which situations technology will assist in making decisions, in controlling, directing, and organizing genetic resources. As the crop breeder learns to shift more responsibility to systematic methods of operation and decision making, the overall task may not become easier. Instead, the crop breeder's attention needs to be directed to planning, reviewing progress, and improving compatibility between selection criteria and future needs.

Jensen (1967) argues for organized problem solving, and the use of a systems analysis in developing better plants for the future. It may take 10 to 15 or more years to solve a problem through breeding. Why not tailor the potential new cultivar to the most likely contingency in the future? We know that fossil energy will be more limited and more expensive. Cultivars will be needed that respond well at low fertility levels, or that produce very efficiently in response to high fertility. Pest resistant plants will be needed as an alternative to pesticide use. Plants will be needed that are competitive with weeds in low-tillage situations. Future cultivars more like those our ancestors used may be needed, with many farmers living on small farms with little access to modern processing and handling facilities. Vast agricultural areas may be abandoned, for example, the irrigated deserts of the southwestern USA. The shortgrass prairies of the Great Plains may be replanted to improved range grasses instead of wheat. And as all non-renewable resources become more limiting, tomorrow's society will be looking to the crop breeder for better ways to capture that ultimate resource—the sun. These are just a few of the possible external factors that could determine the appropriateness of breeding technology

At the same time the crop breeder is confronted with new and changing ideas for diversifying and selecting germplasm. Some early proponents of genetic engineering suggested that recombinant DNA technology would be used to introduce nitrogen-fixing genes into the maize plant. How does the crop breeder respond to such challenging concepts? As if crop breeding were not already a high-risk undertaking, many of today's investors expect test tube miracles. It is thus more important for crop breeders to know the technological context of crop breeding, as well as goals, costs, and risks. Questions may be asked about the application of new technologies. It is important for crop breeders to have an overall plan in order to calculate how new technologies might or might not yield a return.

In the last years of the 20th century, scientists are being asked to be increasingly accountable. Fewer resources are being invested in research and development for the future. Crop breeders must make those commitments count for the farmer and for the consumer. The benefits of better plants have traditionally justified the work of crop breeders. One study reported returns of 300 times the amount of the investment for research to incorporate resistance to diseases (Pardee and Gracen, 1976). Agricultural research has provided generally high rates of return on investment, on the order of 50% annually (Evenson et al., 1979).

Crop breeding has contributed the major part of the yield advance of maize since 1930, estimated to be 60 to 80% of the total yield advance from all research (Sprague et al., 1980). Crop breeding gains have not reached a plateau, not even for the major crops that have been under development for 8,000 years. With increased knowledge, and more organized activity, rapid gains can be made in many crops. But can these efforts be justified in the future? With inflation raging in the world economy, who will underwrite a 10 or 20-year investment in a better cultivar? Thus, in dealing with technological problems, management may be considered a connection to link the crop breeder with the needs of tomorrow. It is clearly not a question of quantity of grain per hectare, but of the quality of crop breeders' solutions and the adequacy of return on investment in their work.

LITERATURE CITED

Burton, G. W. 1974. Recurrent restricted phenotypic selection increases forage yields of Pensacola bahiagrass. Crop Sci. 14:831–835.

———. 1979. Handling cross-pollinated germplasm efficiently. Crop Sci. 19:685–690.

Busey, P. 1978. GENOS: A data system for plant breeders. Agron. Abstr. p. 108.

Eberhart, S. A., and W. A. Russell. 1966. Stability parameters for comparing varieties. Crop Sci. 6:36–40.

Evenson, R. E., P. E. Waggoner, and V. W. Ruttan. 1979. Economic benefits from research: An example from agriculture. Science 205:1101–1107.

Gannon, M. J. 1977. Management: An organizational perspective. Little, Brown and Company, Boston.

Geng, S., and F. J. Hills. 1978. A procedure for determining numbers of experimental and sampling units. Agron. J. 70:441–444.

Gorz, H. J., G. R. Manglitz, and F. A. Haskins. 1979. Selection for yellow clover aphid and pea aphid resistance in red clover. Crop Sci. 19:257–260.

Gould, F. 1978. Predicting the future resistance of crop varieties to pest populations: A case study of mites and cucumbers. Environ. Entomol. 7:622–626.

Hallauer, A. R. 1981. Selection and breeding methods. Ch. 1. *In* Frey, K. J. (ed.) Plant Breeding II. Iowa State University Press, Ames.

Hill, R. R., Jr., and R. A. Byers. 1979. Allocation of resources in selection for resistance to alfalfa blotch leafminer in alfalfa. Crop Sci. 19:253–257.

Jensen, N. F. 1967. Agrobiology: Specialization or systems analysis? Science 157: 1405–1409.

———. 1978. Composite breeding methods and the DSM system in cereals. Crop Sci. 18:622–626.

Koch, E. J., and J. A. Rigney. 1951. A method of estimating optimum plot size from experimental data. Agron. J. 43:17–21.

LeClerg, E. L. 1966. Significance of experimental design in plant breeding. Ch. 7. *In* Frey, K. J. (ed.) Plant Breeding. Iowa State University Press, Ames.

Lyrene, P. M. 1978. A simulated selection experiment in sugarcane. Crop Sci. 18: 971–974.

Mak, C., B. L. Harvey, and J. D. Berdahl. 1978. An evaluation of control plots and moving means for error control in barley nurseries. Crop Sci. 18:870–873.

O'Toole, J. C., and M. A. Maguling. 1981. Greenhouse selection for drought resistance in rice. Crop Sci. 21:325–327.

Pardee, W. D., and V. E. Gracen. 1976. Breeding resistance into crop plants. Crops Soils 28(8):15–16.

Pedersen, A. R., E. H. Everson, and J. E. Grafius. 1978. The gene pool concept as a basis for cultivar selection and recommendation. Crop Sci. 18:883–886.

Quisenberry, J. E., B. Roark, D. W. Fryrear, and R. J. Kohel. 1980. Effectiveness of selection in upland cotton in stress environments. Crop Sci. 20:450–453.

Rosielle, A. A., and J. Hamblin. 1981. Theoretical aspects of selection for yield in stress and non-stress environments. Crop Sci. 21:943–946.

Simmonds, N. W. 1979. Principles of crop improvement. Longman. London.

Soplin, H., H. D. Gross, and J. O. Rawlings. 1975. Optimum size of sampling unit to estimate Coastal bermudagrass yield. Agron. J. 67:533–537.

Sprague, G. F., D. E. Alexander, and J. W. Dudley. 1980. Plant breeding and genetic engineering: A perspective. BioScience 30:17–21.

Tai, G. C. C. 1979. Analysis of genotype-environment interactions of potato yield. Crop Sci. 19:434–438.

Webel, O. D., and J. H. Lonnquist. 1967. An evaluation of modified ear-to-row selection in a population of corn (*Zea mays* L.). Crop Sci. 7:651–655.

Yates, F., and W. G. Cochran. 1938. The analysis of groups of experiments. J. Agric. Sci. 28:556–580.

SUGGESTED READING

Morris, W. T. 1977. Decision analysis. Grid, Inc. Columbus.

Pendleton, J. W. 1974. Philosophical and practical considerations in field research. J. Agron. Educ. 3:58–65.

Render, B., and R. M. Stair, Jr. 1978. Management: A self-correcting approach. Allyn and Bacon, Inc. Boston.

Tversky, A., and D. Kahneman. 1981. The framing of decisions and the psychology of choice. Science 211:453–458.

Chapter 4

Plant Population Management and Breeding[1]

K. J. FREY[2]

Iowa State University
Ames, Iowa

Plant breeders must be concerned with populations of plants from two different points of view. In the first instance, there is the cultivar that is used in crop production. In reality, a cultivar of a crop growing in a field for agricultural production is a population of plants growing together in a community with individual plants interacting with one another at all stages of growth. In the second instance, genetically heterogeneous populations of plants are used with a number of plant breeding methods. These methods are referred to as "population breeding" methods.

CULTIVAR TYPES

Types of plant cultivars are: (1) pure lines, (2) hybrids, (3) multilines, and (4) synthetics.

A breeder or agriculturalist is interested in these kinds of populations for (1) total production per hectare and (2) response and stability of production over environments.

Basically, four mechanisms in plants and plant populations affect the response and stability of production of a cultivar across variable environments. They are: (1) phenotypic plasticity, (2) heterozygosity, (3) heterogeneity, and (4) polyploidy. Mechanisms 1, 2, and 4 are features of a genotype, whereas mechanism 3 is a feature of populations. A given cultivar may have only one mechanism, i.e., a pure-line cultivar of barley, or it may have all four, i.e., a cultivar of alfalfa. In this chapter I assess different cultivar types in terms of productive capacity and production response and stability.

[1] Journal Paper No. J-9873 of the Iowa Agric. and Home Economics Exp. Stn., Ames, Iowa. Project 1752.

[2] C. F. Curtiss distinguished professor in Agriculture, Iowa State Univ., Ames, IA 50011.

Pure Lines

A pure line, from the practical viewpoint, is the bulk progeny that originates from a homozygous or near-homozygous plant. Pure-line cultivars were tried for both outcrossing and self-pollinating crop species shortly after the rediscovery of Mendel's Laws. As agricultural cultivars for outcrossing species, such as maize, pure lines were unsuccessful because inbreeding depression was deleterious to plant vigor.

Pure-line cultivars have proved useful in agriculture, especially where certain desirable traits, such as the quality factors that affect consumer acceptance, are demanded. Such traits usually are not understood genetically; thus, when desirable traits are achieved in a genotype, plant breeders are reluctant to break up the unique genetic combination.

How pure is a pure line? Usually, breeders assume that a pure-line cultivar is somewhat heterogeneous. Briggs and Allard (1953), when discussing the backcross method, suggested that many plants from the recurrent parent pure line should be used as pollen parents in each generation to insure that the backcross-derived cultivar will have the same "cryptic heterogeneity" as the recurrent parent. Wheat and Frey (1961) analyzed three cultivars of hexaploid oats for heterogeneity and found only one plant progeny that differed from the cultivar mean, and no cultivar that showed a significant interaction for individual plant progenies with environments. So this study does not support the hypothesis that genetic heterogeneity within pure-line cultivars of oats contributes to production stability. Frey and Chandhanamutta (1975) showed, however, that mutation rates for di, tetra, and hexaploid oat cultivars were 1.20, 0.51, and 0.32, respectively, per 100 gametes, but no mutant affected yield. Their studies show that pure-line cultivars of oats cannot be kept genetically pure because there is such a high mutation rate. Thus, what plant breeders call pure-line cultivars of self-pollinating species probably are not really pure lines but rather they are "near" pure lines.

Phenotypic Plasticity

Even though pure-line cultivars have some impurity, they are dependent almost exclusively upon phenotypic plasticity for production response and stability across environments. This is the physiological resilience that a genotype displays to environmental stimuli. Phenotypic plasticity will be illustrated with oats. Three cultivars of oats, 'Marion', 'Clintland', and 'Bond,' were evaluated for grain yield at four nitrogen levels, 0, 20, 40, and 80 kg/ha (Fig. 4.1). Bond showed little yield increase at any level of N fertilization, whereas Clintland and Marion increased sharply at each higher fertilization level. Responses of these three cultivars in panicles per plant, seeds per panicle, and seed weight (the three components of grain yield) showed interesting patterns of phenotypic plasticity. For panicles per plant (i.e., tillering), Bond and Clintland did not respond to N fertilization, but Marion did. For seeds per panicle, Clintland and Marion both increased more than 100%. For seed weight, no cultivar responded.

The different phenotypic plasticities of Bond, Clintland, and Marion can be explained by their respective genetic capacities to respond to N fertilization during the three stages of growth and differentiation for oats. Tiller production is determined within 14 days after seedling emergence, so this represents the first critical period during which differential genetic capacities of genotypes to environmental stimuli are expressed. The second critical period involves panicle, spikelet, and floret differentiation, and this period, which lasts for 20 days (from day 21 to day 42 in the oat growth cycle), determines the maximum number of seeds that a panicle can bear. The third critical period is during grain filling when weight per seed is determined. Described in terms of their genetic capacities for response to N fertilization: (1) Bond does not have the ability to respond to N in any stage of growth, (2) Clintland has the genetic capacity to respond in stage 2 (i.e., panicle, spikelet, and floret differentiation), and (3) Marion has the genetic capacity to respond in both stages 1 and 2 (i.e., tillering and panicle, spikelet and floret differentiation).

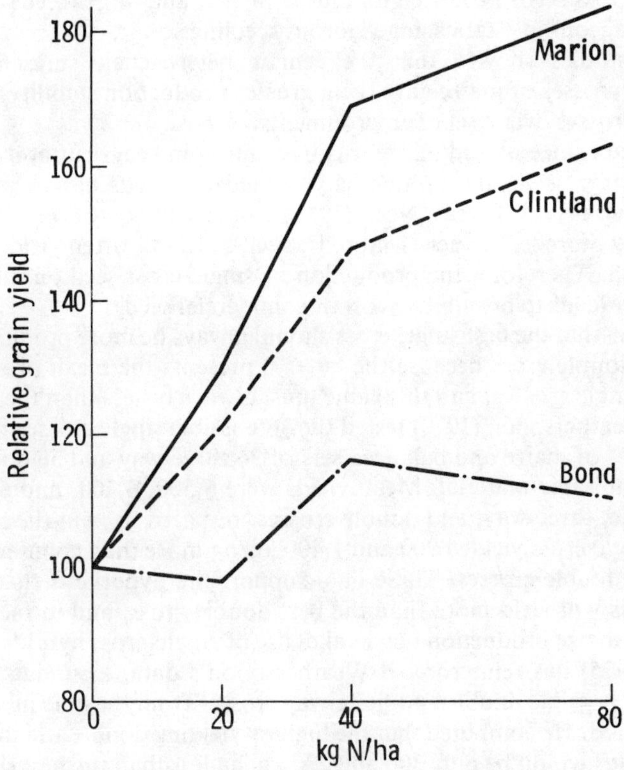

Fig. 4.1. Relative grain yields for three oat cultivars tested at four rates of nitrogen fertilization (Frey, 1959).

Hybrids

Hybrid cultivars are derived by mating unlike genotypes. This type of cultivar generally is more productive than the parents in the first generation, but productivity is reduced in subsequent generations. Yield declines because of the way in which the breeding system is manipulated.

Neal (1935) grew the first and second generations of single, three-way, and double crosses of maize and found that the reductions in yield in these hybrid types closely paralleled their reductions in heterozygosity. Heterozygosity reductions were 50, 33, and 25% for single, three-way, and double crosses, respectively, and the reductions in yield were 48, 37, and 26%, respectively.

When hybrid maize was first used in agricultural production, double crosses were employed for three reasons:

1. Seed production on inbred plants made the cost of single-cross seed prohibitive. By producing double crosses from highly productive single crosses, the cost of seed was acceptable.
2. Double-cross seeds produced on single-cross plants were larger than single-cross seeds produced on inbred plants, and larger seeds were advantageous in establishing vigorous seedlings.
3. Early studies showed that the genetic heterogeneity present in double crosses of maize gave them greater production stability than single crosses over a set of environments.

Over the 40 years since hybrid maize was introduced into agricultural production on a large scale, the productivity of maize inbreds has been improved substantially. Whereas Neal (1935) got 1,200 kg of grain per hectare on early inbreds, the newer inbreds, such as 'B73', often yield four times that much. Therefore, the production of single-cross seed on inbred plants no longer leads to prohibitive cost of commercial seed.

Theory says that the best single cross should always be more productive than the best double cross because the latter represents the mean productivity of four single crosses, and this value must always be less than the best single cross. Weatherspoon (1970) tested the 36 possible single crosses from nine inbred lines of maize and balanced sets of 36 three-way and 36 double crosses from the same material. Mean yields were 6,500, 6,200, and 6,000 kg/ha for single, three-way, and double crosses, respectively, but the highest-yielding single cross yielded 900 and 1,400 kg/ha more than counterpart three-way and double crosses. These data support the hypothesis that the best single cross will yield more than the best double cross, and in fact, in the USA, most maize production now makes use of single-cross hybrids.

Schnell (1975) has reinterpreted Weatherspoon's data, assuming that the yields of all possible double and three-way crosses from the nine inbreds could be predicted. He computed that the highest-yielding double and three-way cross hybrids would be only 300 and 200 kg/ha less than the best single cross. On the basis of these computations, the best single crosses would not be expected to yield much better than the best three-way or double cross.

Stability of Hybrids

A supposed advantage for double-cross hybrids is their intracultivar genetic heterogeneity, which would give them greater stability of production over environments. When Sprague and Federer (1951) combined hybrid data across testing sites, single crosses had a variance component for "hybrids × environments" three times larger than the comparable component for double crosses. When combined across years, the single-cross variance component was 4.5 times larger than the hybrids × environments value for double crosses. These results were interpreted to show that double crosses were more stable (i.e., have lower interactions) than single crosses when grown across variable environments.

The Sprague and Federer (1951) results on the relative stability of single- and double-cross hybrids of maize have not fit with the advances in agricultural practice, however. Single crosses now account for most maize production, and these commercially acceptable single crosses have stability of production across environments comparable to that of double crosses. Eberhart and Russell (1969) reinvestigated the subject, using an analysis developed by Finlay and Wilkinson (1963) and themselves (Eberhart and Russell, 1966). The regression analysis gives two statistics to measure response and stability reactions: (1) the regression value, b, which measures the degree to which yields of an individual hybrid are proportional to the inherent productivities of the test environments and (2) the degree to which yields in individual environments deviate from the regression lines. They found that a number of single crosses had higher yields than the best double cross and that they were more responsive than double crosses to improved environments. Furthermore, two single crosses had deviation mean squares as low as the best double cross. For all 45 single and all 45 double crosses, mean yields in low-, medium-, and high-productivity environments and the means for b-values were nearly identical, whereas the means for deviations mean square were 55 for single crosses and only 34 for double crosses. The ranges for deviation mean squares, however, were 12 to 226 for single crosses and 16 to 84 for double crosses. The analysis showed that although on the average double-cross hybrids of maize tend to be more stable than single crosses over a series of environments, certain single crosses are just as stable as the most stable double cross. Presumably, the greater average stability of double crosses over single crosses may be attributed to the genetic heterogeneity present in the double-cross cultivars.

Hybrid cultivars also may be more stable than pure-line cultivars because of heterozygosis. This mechanism was illustrated in the field by Adams and Shank (1959). They created populations of maize with eight levels of heterozygosity, from none to complete. These were tested in four environments, and intraplot standard deviations were used to compute coefficients of variations (CV). For ear height and ear weight, they found a general decline in CV as the heterozygosity increased, showing that maize single crosses possess two of the stability mechanisms, phenotypic plasticity and heterozygosity, whereas double crosses possess heterogeneity in addition to these two.

Multilines

A multiline is a cultivar type used primarily for self-pollinating species. It is a mixture of isolines or near isolines, i.e., identical genotypes that differ for a single gene or a defined set of genes. To date multiline cultivars have been composed of near-isolines, each of which carries a unique and distinct source of pest resistance, and they have been used specifically to enhance control of pests through genetic resistance. It is conceivable, however, that multiline cultivars could be useful for coping with other environmental stresses as well.

Natural populations and landraces of most crop species usually are heterogeneous for disease resistance as a result of the co-evolution of hosts and pathogens. Plant selectionists and plant breeders, in attempts to improve the host-resistance system, carried out a stepwise homogenization of host resistance until one resistance gene would be used in cultivars grown on a large portion of the area of that crop over a whole continent. To provide agronomic uniformity and heterogeneity for disease resistance in a common cultivar, Jensen (1952) proposed a multiline cultivar.

Isolines, to be used in a multiline cultivar, are developed via backcrossing (Fig. 4.2). The recurrent parent is crossed to a number of donor lines, each of which possesses a unique and useful gene for disease resistance. For each line of descent (i.e., involving one donor and one recurrent parent), the crossing program is carried through the original cross and five backcrosses. The Bc_5F_3 progenies that are homogeneous for resistance and conform to the agronomic type of the recurrent parent are bulked to form one isoline. Donor parents used as sources of resistance genes may be other adapted cultivars, unadapted cultivars, or even wild or weedy related species.

Isoline Composition

Multiline cultivars of wheat with stem rust (*Puccinia graminis tritici*) resistance were developed by Borlaug (1959) in Mexico. Multiline cultivars of oats were developed and released in Iowa in 1968 and they have been highly successful in controlling crown rust disease (*Puccinia coronata avenae*). To determine the isoline composition of a specific multiline cultivar of oats, three sets of information are used (Frey et al., 1977):

1. Agronomic performance of isolines. Crown rust epidemics do not develop every year, and consequently, a multiline cultivar must have a good yield capacity in rust-free as well as in rust-prevalent environments.
2. Adult-plant reactions of the isolines to specific crown-rust races. All isolines are assayed for reactions in single-race crown-rust nurseries in the field.
3. Trends in race composition of the crown-rust spore population. These data are collected by the USDA.

Multiline cultivars of a crop aid in controlling a disease in two ways:

1. Patterns of virulence genes in the pathogen population should be stabilized by multiline cultivars. Many races of the pathogen have an opportunity to survive. If the race structure of the pathogen is stabilized, the genes for resistance should retain their value in protecting the crop for many years. Multiline cultivars of oats have been produced in Iowa for 13 years with no detectable increase in susceptibility to natural populations of the crown-rust pathogen.
2. Multiline cultivars delay intrafield buildup of the pathogen. This effect is called "population resistance."

Seemingly, multiline oat cultivars reduce disease development of crown rust by "spore trapping." That is, in the initial spore shower that inoculates a field of oats, a spore that lands on a resistant plant cannot reproduce; likewise, progeny spores from the initial infections on susceptible plants of a multiline cultivar cannot infect and reproduce if they land on resistant plants. This mechanism reduces the "effective" inoculum load in each generation.

Multiline cultivars of agricultural crops are designed specifically to give disease control. They are most effective for those pathogens that are airborne, possess physiological races, and are explosive in reproduction. The extent to which multiline cultivars are useful in a given crop production area will depend upon the frequency and severity of disease.

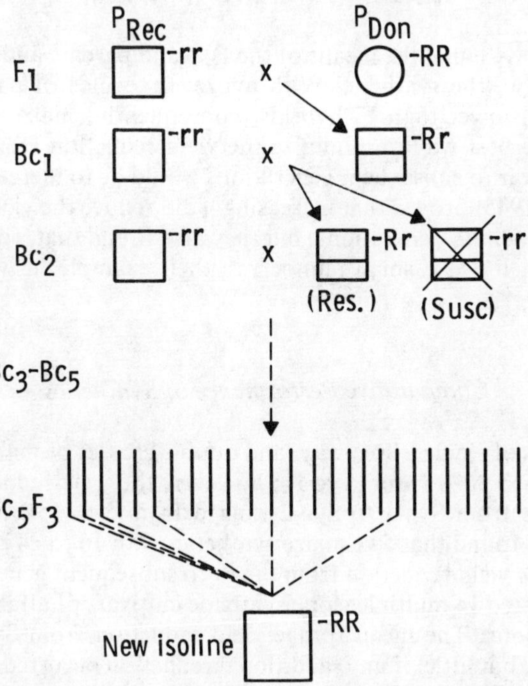

Fig. 4.2. Backcrossing procedure used to develop an isoline. P_{rec} = recurrent parent; P_{don} = donor parent; R = resistant, r = susceptible.

Synthetics

Synthetics are used as cultivars in outcrossing species of crop plants and as gene pools in breeding programs. A famous synthetic that has been used as a breeding population is the Iowa Stiff Stalk Synthetic (BSSS) of maize. The major difference between a natural and a synthetic population is that the latter is composed of selected parental stocks.

Synthetic cultivars of maize are used for grain production in many countries of the world. Most improved forage crop cultivars are synthetics as well. Synthetic cultivars have several desirable features:

1. Yield reduction in advanced generations is less than with a single or double cross, so it is unnecessary to obtain new seed of the cultivar for each production season.
2. By producing successive generations in the habitat of use, the synthetic cultivar may become better adapted to the local environment over time.
3. The cultivar is heterogeneous, which provides for production stability over environments.

The formula for calculating the expected yield of the F_2 generations of a synthetic cultivar is:

$$F_2 \text{ yield} = \overline{X}_{F_1} - [(\overline{X}_{F_1} - \overline{X}_P)/n],$$

where $\overline{X}_{F_1}$ and $\overline{X}_P$ equal the means of the F_1's and parents and n is the number of parents in the synthetic. With maize, the yields of parental inbred lines are much lower than F_1 hybrids; consequently, maize breeders suggested that the best way to minimize the yield reduction from the F_1 of a synthetic cultivar to subsequent generations would be to increase n. Kinman and Sprague (1945) proved that increasing n did reduce the yield differential between the F_1 and F_2 generations, but they also found that, for a maximum yield in thc F_2, n was a small number; in their example, it was five or six maize inbreds.

Comparative Performance of Synthetics

The yields of single, three-way, and double crosses of maize showed reductions of 15 to 30% from F_1 to F_2; however, the yield reductions of synthetic cultivars from Syn-1 to Syn-2 ranged from 5 to 15%. Sprague and Jenkins (1943) found that five maize synthetics with 16 to 24 parent inbreds each gave a 5% yield reduction from the F_1 to subsequent generations. Kehr et al. (1961) tested 14 multiple clone synthetic cultivars of alfalfa in Syn-1 to Syn-4 generations. The mean forage yield reduction from Syn-1 to Syn-2 was about 3%, but little, if any, additional reduction occurred through Syn-4. With five comparable single crosses (i.e., two-clone crosses), the yield reduction was 4 to 5%.

Heterozygosity in Synthetics

The formula for computing the expected productivity of the F_2 generation of a synthetic cultivar assumes:

(1) The species has diploid reproduction

(2) The parents are inbred.

Therefore, this formula, which has proved to be accurate for all maize synthetics, is not appropriate for individual synthetics of forage species that are polyploid and obligate outcrossers. For example, Kehr et al. (1961) found two single crosses of alfalfa that gave reduced yields between Syn-1 and subsequent generations, but one, C17 × C19, gave increased yields. Not only did the yield of the latter cultivar increase, but the increase was gradual and continued through Syn-4. Dudley (1964) was able to explain these opposing yield trends in alfalfa synthetics as being due to the degree of inbreeding to which the parent clone progenies had been carried. This is illustrated in Table 4.1. The percentages of heterozygosity at random mating equilibrium in the five mating types varied from 41 to 88. If yield is a function of heterozygosis, all five F_1 hybrids would regress in yield with advancing generations if the original clones or the S_1 or S_2 clonal progenies were used as parents; however, if highly inbred progenies were used as parents, the yield would increase. Another interesting feature is the effect that allele distribution among the parents can have on heterozygosity. For example, matings $A^2 \times A^0$ and $A^1 \times A^1$ each carry 2A and 6a alleles, but the degree of heterozygosity in the F_1 generation is always higher in the $A^2 \times A^0$ cross, irrespective of the clonal progeny generation used to make the cross.

Because a synthetic cultivar is genetically heterogeneous, natural and artificial selection can modify its genotypic structure. Lonnquist and McGill (1956) practiced selection for increased plant vigor in the Syn-2 of a maize synthetic and obtained an 8 to 11% increase in yield in the Syn-3 and Syn-5 generations.

Table 4.1. Individuals with heterozygous genotypes in the F_1 generation after parents have undergone different degrees of inbreeding, and in random mating populations of an autotetraploid species.†

Original genotypes	Selfed generation of parents when crossed				Random mating
	S_0	S_1	S_2	S_∞	
	%				
$A^4 \times A^3$	50	46	42	25	41
$A^3 \times A^3$	75	71	66	38	68
$A^2 \times A^2$	94	90	86	50	88
$A^2 \times A^0$	83	78	73	50	68
$A^1 \times A^1$	75	71	66	38	68

† Dudley (1964).

‡ A^4 = quadriplex, A^3 = triplex, A^2 = duplex, A^1 = simplex, A^0 = nulliplex.

Table 4.2. Tall, medium, and short alfalfa plants found in seed sources of Ranger grown in different regions.†

Origin of seed	Class of seed	Percent of population‡		
		Tall	Medium	Short
Nebraska	Foundation	14	38	48
Arizona	Certified-1	17	39	44
Arizona	Certified-2	41	36	23

† Canode (1958).
‡ Tall > 14 cm, Short < 6 cm, Medium = 6 to 14 cm.

Natural Selection in Synthetics

The effect of natural selection on synthetic cultivars is of much concern in crops like alfalfa because the seed crop is grown in the western USA which has a different environment than the Midwest where alfalfa is used for forage production. Of primary concern is the potential for losing winterhardiness because the seed production areas have mild winters. Canode (1958) found a significant shift in proportions of tall and short seedlings in 'Ranger' alfalfa when the seed crop was grown in Arizona (Table 4.2). Second generation (certified-2) seed from Arizona had three times as many tall plants and only half as many short ones as did the foundation seed lot from Nebraska. Tall Ranger plants had low winterhardiness, so any shift to tall plants in seed lots grown in Arizona would be detrimental to using the seed in the Midwest where winterhardiness is of paramount importance.

In Indiana, Bula et al. (1969) obtained survival ratings on stands of 'Dollard' red clover cultivar from the original and second- and third-generation seed lots from Prosser, Washington, and Tehachapi and Shafter, California. Survival ratings for second-generation seed lots were not changed significantly in any production environment, but stands from third-generation seed lots from Prosser had increased winterhardiness and those from Shafter had reduced it.

Thus, natural selection can modify synthetic cultivars of plants. Such modification may be beneficial when the seed-production and crop-use environments coincide, but it can be detrimental if they do not. Because seed of most perennial forage species used in the northern agricultural production areas is produced in the western and southwestern parts of the USA, elaborate regulations and procedures have been recommended relative to seed production of these crops. For example, it is recommended that all fields used for producing certified seed to be used in the Midwest should be sown with foundation seed from a midwestern source. And, certified seed should be harvested from an original stand for no longer than 3 years because volunteer plants from shattered seed may cause a shift in the original cultivar.

Synthetic cultivars of outcrossing crops generally are slightly less productive than narrow-base hybrids, but their reduction in yield from inbreeding depression is less. Farmers can use their own seed of Syn-2 and subsequent generations of synthetic cultivars because the advanced generations show only small yield reductions from Syn-1. Natural selection can and often does change the genotypic composition of synthetic cultivars. In some cases, special procedures need to be invoked to preclude such changes.

METHODS FOR POPULATION DEVELOPMENT

Plant breeding consists of three phases: (1) creation of genotypic variation in a population of plants, (2) selection for genotypes that possess the desired combination of genes, and (3) release of the best cultivars for agricultural production.

Creating Genotypic Variation

Genotypic variation in natural breeding populations of plants originates from migration, mutation, and recombination (Falconer, 1960). Genotypic variation available in the artificial breeding populations created by plant breeders have the same sources. Migration corresponds to what plant breeders call "introduction of natural variation"; mutation occurs with a low, but consistent, frequency in artificial breeding populations; and recombination of genes is promoted by crossing.

Natural genotypic variation (including mutation) was the primary source of variability among plants when artificial selection was practiced in pre-Mendelian times. Famous plant selectors—like Vilmorin of France, Nilsson-Ehle of Sweden, Hays, Pringle, and Beal of the USA, and Farrar of Australia—were successful in obtaining new crop cultivars by selecting among natural variants (USDA, 1936).

With the rediscovery of Mendel's Laws in 1900 and the understanding they brought about the particulate nature of "genes," there was a gradual move in the period 1900–1920 to artificial hybridization for creating combinations of genes not available in naturally occurring populations. Most hybridizations were single crosses, which permitted the recombination of genes from two parents. With limited numbers of hybridizations and samples of segregates per hybridization, the pedigree method of plant breeding was used as a selection procedure.

Composite Crosses

In the late 1920's, Harlan et al. (1940) developed a concept of composite crosses for the use of multiparent crosses. This concept changed the breeding methodologies for self-fertilizing species. The composite cross (CC) hybridization scheme is illustrated in Fig. 4.3. This is a mating design whereby 16 or 32 parents are crossed in successive generations into single crosses, double crosses, octuple crosses, etc., until the final hybrid involves all parents. A somewhat similar objective was met by Sprague (1946), who

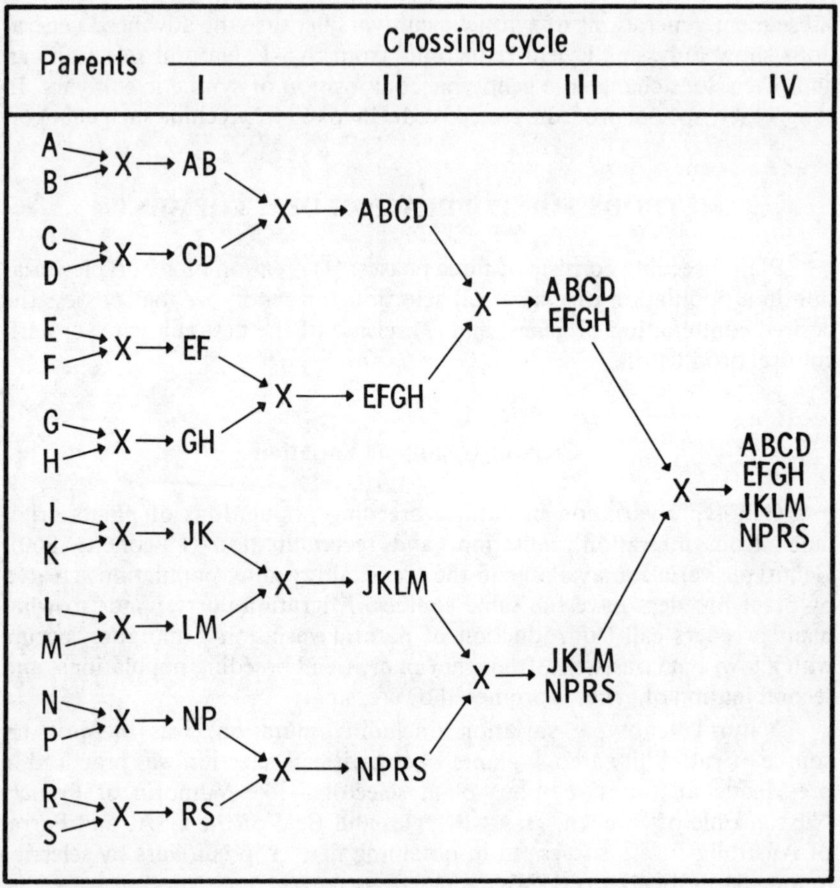

Fig. 4.3. Diagram of a multiple cross involving 16 parents (Harlan et al., 1940).

created the Iowa Stiff Stalk Synthetic (BSSS) of maize, except that, because maize is an outcrossing species, BSSS was simply allowed to random mate in isolation after all single crosses were included in the synthetic. The CC or synthetic (Syn) provided the first purposeful attempt to maximize multiple-parent recombination of genes, especially for selfing species. The physical limitation on making crosses actually places a restriction on how much genetic recombination will take place in a finite population. For example, Ackerman and MacKey (1948), assuming a CC of 16 parents with each carrying a unique desirable gene, calculated the number of hybrid seeds needed in each generation of the crossing program to insure retaining all 16 genes in one plant in the final crossing generation. The numbers of crosses were 8, 64, 131,072, and 65,536[4] in the first, second, third, and fourth crossing generations, respectively. Even with a male-sterility system or with an outcrosser, the numbers of hybrid seeds in generations 3 and 4 would never be attained. Nevertheless, the CC and Syn do provide a genetic matrix for simultaneous recombination of genes from several sources.

MacKey (1954) commented that using 16, 32, or even more parents in a CC would force the inclusion of an unduly large number of unadapted parental lines, which might undo the adapted genetic background. As an alternative, he suggested using a modified backcrossing program to obtain optimal parental materials for the CC crossing scheme. With this modification, each unadapted strain would be crossed, and perhaps backcrossed, to an adapted strain of the species before it was used in the CC. In this scheme, several unadapted strains could be combined with more than one cultivar. The proportion of unadapted germplasm that would be desirable in a CC would depend upon the degree of "unadaptiveness" of the unadapted germplasm. If the unadapted strains were introductions of the same cultivated species adapted to the same latitude, probably single crosses would be adequate. A good example of this level of introgression was CC XII of barley in which the hybrid of 26 more or less unadapted cultivars created by successive generations of crossing was hybridized with the F_1 of 'Atlas' × 'Vaughn,' both of which were adapted to California where this CC was made (Suneson, 1956). On the other hand, Lawrence and Frey (1976), who introgressed germplasm from *Avena sterilis* L., a weedy oat from the Mediterranean Sea area, into cultivated oats, found it necessary to backcross three to five times before the base of "adapted germplasm" was appropriate for a practical breeding program. With these levels of backcrossing, the expected proportions of adapted germplasm ranged from 94 to 98%.

Selection of Desired Genotypes

There are several different "population methods" of plant breeding: (1) Bulk, (2) Mass selection, (3) Recurrent selection, and (4) Population improvement.

I will assess the different effects of these methods in terms of selection efficiency and the choice of appropriate methods for the selection of desired genotypes in segregating populations.

Bulk Method of Breeding

With CC and Syn schemes, new dimensions occurred for obtaining genotypic recombinants. The genotypic recombination possible with these mass methods of hybridization among many parents has required development of new methods for managing large populations of segregates.

One such attempt was the "bulk method" of breeding proposed by Harlan et al. (1940). This method involved yield testing the F_2 bulk progenies from crosses and discarding whole crosses on the basis of the yield data. These researchers made 379 single crosses of barley and tested the F_2 bulk progenies for grain yield. On the basis of their F_2 data, Harlan et al. (1940) discarded whole crosses and concentrated selection on those crosses likely to produce high-yielding lines. Smith and Lambert (1968) found that barley crosses that contained many superior lines were those with high mid-parent values and high F_2 bulk yields, and Cregan and Busch (1977) reported highly significant positive correlations between F_5-derived means and F_1 to F_5 bulk yields for crosses of spring wheat.

Studies that do not support the bulk method have also been published. Kalton (1948) and Fowler and Heyne (1955), using 25 soybean crosses and 45 wheat crosses, respectively, found non-significant intergeneration correlations (between 0.15 and 0.25) for grain yields of bulks in F_2, F_3, F_4, and F_5. Also, Atkins and Murphy (1949) duplicated the barley experiment of Harlan et al. (1940) using oats and found little relationship between the means of yields of F_8-derived lines and the F_2 to F_6 bulk yields of crosses from which they originated.

The original objective for the use of the bulk method according to Harlan et al. (1940) was to stratify crosses for selection of parents based upon yield values. The assumptions were that (1) variation for yield would be similar in all crosses and (2) intergeneration yields would be subject to little or no genotype-environment interaction. Since neither of these assumptions have held true consistently, the bulk method of plant breeding has become simply a procedure for propagating large populations of self-pollinating plants from crosses to establish homozygosity from F_2 onward.

The bulk method is an inexpensive way to carry large numbers of plants. However, it is useful only if "desirable" plants survive through the generations of propagation. The differential survival of genotypes in heterogeneous populations of plants is known as "natural selection." From the viewpoint of crop breeding, evidence for the importance of natural selection in bulk populations comes from (1) studies on the differential survival among pure-line genotypes in designed mixtures, (2) survival of different plant types in hybrid-derived populations, and (3) experiments designed to measure the competitive abilities of genotypes.

A classical experiment of differential survival of pure-line genotypes in mixtures was reported by Suneson (1949) more than 30 years ago. He mixed four adapted barley cultivars, 'Atlas,' 'Club Mariout,' 'Hero,' and 'Vaughn,' in equal quantities and grew the mixture for 16 consecutive years in California. By 1948, Atlas comprised 88% of the mixture, and Vaughn was eliminated completely (Table 4.3). When tested in pure stands, the grain yields of Atlas and Vaughn were low and high, respectively. Lee (1960) found that the competitive advantage of Atlas was attributable to the fact that it initiated root proliferation in the spring about 1 week earlier than

Table 4.3. Analysis of barley cultivars remaining in a mixture propagated for 16 consecutive years.†

Cultivar	Year of analysis 1933	1940	1948
	%		
Atlas	25	63	88
Club Mariout	25	17	11
Hero	25	8	1
Vaughn	25	12	0

† Suneson (1949).

Vaughn. When moisture (and perhaps nutrients) was limiting, Atlas's larger root system gave it an initial and continuing advantage in plant size and development, and an Atlas plant in a mixed stand could command more moisture and nutrients than one in pure stand where its competitors would be other Atlas plants. These unseen effects of root growth initiation and proliferation were manifested in tiller and spike production.

Competition in Soybeans

In a soybean mixture of three cultivars, Mumaw and Weber (1957) showed that within 5 years of propagation, 'Bavender Special,' an unadapted, profusely branching type, constituted 70% of the mixture, whereas 'Adams,' a well-adapted cultivar, had been eliminated.

Competition in Rice

Jennings and de Jesus (1968) studied the survival of five rice cultivars in a mixture and found that 'TN1,' 'Ch 242', and 'M6,' non-tillering and short cultivars, were practically eliminated from the mixture in 1 year of propagation, whereas 'BJ' and 'MTV,' tall leafy cultivars, dominated. Yield abilities of these five cultivars in pure stands were exactly inverse to their capabilities to survive in mixtures. Plant height and tillering ability were two obvious traits of importance to survival of genotypes of rice cultivars in mixtures.

What is most obvious from the study of survival of cultivars in mixtures is the lack of a positive relationship between yield and ability to survive.

In studies with cultivar mixtures of self-fertilizing crops, the interactions are simple compared to those in a segregating population of plants.

Jennings and Herrera (1968) studied a segregating rice population from the cross of 'Peta' (P), a tall tropical cultivar, and 'Taichung Native 1' (TN1), a short cultivar. Plant height in this cross segregated on a mono-factorial basis. Whether plants were sown in a grid of 30 × 30 cm or 30 × 15 cm, short plants were consistently eliminated from the population. The relative reproductiveness of short plants (where tall plants = 1.0) was 0.68 in the zero-nitrogen environment and 0.36 in the high-nitrogen one. However, the short F_6-derived lines were much superior to the tall ones for grain yield in pure stands. Obviously, natural selection in the segregating rice population was counterproductive to the goal of developing a high-yielding rice cultivar. Jennings and Aquino (1968) determined that, even though competition between tall and short rice plants did not begin until 50 to 60 days after germination, tall plants had vigorous vegetation, a decided competitive advantage for light interception.

The studies with cultivar mixtures and bulk segregating populations lead to the same conclusion. Whether natural selection operates beneficially or detrimentally on a heterogeneous population of plants is highly specific. The results depend upon the interaction system in the population, the propagation environment, and the breeders' goals. Seldom will natural selection be neutral.

Survival value of a specific genotype is described as a measure of its competitive ability. Sakai (1955) described a procedure for measuring competitive ability among plants. Plants of the genotype to be evaluated for competitive ability are sown in the center of two paired plots. In one, the center plant (O^T) is surrounded by six plants of its own genotype (O), and in the other, the center plant (O^A) is surrounded by six plants of a tester genotype (X). The ratio of O^A/O^T is Sakai's index of competitive ability. He and his colleagues (Sakai and Suzuki, 1955a, 1955b; Sakai and Gotoh, 1955; Sakai, 1956, 1961) measured many crop species for competitive ability and concluded:

1. In barley and rice, cultivars showed differential competitive abilities.
2. In barley, competitive ability was not associated with any other measured trait.
3. F_1's were vigorous but poor competitors.
4. Autotetraploids were poor competitors, but allotetraploids were good ones.
5. Competitive ability in a diploid was reflected in its autotetraploid.
6. Competitive ability was an inherited trait.

Types of Intergenotypic Competition

Schutz and Brim (1967) used a variation of the Sakai method to test competition in soybeans tested in hills or in rows. They found the competitive relationships between pairs of genotypes to fit four patterns:

1. Complementary = Genotype A shows an increase in yield and genotype B shows an equal decrease.
2. Neutral = The yield of neither genotype A nor genotype B is affected by its counterpart.
3. Undercompensatory = Genotype A shows no change in yield, whereas that of genotype B decreases.
4. Overcompensatory = Genotype A shows no change in yield, whereas that of genotype B increases.

Allard and Adams (1969) concluded that these four intergenotypic interactions would have the following consequences in a bulk population of self-pollinating plants:

1. Complementary interaction leads to the loss of some components, but many genotypes will persist.
2. The neutral pattern will result in maintaining the initial genotype frequencies.
3. With undercompensation, the genotype with the highest initial frequency will predominate.
4. Overcompensation leads to equilibrium of frequencies of all genotypes, and the rapidity of reaching equilibrium depends on the degrees of overcompensation.

Actually, no set of genotypes would show a constant competitive pattern over a set of consecutive propagation environments, and neither would a population contain a single competitive system. These complexities lead to an undefinable situation in a population of plants relative to the effect of genotypic interaction systems upon survival.

As already mentioned, the bulk method is an inexpensive way to grow very large numbers of plants. However, a breeder must expect the genotypic frequencies in a bulk population to change during the propagation period due to natural selection (Suneson and Stevens, 1953). Shifts in genotypic frequencies usually are not predictable, and often they may be opposite to breeding goals. To eliminate shifts in genotypic frequencies in bulk populations, Brim (1966) proposed using the "single-seed descent" method. With this method, one seed is picked from every plant to provide the bulk with which to sow the next generation. It should maintain the genotypic array, but it adds greatly to the expense of using the bulk method.

Mass Selection

Whereas genotypic changes in heterogeneous plant populations by natural selection tend to be unpredictable, mass selection permits a breeder to shift gene and genotypic frequencies in a desired direction.

Poehlman (1979) said mass selection is "a system of breeding in which seeds from individuals selected on the basis of phenotype are composited and used to grow the next generation." Note that (1) evaluation and subsequent selection of certain plants is based only on their phenotypes and (2) progenies from selected plants are grown in bulk.

Mass selection is the oldest breeding method for plant improvement. It was applied by the early farmers during the development of cultivated species from their ancestral forms. With the rediscovery of Mendel's Laws in 1901, there was a shift away from mass selection. Sprague (1955) attributed the lack of control of the male parent and the confounding effects of soil variability on the phenotypes of selected plants as reasons why mass selection became unpopular with maize breeders. Other negative factors were that:

1. Data were collected from outcrossing species that were subject to inbreeding depression.
2. Grain yield was the selected trait and was subject to inbreeding depression.
3. The evaluation often was subjective.

Mass selection should be successful if:

1. Gene action for the selected trait is primarily additive in the population under selection.
2. Heritability of the trait, either natural or manipulated, is high. (Genotype × environment interaction must be eliminated.)
3. Traits selected and traits to be improved are genotypically correlated. (This could be the same trait in two environments.)
4. In outcrossing species, the parental sample size is large enough to preclude inbreeding depression.

Using mass selection, a plant breeder selects on the basis of the phenotype of an organism, but really selects a genotype. Heritability is a measure of the degree of correspondence between phenotype and genotype. Only certain traits with a high heritability, e.g., plant maturity, respond to selection via single plants.

Mass selection also may be indirect, whereby a plant breeder improves one trait via selection for a second. Greatest success from indirect selection occurs when either (1) the two traits result from pleiotropy of genes or (2) the selected trait is a component of the trait to be improved. So far, the literature does not verify that pleiotropy was responsible for the success of indirect mass selection. A case of improving a trait by selecting for one of its components was reported by Frey (1967). Selection was practiced for seed width by passing large numbers of oat seeds over a screen that had 12.7 × 2.4 mm slots. Seeds too wide to go through the slots were bulked and used to propagate the F_3 through F_7 generations. Mean weight per 100 seeds was increased 0.1 g per generation by selecting for seed width via this procedure. Indirect mass selection has been used extensively to select soybean bulk populations for high seed protein or high seed oil by using density separation of the seed (Hartwig and Collins, 1962).

Heritability, as it applies to mass selection with self-fertilizing species or to outcrossers when selection is practiced before anthesis, is computed according to the formula:

$$h^2 = \sigma_a^2/(\sigma_g^2 + \sigma_{gE}^2 + \sigma_e^2),$$

where σ_a^2, σ_g^2, σ_{gE}^2, and σ_e^2 are variances from additive genetic effects, genotype, genotype × environment interaction, and error, respectively. (For the case where mass selection is practiced on out-crossing species after anthesis, the numerator would be $\sigma_a^2/2$.) This formula represents the real progress that will be made with mass selection. In practice, the predicted progress often uses $\sigma_g^2 + \sigma_{gE}^2$ as a substitute for σ_a^2. Thus, the similarity of actual and predicted gains from mass selection will depend upon the similarity of σ_a^2 to σ_g^2 and the magnitude of σ_{gE}^2.

Gardner (1961, 1969) and Harris et al. (1972) have presented an interesting manipulation of heritability in mass selection for grain yield of maize. Mass selection, as practiced in the early days, was to harvest a plot, say of 0.25 ha, and after the maize was in a bin, ears would be weighed, and the heaviest ones saved. No progress in improving yield by this method was ever shown, probably because the environmental variation was so large relative to the genotypic variation that heritability for yield was near zero. Gardner subdivided the field into blocks, each with a more or less uniform environment. He used 40-plant blocks and saved the upper 10% for yield from each block, so he was dividing σ_e^2 into two parts, σ_b^2 and σ_w^2; i.e., variances between and within blocks, respectively. His heritability formula became:

$$h^2 = (\sigma_g^2 + \sigma_{gE}^2)/(\sigma_g^2 + \sigma_{gE}^2 + \sigma_w^2).$$

The denominator of the equation will be reduced if σ_w^2 is smaller than σ_e^2. Although Gardner had no direct measurements of the values of σ_b^2 and σ_w^2, he increased yield of 'Hays Golden' open-pollinated cultivar by 2.7% per cycle through 11 cycles, so it appears that σ_w^2 was much smaller than σ_e^2.

Hallauer and Sears (1969) used Gardner's method on 'Krug' synthetic and 'Iowa Ideal' open-pollinated cultivars. They found no evidence for significant yield improvement in either source after 5 and 6 cycles, respectively, but they noted that, very likely, their failure was due to σ_{gE}^2 being a confounding factor of σ_g^2. For selection, Gardner used a uniform soil area, applied irrigation when needed, and did all evaluations at the same site where selection was practiced. Hallauer and Sears used the natural climate (including drought) at Ames, Iowa, for selection, and evaluated the progress from selection in one to four Iowa locations away from Ames. In Gardner's study, little genotype × environment interaction for yield occurred because of the frequent application of water and having the selection and test environments coincide, whereas in the Hallauer study, there was great opportunity for genotype × environment interaction to occur.

Examples of Mass Selection

Tiyawalee and Frey (1970) caused artificial epidemics of crown rust in heterogeneous oat populations by inoculation of spreader rows. The oats were harvested en masse at maturity and bulk-threshed, and the seed lot was winnowed. Heavy seeds were used for continued propagation of the population through F_{10}. This technique moved the gene frequency for rust resistance from 0.21 in the original population to 0.35 by F_{10}.

A mass-selection project on alfalfa was begun in the mid-1950's, using the complex gene pools A and B that were initiated by Hanson et al. (1972), who intercrossed vigorous plants from 3- and 4-year-old stands. In each cycle of selection, 1,800 to 10,000 plants were evaluated in each pool, and 80 to 500 selected plants were intercrossed to initiate the next cycle. Parent plants were selected for vigor and pest resistance. Crossing was permitted only between resistance plants, so both male and female plants were selected. Eighteen cycles of selection were reported with selection for resistance to four diseases and two insects. Selection for rust resistance was very effective, lowering the mean rust scores from about 6.0 to less than 3.0 by cycle 6, but the resistance plateaued at that level. Resistance to leaf hopper in the germplasm pools after 11 cycles increased, with additional improvement possible. For common leaf spot, three cycles of selection lowered the scores of both A and B pools so they were more resistant than 'Du Puits' cultivar, one of the most resistant known. Resistance to spotted alfalfa aphid in pool A was accomplished in one cycle, but improvement in pool B was more gradual. Likewise, the reactions of the two germplasm pools to selection for resistance to bacterial wilt were different. Pool A responded abruptly, whereas pool B responded gradually. Improvement in anthracnose resistance was not dramatic in either germplasm pool, but it was significant.

In addition to its effectiveness for modifying genotypic and gene frequencies, mass selection may also cause correlated response to selection in

Table 4.4. Genotypic correlations among traits in F_9 oat lines.†

Traits	Height	Yield	Seed width	Seed weight	Seeds per plant
Heading date	0.48	−0.45	0.42	0.61	−0.52
Height		−0.10	0.16	0.44	−0.11
Yield			−0.20	−0.18	0.51
Seed width				0.61	−0.45
Seed weight					−0.67

† Geadelmann (1970).

secondary traits. Some examples in oats reported by Geadelmann (1970) are shown in Table 4.4. Theoretically, such correlated responses are due to pleiotropism or linkage.

Mass selection for improvement of plant populations can be effective and inexpensive. It can be used on heterogeneous populations of both outcrossing and self-pollinating crop species. Its success depends upon the heritability, either inherent or manipulated, of the selected trait, the presence of additive gene action for the selected trait, and minimal confounding by genotype × environment interaction expression by the trait. In some instances, indirect selection for one trait by way of a second trait may be more effective and less costly than direct selection for the first trait. Associated changes between selected and unselected traits can be expected consistently in self-pollinating species.

Recurrent Selection

Recurrent selection is a method of plant breeding for quantitatively inherited traits by which the frequencies of favorable genes are increased in populations of plants. The methodology is cyclic with each cycle encompassing two phases: (1) selection of genotypes that possess favorable genes and (2) crossing among the selected genotypes. It causes a gradual upgrading of the frequencies of desired alleles. Theoretically, an individual plant with all of the favorable genes in a population could be obtained in a single generation if a breeder could work with a population of infinite size. Practically, the procedure is not possible. Recurrent selection was developed as a technique to increase frequencies of favorable genes gradually until there is a reasonable likelihood of obtaining the ultimate genotype in a finite sample. The mean of the trait under selection will improve gradually; and the shift will continue as long as genetic variability exists in the population. The methodology and expected results for recurrent selection are illustrated in Fig. 4.4. Genotypes at the plus end of the frequency distribution are selected and intercrossed. Progenies arising from the intercrosses form the cycle I population from which the plus genotypes are selected again. These are intercrossed to form the cycle II population, and so on through successive crosses. Note that the frequency distribution and mean of each successive generation are shifted in the plus direction, which increases the probability of recovering the desired genotypes.

Recurrent selection can be applied to the improvement of quantitatively inherited traits in either outcrossing or self-pollinating species, regardless of the type of gene action involved in determination of the trait. The tech-

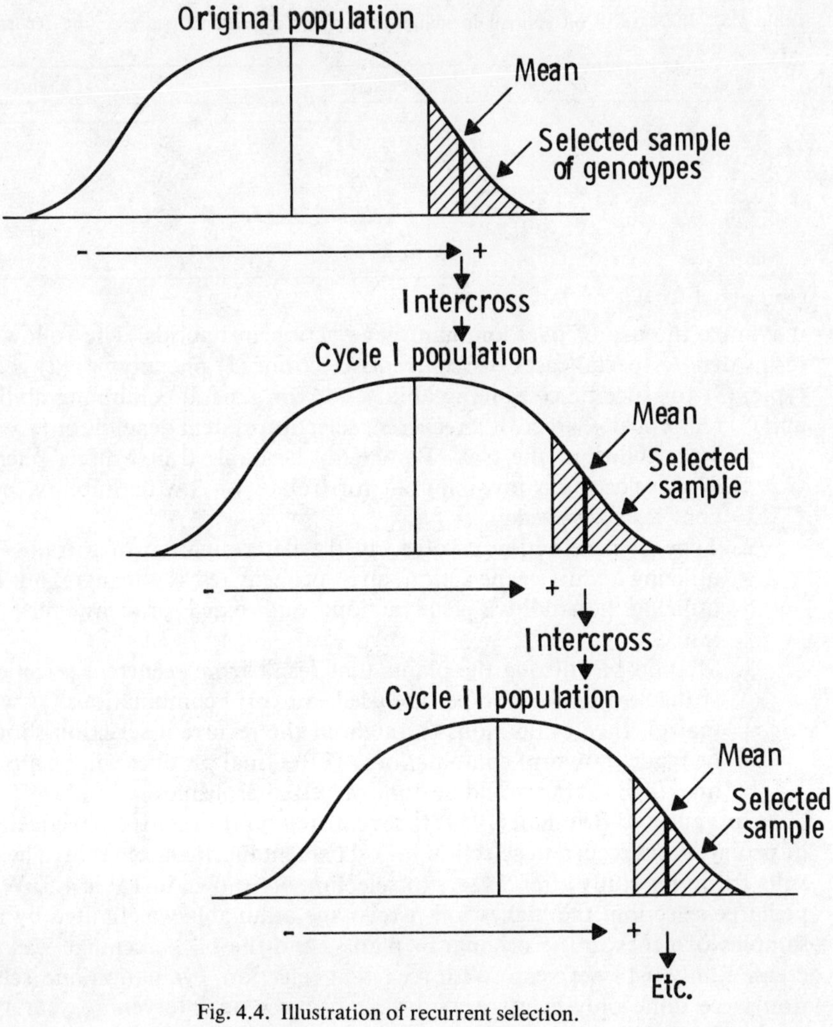

Fig. 4.4. Illustration of recurrent selection.

nique has been applied more to outcrossing species because it is dependent upon massive crossing among selected genotypes in each cycle. The technical problem of obtaining enough crossed seeds for a self-pollinating crop discourages breeders from extensively using recurrent selection. However, male sterility is a tool that could make the technique suitable for self-pollinated crops.

Requirements for Recurrent Selection

Recurrent selection has been practiced ever since crop breeding became a profession, but the term "recurrent selection" was coined by Hull (1945). He proposed its use for developing inbred lines of maize that would

Table 4.5. Increase in oil content in maize grain from 5 years of pedigree and recurrent selection.†

Generation	Pedigree	Recurrent
	%	
S_1	5.0	4.2
S_2	4.6	--
S_3	5.0	5.2
S_4	5.2	--
S_5	5.6	7.0
Gain per year	0.13	0.41

† Sprague et al. (1952).

maximize the use of overdominant gene action in hybrids. The following terms denote special cases of recurrent selection: (1) phenotypic, (2) genotypic, (3) for specific combining ability, (4) for general combining ability, and (5) reciprocal. Choice of a recurrent selection system depends on:

1. Heritability of the trait. For highly heritable traits, direct phenotypic selection is possible, but for traits with low heritability, progeny tests are needed.
2. Type of gene action involved in the determination of a trait. For utilizing additive gene action, an S_1 progeny test is adequate, but for utilizing nonadditive gene action, outcrossed progeny must be tested.
3. Method of utilizing the plants that result from recurrent selection. If the resultant genotypes are used in hybrid combinations (as with maize), then evaluations throughout the recurrent selection should be made in hybrid combinations. If the final product will be a pure line, then testing would be done on selfed progenies.

Sprague and Brimhall (1950) first reported on the relative effectiveness of pedigree and recurrent selection for oil percentage in maize grain. The results from this study after 5 years of selection are shown in Table 4.5. With pedigree selection, the highest oil percentage attainable was limited by the samples of genes in the original S_1 plants, and the oil percentage was increased only 0.13 per year. With recurrent selection, evaluation and selection were done only every second year because an intervening year was needed for crossing among selected genotypes. For three cycles, the oil percentage was increased 0.41 per year, an increase three times greater than the annual progress from pedigree selection. Furthermore, genetic variation among the third cycle genotypes from recurrent selection provided opportunity for additional progress.

Evaluation of Recurrent Selection

The first comprehensive evaluation of recurrent selection was done with sweetclover by Johnson and his coworkers (Johnson and Goforth, 1953; Johnson, 1952, 1956; Johnson and El Banna, 1957). They began the study with 2,000 plants from 'Madrid' cultivar and selected 200 late-flowering plants. Each plant was selfed and allowed to outcross randomly. About one-third of the plants set 10 or more selfed seeds, and these were used to establish S_1 progenies. Open-pollinated seeds from each plant were used to

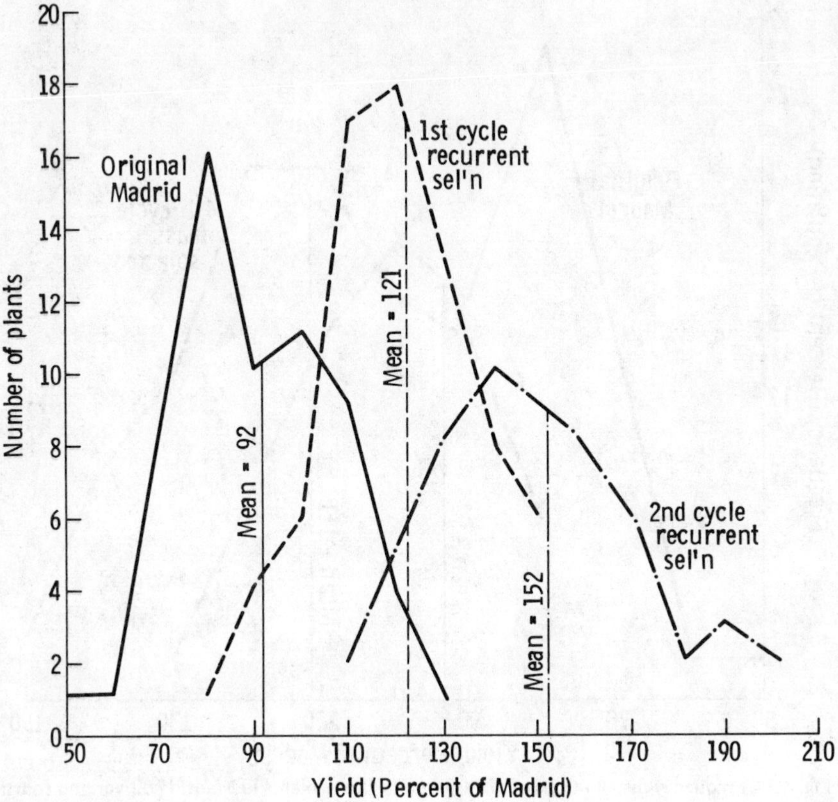

Fig. 4.5. Frequency distributions of yields of sweetclover plants in Madrid cultivar and derived populations after one and two recurrent selection cycles (Johnson, 1956).

establish a yield test in which Madrid was the control. The S_1 progenies from the 10 highest yielding S_0 topcrosses were used to make all possible two-parent crosses. The seed from this intercrossing was bulked to form a Syn-1 population of about 2,000 plants representing cycle-1 from recurrent selection. This process was repeated for four cycles, and the results from the first two are shown in Fig. 4.5. In one cycle of recurrent selection, the yield per plant was increased from 92 to 121% of Madrid cultivar, and a second cycle increased the mean to 152% of Madrid. This rapid advance in yield per plant from recurrent selection was much more astounding than expected. In a companion study using the same initial plants, Johnson and Goforth (1953) did mass selection in which recombination occurred among open-pollinated progenies instead of among S_1 progenies as used for recurrent selection. After three cycles of mass selection, the mean plant yield had been raised from 92 to 111% of Madrid (Fig. 4.6). From three cycles of recurrent selection, the improvement was 60%. This result points out the efficiency of recurrent selection as a plant breeding method.

Recurrent selection should be effective as long as genetic variation remains in the population. With selection, genetic variance in a population should decrease until the selected alleles become fixed. Then, the genetic

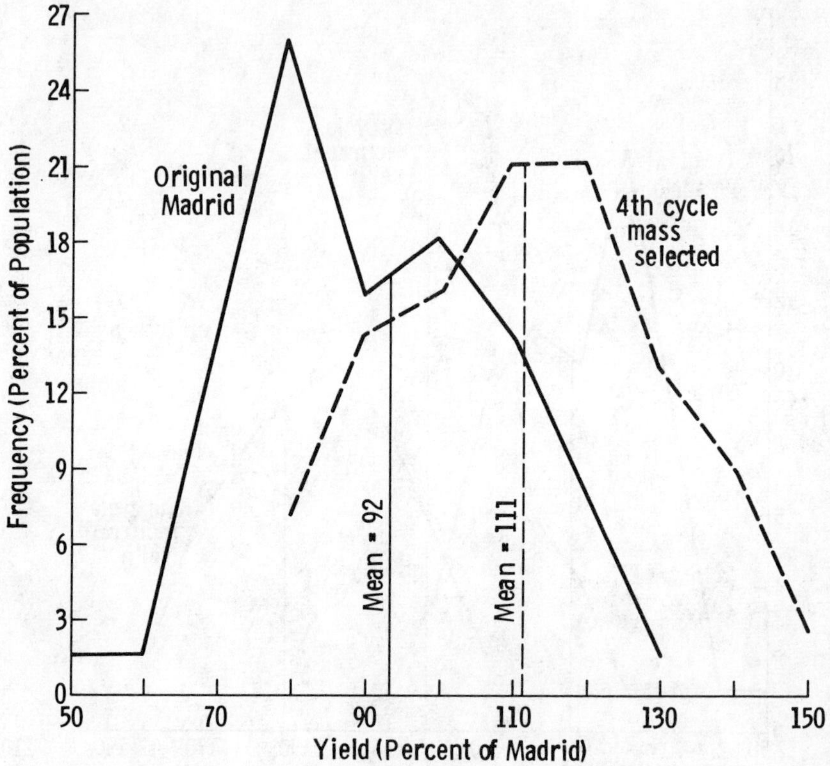

Fig. 4.6. Frequency distributions of yields of sweetclover plants in Madrid cultivar and fourth cycle population from mass selection (Johnson and Goforth, 1953).

variance would become zero and no further progress could be expected. Most traits to be improved through recurrent selection are quantitatively inherited. The large number of loci involved preclude a rapid depletion of genetic variance. Johnson and El Banna (1957) measured the relative genetic variation in the Syn-1 generation of sweetclover populations after 1, 2, and 3 cycles of recurrent selection. Mean yield per plant increased markedly relative to Madrid, and they found no evidence that the recurrent selection for three cycles had caused any reduction in the genetic variance.

Recurrent Selection for Combining Ability

The first positive report of improving yield of maize via recurrent selection was published by Sprague et al. (1959), who showed that hybrids of the inbred 'Hy' with C_2 populations of 'Lancaster' and 'Kolkmeier' cultivars had increases of 9 and 29%, respectively, over hybrids of Hy with their C_0 counterparts. Improvement in yield from recurrent selection in these maize cultivars was for combining ability.

Two other extensive studies on recurrent selection for specific combining ability in maize have been conducted. In Florida, Horner et al. (1976) reported on six cycles of recurrent selection for specific combining ability.

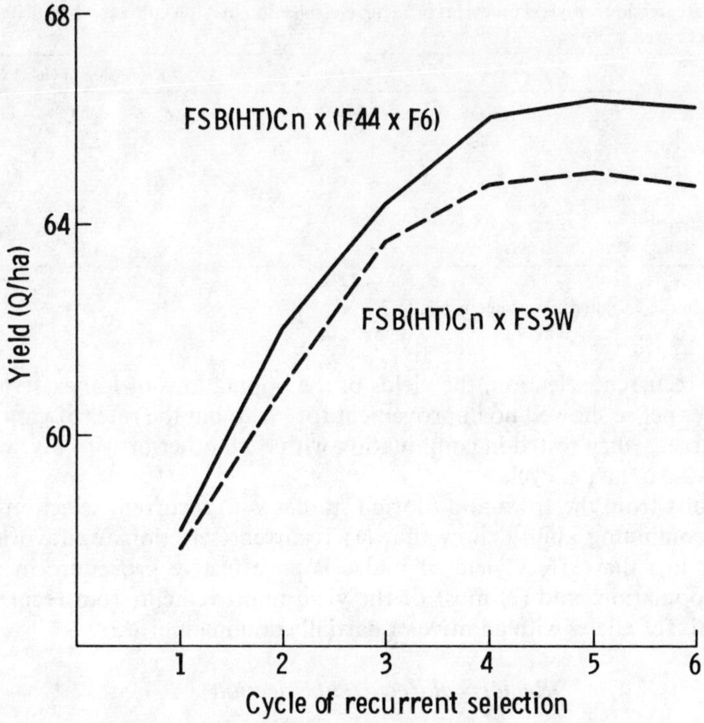

Fig. 4.7. Yield trends of FSB (HT) C_n populations when crossed to a narrow (F44 × F6) and a broad (FS3W) tester (Horner et al., 1976).

Between 114 and 610 S_0 plants of the population, [FSB(HT)], were tested per cycle of recurrent selection with the tester $F_{44} \times F_6$. Selection intensity was about 20%. After six cycles of recurrent selection were completed, Syn-2 populations of each cycle were crossed to $F_{44} \times F_6$ and the unrelated synthetic FS3W; the yield trends are shown in Fig. 4.7. Recurrent selection improved the yield of FSB(HT) Cn about 15% in four cycles, but in cycles 5 and 6, the yield had reached a plateau. This plateau probably resulted because the genetic variation for combining ability in FSB(HT) had been exhausted.

Russell et al. (1973) carried out five cycles of recurrent selection on two populations, 'Alph,' an open-pollinated maize cultivar, and the F_2 of the single cross, WF9 × B7. Gains in yield per cycle of recurrent selection are shown in Table 4.6. This program took 16 years to complete, and gains per year were 1.3% for Alph and 0.6% for WF9 × B7. Obviously, recurrent selection was successful in both sources for increasing the frequencies of the favorable alleles for combining ability. The population crosses (i.e., Alph C_n × (WF9 × B7) C_n) had a rate of gain of 4.1 q/ha per cycle. Alph and WF9 × B7 contributed 64 and 36% of the population × population gain, respectively.

Walejko and Russell (1977) reported on recurrent selection in Lancaster and Kolkmeier maize cultivars with Hy as the tester. After five

Table 4.6. Regressions of yield on cycles of recurrent selection in Alph cultivar when different testers were used.†

Tester	Regression of yield‡
	q/ha
Alph C_n	2.1
B14	3.1
BSBB	3.6
Alph C_0	2.3
(WF9 × B7)C_0	2.6
(WF9 × B7)C_n	4.1

† Russell et al. (1973).
‡ All regressions were significantly greater than zero.

cycles of recurrent selection, the yields of the populations of Lancaster and Kolkmeier per se showed no improvement for yield, but the rates of gain for both sources, when tested in combination with each other or with Hy, were all about 2.5 q/ha per cycle.

Results from the Iowa and Florida studies with recurrent selection for specific combining ability show that (1) recurrent selection for favorable alleles at loci that affect yield of maize is an effective procedure in any source population, and (2) most of the yield improvement from recurrent selection is for alleles with additive or partially dominant effects.

Reciprocal Recurrent Selection

Reciprocal recurrent selection is a procedure in which progressively improved populations of two germplasm pools are used reciprocally as testers. In other words, pool AC_{n-1} is used as a tester for pool BC_n, and pool BC_{n-1} is sued as a tester for pool AC_n.

Moll and Stuber (1971) reported modest increases in 'Jarvis' and 'Indian Chief' maize cultivars for yield per plant by using reciprocal recurrent selection. In Iowa, Eberhart et al. (1973) used reciprocal recurrent selection to improve two synthetic cultivars, BSSS and Iowa Corn Borer Synthetic No. 1 (BSCB1). They carried out five cycles of reciprocal recurrent selection over a 23-year period. Neither BSSS nor BSCB1 populations per se showed significant improvement in yield from reciprocal recurrent selection, but the yield improvement in the cross, BSSSC_n × BSCB1C_n, was significant.

All evidence reported for yield of maize shows that recurrent selection has been effective in increasing favorable alleles in the populations. Generally, improvement has been for general combining ability, regardless of the type of tester used. On the other hand, the reciprocal recurrent selection programs select for a certain degree of specific combining ability between the populations being improved as evidenced by the fact that heterosis in the population crosses becomes greater with advancing cycles. Russell et al. (1973) found that heterosis for the cross of Alph and WF9 × B7 populations was 35% in C_0 generation and 52% in C_5. Eberhart et al. (1973) found 14% heterosis in C_0 × C_0 of BSSS(R) by BSCB1(R) and 35% in C_5 × C_5.

Recurrent Selection for Pest Resistance

Recurrent selection has been used to increase gene frequencies for pest resistance in maize populations. Jinahyon and Russell (1969), in three cycles of recurrent selection for resistance to *Diplodia* stalk rot in Lancaster cultivar, reduced mean rating from 3.7 in C_0 to 1.7 in C_3. Penny et al. (1967) selected for resistance to European corn borer in five maize synthetics via recurrent selection and improved resistance to borer feeding in all five. Response to recurrent selection in these populations was generally linear.

In most cases, populations in which traits are improved by recurrent selection are not used for agricultural production per se. They serve as breeding populations from which to select inbreds, pure-line cultivars, or parents for synthetics and crosses. Recurrent selection is a powerful breeding method for concentrating the frequencies of desirable genes in a cultivar or synthetic population of plants. Shifts in gene frequencies are accomplished gradually within samples of finite size. Also, it has the advantage of permitting greater recombination among linked alleles than would be accomplished in a single selection cycle.

Population Improvement

The success with recurrent selection has led to several "population improvement" schemes. Recurrent selection to date has been applied to "closed" genetic systems; i.e., no new germplasm was added to the population during selection. Potential limitations to progress from recurrent selection in closed populations have led to proposed breeding techniques known collectively as "population improvement." Eberhart et al. (1967) proposed a "comprehensive breeding system" applicable to both outcrossing and self-pollinating species, and Jensen (1970) proposed the "diallel selective mating system," designed primarily for self-pollinating species. With these procedures, populations of plants are dynamic gene pools (1) to which new sources of germplasm are added when feasible, (2) in which the frequencies of favorable alleles are progressively increased via recurrent selection, (3) in which genetic recombination is enhanced by massive hybridization among selected genotypes, and (4) from which cultivars, inbreds, or parental lines can be extracted at any stage.

Population Synthesis

Eberhart et al. (1967) recommended that a breeder who intends to initiate a "population breeding program for corn" should spend 3 to 5 years evaluating and selecting plant materials for inclusion in the population. They stress the need for evaluating both exotic and locally adapted lines and including the best of both in the breeding population. The exotic materials greatly expand genetic variation in the breeding population. The initial population of maize could include (1) sources of resistance to important pests in the production area where the breeding population would be used, (2) a sizable number of lines with good agronomic traits, such as grain quality, lodging resistance, and proper maturity ratings, and (3) more than half of the germplasm would have good productivity potential in the region.

Current Population Improvement Programs

The population generation and improvement system for maize has been adopted by the International Maize and Wheat Improvement Center (CIMMYT) (1975). CIMMYT has originated 48 maize populations at its Mexican site for distribution to national testing programs around the world. Breeders responsible for national breeding programs are free to practice any type of selection. In most cases, S_1 progenies are evaluated intensively, after which they are recombined.

The first maize population established (other than naturally occurring open-pollinated cultivars) was BSSS. It was originated by compositing hybrid seeds from intercrossing 16 Corn-Belt inbreds chosen for their resistances to stalk lodging (Sprague, 1946). Until 1976, a total of 82 commercially used inbreds have been isolated from BSSS. These inbreds were selected in 13 states of the USA and three provinces of Canada. An Iowa inbred from BSSS, B73, is now a parent in single-cross hybrids grown on 35% of the U.S. maize hectarage.

Several populations of sorghum have been originated and released for use by breeders in the USA. Nordquist et al. (1973) reported three, NP1BR, NP2B, and NP3R. NP1BR had 21 germplasm sources, and NP3R had 30 germplasm sources. Sorghum populations are being developed extensively at the International Crops Research Institute for the Semi-Arid Tropics for distribution to national programs in developing countries.

Except for the initiation of several second-generation synthetics by Hallauer et al. (1974) in maize, and perhaps the system used by CIMMYT, little attempt has been made to add new germplasm to breeding populations on a regular basis. Therefore this underlying principle of population improvement generally has not been utilized.

Population Improvement for Multiple Traits

Selection systems to improve breeding populations of plants have been used in tandem, such as those used by Hanson et al. (1972) in the alfalfa studies mentioned earlier, or they have been intensive for improving a primary trait and mild for several secondary traits. In the alfalfa study, the intense selection for pest resistances was done in tandem order in the following phases:

1. Selection was practiced simultaneously for resistance to leafhopper yellowing and rust for eight cycles.
2. The population from phase 1 was selected subsequently for resistance to leaf-hopper yellowing for three cycles.

3. The population from phase 2 was selected for resistance to common leaf spot for three cycles.
4. The population from phase 3 was selected for resistance to bacterial wilt for two cycles.
5. The population from phase 4 was selected for resistance to anthracnose for two cycles.

Mild selection was practiced among resistant plants in each of the 18 cycles for plant vigor. From this massive population improvement program, there resulted a germplasm pool with a high level of multiple pest resistance and with vigor equal to the original pools A and B.

Generally, improvement of populations of maize has been concentrated on improving a single trait in a single population, and most often, that trait has been yield. Lines used to constitute BSCB1 were highly resistant to the European corn borer, but all selection in it has been for increasing yield. Thus BSCB1(R)C6 should be corn-borer resistant, owing to its origin, and high yielding, owing to its recurrent selection history. Generally, in the population improvement work with maize, mild selection intensity has been practiced for lodging resistance and appropriate maturity.

Population Release

An important aspect of population improvement is the potential for releasing cultivars and inbreds from the population while it is undergoing improvement. As pointed out, several maize inbreds were selected from BSSS as it was being improved via recurrent selection. Also, in alfalfa, pool B was released as the cultivar 'Cherokee' after eight cycles of selection for resistance to leaf-hopper yellowing and rust (Dudley, 1963). One germplasm pool begun in North Carolina, after being subjected to mass selection for six cycles for resistance to alfalfa weevil, was released as 'Team' cultivar (Barnes et al., 1973). After three additional cycles of mass selection for resistance to bacterial wilt and anthracnose, it was released as 'Arc' cultivar (Devine et al., 1975).

Diallel Selective Mating System

The diallel selective mating system was specifically designed for self-pollinating crop plants (Jensen, 1970). It provides for a conventional bulk population from a series of single crosses and for multiple-parent crosses with opportunities for recombination among alleles from more than two parents. Selection is applied to upgrade each population series (mass selection), and cultivars can be selected at any stage and from any populations. Further, there is opportunity for using new germplasm from outside the population at any time. Jensen (1970) suggested having several populations, each with a rather narrow objective in terms of crop improvement. However, the diallel mating system could accommodate a population with multiple breeding objectives.

Redden and Jensen (1974) published results on use of this system. Mass selection for increased tillering in barley and wheat gave improvements of 6.3 and 10.3% per cycle, respectively. For both crops, selection was more effective in hybridized (hybridization consisted of random mating among selected plants) than in selfed series. With barley, selection response was 9.0% in the selfing and 15.5% in the hybridized series in two cycles, and with wheat, the respective responses were 18.5 and 22.6%. These 4.1 and 6.5% differentials for wheat and barley, respectively, were attributed to genetic recombination that occurred from hybridization.

Population improvement is a breeding system that involves recurrent selection as the focal procedure. The breeding population being improved is dynamic as the infusion of new germplasm adds new alleles, selection increases the frequencies of favorable alleles, opportunity for recombination is maximized, and useful agricultural cultivars can be extracted at any stage in its evolution. Population improvement is applicable to both outcrossing and self-pollinating species, and it is a powerful procedure for breeding programs.

LITERATURE CITED

Adams, M. W., and D. B. Shank. 1959. The relationship of heterozygosity to homeostasis in maize hybrids. Genetics 44:777–787.

Ackerman, A., and J. MacKey. 1948. The breeding of self-fertilized plants by crossing. p. 46–71. *In* A. Ackerman, O. Tedin, and K. Froier (ed.) Plant breeding, Svalof 1896–1946. K. Bloms Boktr, Lund.

Allard, R. W., and J. Adams. 1969. The role of intergenotypic interactions in plant breeding. 3:349–370. *In* C. Oshima (ed.) Proc. XII Int. Cong. Genet. Tokyo, Japan, 19–28 Aug. 1968. Dai Nippon Printing Co. Ltd., Tokyo.

Atkins, R. E., and H. C. Murphy. 1949. Evaluation of yield potentialities of oat crosses from bulk hybrid tests. Agron. J. 41:41–45.

Barnes, D. K., C. H. Hansen, R. H. Ratcliffe, T. H. Busbice, J. A. Schillinger, and G. R. Buss. 1973. Registration of Team alfalfa. Crop Sci. 13:769.

Borlaug, N. E. 1959. The use of multilincal or composite varieties to control airborne epidemic diseases of self-pollinated crop plants. p. 12–26. *In* Proc. First Int. Wheat Genet. Symp. Univ. of Manitoba, Winnipeg. 11–15 Aug. 1958. Public Press Ltd., Winnipeg.

Briggs, F. N., and R. W. Allard. 1953. The current status of the backcross method of plant breeding. Agron. J. 45:131–138.

Brim, C. A. 1966. A modified pedigree method of selection in soybeans. Crop Sci. 6: 220.

Bula, R. J., R. G. May, C. S. Garrison, C. M. Rincker, and J. G. Dean. 1969. Floral response, winter survival, and leaf mark frequency of advanced generation seed increases of Dollard red clover, *Trifolium pratense* L. Crop Sci. 9:181–184.

Canode, C. L. 1958. Natural selection within Ranger alfalfa. Univ. Idaho Agric. Exp. Stn. Res. Bull. 39.

Cregan, P. B., and R. H. Busch. 1977. Early generation bulk hybrid yield testing of adapted hard red spring wheat crosses. Crop Sci. 17:887–891.

Devine, T. E., R. H. Ratcliffe, C. M. Rincker, D. K. Barnes, S. A. Ostazeski, T. H. Busbice, C. H. Hanson, J. A. Schillinger, G. R. Buss, and R. W. Cleveland. 1975. Registration of Arc alfalfa. Crop Sci. 15:97.

Dudley, J. W. 1963. Registration of Cherokee alfalfa. Crop Sci. 3:458–459.

———. 1964. A genetic evaluation of methods of utilizing heterozygosis and dominance in autotetraploids. Crop Sci. 4:410–413.

Eberhart, S. A., S. Debela, and A. R. Hallauer. 1973. Reciprocal recurrent selection in the BSSS and BSCB1 maize populations and half-sib selection in BSSS. Crop Sci. 13:451–456.

———, M. N. Harrison, and F. Ogada. 1967. A comprehensive breeding system. Der Züchter 37:169–174.

———, and W. A. Russell. 1966. Stability parameters for comparing varieties. Crop Sci. 6:36–40.

———, and ———. 1969. Yield and stability for a 10-line diallel of single-cross and double-cross maize hybrids. Crop Sci. 9:357–361.

Falconer, D. S. 1960. Introduction to quantitative genetics. The Ronald Press Co. New York.

Finlay, K. W., and G. N. Wilkinson. 1963. An analysis of adaptation in a plant-breeding programme. Aust. J. Agric. Res. 14:742–754.

Fowler, W. L., and E. G. Heyne. 1955. Evaluation of bulk hybrid tests for predicting performance of pure line selections of hard red winter wheat. Agron. J. 47: 430–434.

Frey, K. J. 1959. Yield components of oats: II. Effect of nitrogen fertilization. Agron. J. 51:605–608.

———. 1967. Mass selection for seed width in oat populations. Euphytica 16:341–349.

———, J. A. Browning, and M. D. Simons. 1977. Management systems for host genes to control disease loss. P. R. Day (ed.) The genetic basis of epidemics in agriculture. Ann. N.Y. Acad. Sci. 287:255–274.

———, and P. Chandhanamutta. 1975. Spontaneous mutations as a source of variation in diploid, tetraploid, and hexaploid oats (*Avena* spp.). Egypt. J. Genet. Cytol. 4:238–249.

Gardner, C. O. 1961. An evaluation of effects of mass selection and seed irradiation with thermal neutrons on yield of corn. Crop Sci. 1:241–245.

———. 1969. Genetic variation in irradiated and control populations of corn after ten cycles of mass selection for high grain yields. p. 469–477. *In* Induced mutations in plants. Int. Atomic Energy Agency, Vienna.

Geadelmann, J. L. 1970. Estimation of mass selection parameters from a heterogeneous oat population. Iowa State Univ. Publication No. 71-14226. 93 p. Ph.D. Dissertation (Diss. Abstr. 31/12B:7045).

Hallauer, A. R., S. A. Eberhart, and W. A. Russell. 1974. Registration of maize germplasm. Crop Sci. 14:341–342.

———, and J. H. Sears. 1969. Mass selection for yield in two varieties of maize. Crop Sci. 9:47–50.

Hanson, C. H., T. H. Busbice, R. R. Hill, Jr., O. J. Hunt, and A. J. Oakes. 1972. Directed mass selection for developing multiple pest resistance in conserving germplasm in alfalfa. J. Environ. Qual. 1:105–111.

Harlan, H. V., M. L. Martini, and Harland Stevens. 1940. A study of methods in barley breeding. USDA Tech. Bull. 720.

Harris, R. E., C. O. Gardner, and W. A. Compton. 1972. Effects of mass selection and irradiation in corn measured by random S_1 lines and their topcrosses. Crop Sci. 12:594–598.

Hartwig, E. E., and F. I. Collins. 1962. Evaluation of density classification as a selection technique in breeding soybeans for protein or oil. Crop Sci. 2:159–162.

Horner, E. S., M. C. Lutrick, W. H. Chapman, and F. G. Martin. 1976. Effect of recurrent selection for combining ability with a single-cross tester in maize. Crop Sci. 16:5–9.

Hull, F. H. 1945. Recurrent selection for specific combining ability in corn. J. Am. Soc. Agron. 37:134–145.

International Maize and Wheat Improvement Center. 1975. Preliminary report of international progeny testing trials and experimental variety trials. Int. Maize and Wheat Imp. Center, Mexico.

Jennings, P. R., and R. C. Aquino. 1968. Studies on competition in rice. III. The mechanism of competition among genotypes. Evolution 22:529–542.

———, and J. de Jesus. 1968. Competition studies in rice. I. Competition in mixtures of varieties. Evolution 22:119–124.

———, and R. M. Herrera. 1968. Studies on competition in rice. II. Competition in segregating populations. Evolution 22:232–236.

Jensen, N. F. 1952. Intra-varietal diversification in oat breeding. Agron. J. 44:30–34.

———. 1970. A diallel selective mating system for cereal breeding. Crop Sci. 10: 629–635.

Jinahyon, S., and W. A. Russell. 1969. Evaluation of recurrent selection for stalk-rot resistance in an open-pollinated variety of maize. Iowa State J. Sci. 43:229–237.

Johnson, I. J. 1952. Effectiveness of recurrent selection for general combining ability in sweetclover, *Melilotus officinalis*. Agron. J. 44:476–481.

———. 1956. Further progress in recurrent selection for general combining ability in sweetclover. Agron. J. 48:242–243.

———, and A. S. El Banna. 1957. Effectiveness of successive cycles of phenotypic recurrent selection in sweetclover. Agron. J. 49:120–125.

———, and F. Goforth. 1953. Comparison of controlled mass selection and recurrent selection in sweetclover, *Melilotus officinalis*. Agron. J. 45:535–539.

Kalton, R. R. 1948. Breeding behavior at successive generations following hybridization in soybeans. Iowa Agric. Exp. Stn. Res. Bull. 358.

Kehr, W. R., H. O. Graumann, C. C. Lowe, and C. O. Gardner. 1961. The performance of alfalfa synthetics in the first and advanced generations. Nebraska Agric. Exp. Stn. Res. Bull. 200.

Kinman, M. L., and G. F. Sprague. 1945. Relation between number of parental lines and theoretical performance of synthetic varieties of corn. J. Am. Soc. Agron. 37:341–351.

Lawrence, P. K., and K. J. Frey. 1976. Inheritance of grain yield in oat species crosses. Egypt. J. Genet. Cytol. 5:400–409.

Lee, J. A. 1960. A study of plant competition in relation to development. Evolution 14:18–28.

Lonnquist, J. H., and D. P. McGill. 1956. The performance of corn synthetics in advanced generations of synthesis and after two cycles of recurrent selection. Agron. J. 48:249–253.

MacKey, J. 1954. Breeding of oats. Handbüch fur Pflanzenzüchtung 2:512–517.

Moll, R. H., and C. W. Stuber. 1971. Comparison of response to alternate selection procedures initiated in two populations of maize (*Zea mays*). Crop Sci. 11:706–711.

Mumaw, C. R., and C. R. Weber. 1957. Competition and natural selection in soybean varietal composites. Agron. J. 49:154–160.

Neal, N. P. 1935. The decrease in yielding capacity in advanced generations of hybrid corn. J. Am. Soc. Agron. 27:666–670.

Nordquist, P. T., O. J. Webster, C. O. Gardner, and W. M. Ross. 1973. Registration of three sorghum germplasm random-mating populations. Crop Sci. 13: 132.

Penny, L. H., G. E. Scott, and W. D. Guthrie. 1967. Recurrent selection for European corn borer resistance in maize. Crop Sci. 7:407–409.

Poehlman, J. M. 1979. Breeding field crops. 2nd ed. AVI Publishing Co., Inc. Westport, Conn.

Redden, R. J., and N. F. Jensen. 1974. Mass selection and mating systems in cereals. Crop Sci. 14:345–350.

Russell, W. A., S. A. Eberhart, and U. A. Vega. 1973. Recurrent selection for specific combining ability for yield in two maize populations. Crop Sci. 13: 257–261.

Sakai, K. I. 1955. Competition in plants and its relation to selection. Cold Spring Harbor Symp. Quant. Biol. 20:137–157.

———. 1956. Studies on competition among plants. VI. Competition between autotetraploids and their diploid prototypes in *Nicotiana tabacum* L. Cytologia 21:153–156.

———. 1961. Competitive ability in plants: Its inheritance and some related problems. Symp. Soc. Exp. Biol. 15:245–263.

———, and K. Gotoh. 1955. Studies on competition in plants. IV. Competitive ability of F_1 hybrids in barley. J. Hered. 46:136–143.

———, and Y. Suzuki. 1955a. Studies on competition in plants. II. Competition between diploid and autotetraploid plants of barley. J. Genet. 53:11–20.

———, and Y. Suzuki. 1955b. Studies on competition in plants. V. Competition between allopolyploids and their diploid parents. J. Genet. 53:585–590.

Schnell, F. W. 1975. Type of variety and average performance in hybrid maize. Z. Pflanzenzucht. 74:177–184.

Schutz, W. M., and C. A. Brim. 1967. Intergenotypic competition in soybeans. I. Evaluation of effects and proposed field design. Crop Sci. 7:371–376.

Smith, E. L., and J. W. Lambert. 1968. Evaluation of early generation testing in spring barley. Crop Sci. 8:490–493.

Sprague, G. F. 1946. Early testing on inbred lines of corn. J. Am. Soc. Agron. 38: 108–117.

———. 1955. Problems in the estimation and utilization of genetic variability. Cold Spring Harbor Symp. Quant. Biol. 20:87–92.

———, and B. Brimhall. 1950. Relative effectiveness of two systems of selection for oil content of the corn kernel. Agron. J. 42:83–88.

———, and W. T. Federer. 1951. A comparison of variance components in corn yield trials: II. Error, year × variety, location × variety and variety components. Agron. J. 43:535–541.

———, and M. T. Jenkins. 1943. A comparison of synthetic varieties, multiple crosses and double crosses in corn. J. Am. Soc. Agron. 35:137–147.

———, P. A. Miller, and B. Brimhall. 1952. Additional studies of the relative effectiveness of two systems of selection for oil content of the corn kernel. Agron. J. 44:329–331.

———, W. A. Russell, and L. H. Penny. 1959. Recurrent selection for specific combining ability and type of gene action involved in yield heterosis in corn. Agron. J. 51:392–394.

Suneson, C. A. 1949. Survival of four barley varieties in a mixture. Agron. J. 41: 459–461.

———. 1956. Evolutionary plant breeding. Agron. J. 48:188–190.

———, and H. Stevens. 1953. Studies with bulked hybrid populations of barley. USDA Tech. Bull. 1067.

Tiyawalee, D., and K. J. Frey. 1970. Mass selection for crown rust resistance in an oat population. Iowa State J. Sci. 45:217–231.

U.S. Department of Agriculture. 1936. Yearbook of Agriculture. USDA, U.S. Government Printing Office, Washington, D.C.

Walejko, R. N., and W. A. Russell. 1977. Evaluation of recurrent selection for specific combining ability in two open-pollinated maize varieties. Crop Sci. 17: 647–651.

Weatherspoon, J. H. 1970. Comparative yields of single, three-way, and double crosses of maize. Crop Sci. 10:157–159.

Wheat, J. G., and K. J. Frey. 1961. Number of lines needed in oat-variety purification. Agron. J. 53:39–41.

SUGGESTED READING

Allard, R. W., and J. Adams. 1969. Population studies in predominantly self-pollinating species. XII. Intergenotypic competition and population structure in barley and wheat. Am. Nat. 103:621–645.

Sprague, G. F. 1967. Plant breeding. Annu. Rev. Genet. 1:269–294.

Chapter 5

Utilization of Hybrid Vigor

G. W. BURTON
Coastal Plain Station
Tifton, Georgia

Hybrid vigor is an excess in vigor of a hybrid over the midpoint between its parents. The percent hybrid vigor is calculated as follows:

$$\text{Percent hybrid vigor} = \left[\frac{F_1 - (P_1 + P_2)/2}{(P_1 + P_2)/2}\right] 100$$

Only when the offspring (F_1) exceeds the better parent is hybrid vigor of practical significance.

As early as 1716, Cotton Mather recorded the effects of cross-fertilization in plants, ascribing the different-colored grains on an ear of maize to a wind-borne intermixture of cultivars. During the next half century, others made and described hybrids in plants, usually maize. From 1761 to 1766, the German botanist Koelreuter carried out the first systematic studies on plant hybridization (Koelreuter, 1766). He noted that interspecific hybrids were difficult to make and often were sterile. A description of the increased vigor in some of his hybrids is the first published record of hybrid vigor.

During the next century, plant hybridization was routinely practiced by many biologists and farmers. In his classical paper, "Versuche uber Pflanzen—Hybriden," Mendel (1865) wrote, "In repeated experiments, stems of 1 foot and 6 feet in length yielded, without exception, hybrids which varied in length between 6 and 7½ feet." Darwin (1877) carefully measured the amount of hybrid vigor in many hybrids and published his data in *The Effects of Cross and Self Fertilization in the Vegetable Kingdom.* In the first sentence of his last chapter, Darwin wrote, "The first and most important conclusion which may be drawn from the observations in this volume is that cross-fertilization is generally beneficial and self-fertilization injurious."

Although many scientists had studied and reported on hybridization, inbreeding depression, and hybrid vigor, Shull (1908) deserves credit for the modern concept of hybrid vigor. In his publication, *The Composition of a Field of Maize,* Shull (1908) suggested that the increased vigor in hybrids was due to the "unlikeness" in the genetic constitutions of the uniting parental gametes. He later coined the word "heterosis" to describe the stimulation resulting from increased heterozygosity. Today hybrid vigor and heterosis are usually used synonymously to describe the beneficial effects of hybridization.

The widespread use of hybrid vigor in plant improvement today must be credited to the success of hybrid maize. By 1910, Shull and others had provided the essential framework for the commercial production of hybrid maize. Actual use, however, was delayed for 20-odd years by a lack of suitable inbred lines and by skepticism about the feasibility of commercial production of hybrid seed. The second of these objections was removed when Jones (1918) developed the double-cross procedure. The first, lack of suitable lines, remained a limiting factor until the late 1920's. Across the years, many thousands of lines were started and carried to various stages of homozygosity. Most of these lines were discarded because they were inherently weak and performed poorly in hybrid combinations. The relatively few surviving lines of maize in extensive commercial use today have been improved to the extent that single-cross seed can be produced commercially.

In 1933, 0.1% of the U.S. maize crop was planted as hybrid seed. Ten years later, it had increased to 51.6%. Today, hybrids make up nearly all of the crop in the USA.

GENETIC BASIS FOR HYBRID VIGOR

A genetic explanation for hybrid vigor must account for the inbreeding depression in outcrossing populations and the increased vigor in F_1 hybrids between unrelated lines. Two main hypotheses, the dominance hypothesis and the overdominance hypothesis, have been the subject of research and debate among geneticists for more than 60 years. Evidence supports both hypotheses and there is good reason to believe that both function.

Dominance Hypothesis

The dominance hypothesis, suggested first by Davenport (1908), states that hybrid vigor results from the action and interaction of favorable dominant genes. According to this hypothesis, inbreeding depression results when deleterious recessive genes, hidden in the heterozygous condition in the cross-pollinated plants, become homozygous with inbreeding. The crossing of unrelated homozygous lines obscures the deleterious recessives and restores vigor. Vigor of the hybrid may exceed that of the non-inbred parents, if they possess different dominant genes that are homozygous. Theoretically, plants homozygous for all favorable genes could be developed that would perform as well as those heterozygous for these genes. The failure to produce such plants was considered a criticism of the dominance hypothesis until it was recognized that their frequency, 1 in 4^n, if independently inherited, would be astronomical if n (the number of genes) were 25 to 30.

The F_2 distributions from F_1 hybrids that involve metrical characters, such as yield, tend to be symmetrical. If such characters were controlled by five dominant genes, the frequency of the six phenotypic F_2 classes would be asymmetrical (0.1% + 1.5% + 8.8% + 26.4% + 39.5% + 23.7%). This asymmetry has been considered one objection to the dominance theory. Jones (1918) pointed out that linkage would reduce asymmetry. Increased gene number, incomplete dominance, and environmental effects also would help to reduce asymmetry in the F_2 population.

A linear relationship between heterosis and heterozygosity indicates dominant gene action, with many findings that F_2 grain yields generally fall midway between yields of the parents and those of the F_1's. Such studies have usually involved inbred lines that produce less grain under adversity than the plant size would seem to support. To circumvent this problem, Burton (1968a) measured total above-ground plant yield of pearl millet inbred lines (A, B, C, and D), their F_1 single crosses, F_2 single crosses, backcrosses, (A × B) (A × C) crosses, three-way crosses, and double crosses. He found actual yields agreed well with yields projected from the assumption of a linear relationship between yield and 0, 50, 75, and 100% heterozygosity of the material.

Epistasis, the interaction between non-alleles, is a type of gene action that may affect hybrid vigor. Bauman (1959), Sprague et al. (1962), and others have shown that epistasis affects grain yields in maize. The nature and magnitude of the effect of epistasis on yield is difficult to assess. Bauman (1959) obtained both positive and negative epistatic deviations for grain yield in maize. Burton (1968b) suggested that epistasis tended to reduce forage yields of pearl millet hybrids. Both Bauman and Burton found that quantitative epistatic effects were greatly influenced by environment.

Castro et al. (1968), Eberhart and Hallauer (1968), and others used mathematical models with and without epistasis to analyze grain yield for parental generations, F_1's, F_2's, and backcrosses from open-pollinated maize cultivars. The models fit the results well enough to suggest that epistasis is relatively unimportant in such material. Studies conducted to date suggest that epistasis is a relatively unimportant source of genetic variation, but is more important in hybrids between selected inbred lines than between hybrids involving open-pollinated cultivars.

Overdominance Hypothesis

The overdominance hypothesis, proposed first by Shull (1908), states that the heterozygote (*Aa*) of a gene is more vigorous than either homozygote (*AA* or *aa*). It recognizes the effects of partial to complete dominance but states that these effects alone cannot account for all of the vigor observed in many hybrids. Overdominance has been demonstrated in several traits that are controlled by a single gene. Indisputable evidence of its existence in polygenic characters, such as yield, is much more difficult to find. Hull (1952) used regression of progenies on their parents to show overdominance in diallel crosses of maize. Other researchers, such as Rumbaugh and Lonnquist (1959), concluded from their studies of diallel crosses in maize that most of the genetic effects were additive and that little, if any, overdominance was evident.

Numerous theories have been proposed to explain overdominance. Brewbaker (1964) describes four of them: (1) supplementary allelic action, (2) alternative pathways, (3) optimal amount, and (4) hybrid substance. His concluding statement on the subject was: "The tenuous evidence for overdominance seems to encourage the view that the dominance theory of heterosis is, as biological theories go, an uncommonly reliable one for applied research."

UTILIZATION OF HYBRID VIGOR

Success of hybrid maize motivated plant breeders to explore the existence and magnitude of hybrid vigor in many other species with economic importance. The general conclusion from such research is that hybrid vigor, a common phenomenon in plants, is of sufficient magnitude to warrant commercial exploitation if appropriate techniques can be devised.

The economic significance of hybrid vigor has been demonstrated by hybrid maize. From 1928 to 1932 (the period just before hybrids) the USA averaged 1,250 kg/ha on 42 million ha. In 1972 the average was 4,850 kg/ha on 23 million ha. Hybrids, and associated production practices, quadrupled yield and made it possible to nearly double maize production and to free 18.6 million ha for other crops. Hybrids have been credited for as much as 60% of this increase. Hybrids without fertilization and the other production practices could not have produced these yields but neither could these yields have been realized without hybrids.

Hybrids increase yields. They are usually more efficient in use of growth factors and give a greater return per unit for the growth factors such as water and fertilizer. Under stress, F_1 hybrids are generally superior to parental cultivars, with a more stable performance over a wide range of environments. With hybrids, there is uniformity in product and maturity that often facilitates harvest and increases the value of the product in the market place. The F_1 hybrid may combine characters that are difficult or impossible to combine in other ways. This is particularly true of many interspecific and intergeneric hybrids. Often a superior hybrid genotype with economic value can be produced with less development time than a superior cultivar.

Economic considerations usually determine the use of hybrid vigor. Usually it costs more to produce hybrid seed. The superiority of the hybrid must be such that the grower can afford the extra cost of the seed. Also, production techniques that lower propagation costs can expand the use of hybrid vigor and the subsequent benefits.

The commercial applications of hybrid vigor may be classified into two broad groups, dependent on whether the F_1 hybrid is propagated by sexual or asexual means. Opportunities and techniques for the use of hybrid vigor will differ in these two groups, which require different breeding procedures.

Whereas there may be reason for inbreeding in the development of sexually propagated hybrids, there appears to be little need for inbreeding in the production of asexually propagated hybrids.

Asexual Propagation

Vegetative Propagation

Vegetative propagation offers many advantages to the plant breeder. Only one superior plant must be created. The long task of progeny testing can be eliminated. Heterosis, disease resistance, and other desirable traits found in the superior selection can be maintained. If the commercial product is a plant part, such as a tuber, root, stem, or forage, sterile hybrids can be used. Wide crosses to combine the desired traits of different species or genera can be attempted with the assurance that good hybrids can find commercial use even though sterile. Sterile cultivars can be eradicated easily and are less likely to become serious weeds than those which produce seeds. Vegetative propagation helps to identify new cultivars and to make them available to small growers who have more labor than money. Vegetative propagation carries so much potential that plant breeders can afford to use time and resources to develop new methods that may be economically feasible.

Horticulturists have long been able to exploit hybrid vigor for commercial propagation of superior genotypes by grafting, budding, layering, rooting stem cuttings, planting tuber sections, etc. Many superior nut and fruit tree cultivars in use today were seedlings or bud mutants observed by a farmer and sent to a research station for evaluation. The populations of seedlings from which these superior cultivars originated were probably much larger than breeders of these crops could maintain today.

Most of the world's sugar is produced by hybrid canes propagated by planting short stem sections. Elephantgrass and guineagrass are bunch grasses propagated vegetatively in the subtropics from stem or crown cuttings. Rhizomes and stolons have been used to plant over 4.5 million ha of the F_1 forage hybrid, 'Coastal' bermudagrass. Sterile triploid turf-bermudagrass hybrids such as 'Tifgreen', 'Tifway', and 'Tifdwarf' have been planted vegetatively on many golf courses and thousands of lawns. Stoloniferous forage crosses such as pangolagrass and 'Coastcross-1' bermudagrass, which can be planted by broadcasting and disking freshly mowed stems into moist soil, can be established at less cost than many seed-propagated cultivars.

Many vegetatively propagated cultivars arose as natural hybrids. Further improvement in these has usually been attempted by intermating cultivars carrying desirable complementary traits and screening the F_1 population for the desired combination of characters. Theoretically, careful screening of a large F_1 population of such crosses should give the desired combination of characters and some additional hybrid vigor if the parents were unrelated. In practice, recurrent phenotypic selection to increase the frequency of desired genes may be more productive if the cycles can be completed in a relatively short period of time.

Apomixis

Apomixis is most simply defined as vegetative reproduction through the seed. Although the mechanics of apomixis may be varied and complex, the result is in the reproduction of the female through seed to permit the transmission of heterosis without loss from one seed generation to another. This result occurs in many genera of plants including *Citrus, Allium,* and *Rubus.* Many pasture grasses also exhibit this phenomenon. 'King Ranch' bluestem, 'Argentine' bahiagrass, and 'Tucson' sideoats grama are examples of pasture grass cultivars that reproduce by apomixis.

Some species, such as *Poa pratensis* L., are classed as facultative apomicts that can produce both sexual and apomictic seeds. Sexual plants in these species can be mated with facultative apomictic plants to produce new hybrids.

Other species, such as most of the naturally occurring tetraploid *Paspalum notatum* Flugge ecotypes, are obligate apomicts which rarely produce sexual types. When these plants are used as females in crosses, they produce plants like the female and an occasional hybrid consisting of an unreduced egg plus a reduced pollen gain (Burton, 1948). Such hybrids are less fertile and continue to be apomictic. Reduction division is normal in pollen mother cells, and obligate apomixis can be broken by crossing a sexual female with an apomictic male parent. Sexual cultivars, such as the 'Temple' orange, are used as females in the breeding of new cultivars of obligate apomictic citrus.

Apomixis in *Paspalum notatum,* a tetraploid obligate apomictic, was broken by creating induced tetraploids of 'Pensacola' bahiagrass, a sexual diploid. These sexual tetraploids, used as females, were crossed with obligate apomicts to release greater variability than previously observed in nature (Burton and Forbes, 1960). By selection and recombination, superior sexual tetraploid female plants with recessive white stigmas have been developed. Hybrids (made by mutual pollination with emasculation) between these females and red stigma apomicts can be detected by their red stigmas. The uniformity of a five-seedling progeny from open-pollinated seed of these hybrids proves whether they are apomictic (uniform) or sexual (variable). Seed from the uniform progenies are used to establish replicated plots for further evaluation.

The discovery by Pat Higgins, a Texas rancher, of a sexual plant of buffelgrass (Bashaw, 1962) has permitted the rapid genetic improvement of this species (Taliaferro and Bashaw, 1966) and the release of the true-breeding apomictic hybrid, 'Higgins.'

Apomixis is controlled genetically and it may be used to fix hybrid vigor and other desired traits. Unfortunately, apomixis does not guarantee good seed production or other desired traits. The mutual pollination technique used for making bahiagrass hybrids has yielded approximately 20% hybrids, and less than 15% of these hybrids have been apomictic.

Thus, a great deal of material must be handled to obtain a few apomictic hybrids. A treatment that would make apomictic plants temporarily sexual when hybrids are being made would greatly facilitate genetic improvement. It would also enable hybridization in those species where good sexual plants have not been found.

Environmental change during reproduction has been reported to produce sexuality in some apomictic species. At Tifton, Ga., a single, red leafed, apomictic hybrid of tetraploid bahiagrass that was heterozygous for stigma color, on one occasion, gave rise to a heterozygous sexual progeny that segregated for stigma color. Although attempts to make this clone and several others sexual by greatly altering photoperiod, temperature, soil moisture, and soil fertility have not been successful, the possibility is enough to warrant a search for other methods (perhaps chemical treatments) capable of making apomictic plants temporarily sexual (Burton, 1982a).

Sexual Propagation

Hand Emasculation and Pollination

Hand pollination has long been used to produce hybrid seed of ornamentals and tomatoes. Generally, the cost of producing hybrid seed this way is so great that its use is restricted to high value crops. Much of this seed is planted by the flower fancier and the home gardener, who are willing to pay high prices to add zest to their hobbies.

Hand Emasculation and Natural Pollination

This procedure has been restricted to wind- or insect-pollinated plants, usually species in which the sexes are separated on the same plant, such as maize. For many years, the hybrid seed industry has used this method, employing workers to pull tassels from the plants before they shed pollen to make them female. Wind-carried pollen from male rows pollinate these female plants. Both machine and hand detasseling are used today.

Chemical Emasculation

A chemical treatment capable of temporarily making bisexual plants male-sterile would have great potential in the commercial production of hybrid seed. Desirable hybrid combinations could be made simply by applying the chemical to whichever parent would make the best female. Because the treatment would be temporary, the fertility restoration problems associated with the use of cytoplasmic male-sterility would cease to exist. To date, chemicals capable of emasculation under some circumstances have failed to show the reliability required for hybrid seed production. Continued search for such chemicals and techniques for use may one day result in a reliable means of chemically emasculating plants in a commercial hybrid-seed production program.

Dioecious and Monoecious Types

Some seed plants, such as the willows and buffalograss, produce individuals that are either male or female having only staminate or pistillate flowers. If these dioecious plants are perennials and can be propagated vegetatively, commercial hybrid seed may be harvested from isolated fields in which selected male and female clones have been interplanted. 'Mesa' buffalograss, bred in Oklahoma, is a vigorous F_1 hybrid that may be established from seed. Commercial seed of this hybrid is produced by vegetatively interplanting the dioecious parents in isolated seed fields (Hanson, 1972).

Certain monoecious plants, such as the Cucurbitaceae, produce pistillate and staminate flowers on the same plant. In the cucumber, first flowering nodes bear all male flowers, followed by nodes with mixed sexes, and finally, by nodes bearing all female flowers. Short days and low night temperatures shift sex to femaleness, whereas gibberellins (GA) cause strong shift to maleness. Galun (1973) used these responses and a female plant from his stocks of Japanese 'Fushinari' cucumbers to develop a unique system of producing F_1 hybrid seed. In this system, the Israeli virus-resistant cultivar 'Ellem' was made heterozygous-female through a series of backcrosses. Segregating female plants in this population were treated with GA to make them produce male flowers. They were then self-pollinated. Twenty selfed seeds from each plant were progeny tested, and remnant seeds of all homozygous females were increased in isolated fields. Every third row was sprayed with GA_3 (300 mg/L with detergent) once a week to make them produce male flowers. The F_1 hybrid seed was obtained by harvesting seed from the two-row-strips of female stock planted between single rows of the selected monoecious male. Honeybees from hives placed in fields did the pollinating.

Commercial hybrid seed production of castorbean was made possible in Israel by breeding a female plant to serve as the seed parent of hybrid seeds (Atsmon, 1973). Sections of this female plant, after producing female flowers, revert irreversibly in a chimera-like manner to the normal monoecious condition and produce bisexual flowers. Selection for delayed reversion resulted in practically non-reverting females producing, upon selfing, close to 100% female offspring. Interspersed staminate flowers, essential for selfing these females, develop only in the winter under Israeli conditions. Thus female seed increased by this reversion in the winter produces female plants for commercial hybrid seed production in the summer.

Genetic Male Sterility

Genetic male-sterility occurs in many species. Usually it is controlled by a single recessive *ms* gene that must be homozygous to cause male sterility. Breeders who use genetic male-sterility for hybrid seed production usually develop a phenotypically uniform female line that segregates 1:1 for *Msms* and *msms* individuals. Seed for these lines is increased in isolation by harvesting seed from *msms* plants that are pollinated from *Msms* plants. To produce commercial F_1 hybrid seed with genetic male-sterility, the 50% of fertile *Msms* plants must be rogued from the field as soon as their fertility can be identified. This technique was used in the production of 'Giant African' hybrid marigold seed. The *Msms* plants were rogued from the female rows as soon as the first flower opened. Remaining *msms* plants were fertilized by bees that carried pollen from the fertile male rows planted between the female rows.

Genetic markers linked to the *Ms* gene would assist in the identification and roguing of fertile *Msms* plants. Wiebe (1960) suggested that hybrid barley seed could be produced easily if an *ms* gene could be tightly linked to the dominant *Ddt* gene that causes barley seedlings, sprayed with DDT, to die. Spraying the segregating female rows with DDT would kill all the fertile *Ms-Ddt* plants, leaving only the male-sterile *ms-ddt* plants to produce the F_1 hybrid seed. Wiebe reported no such linkage was available and that the percentage of seed set on male-sterile barley plants was low. The technique was never developed, but is described here to stimulate thinking about similar procedures that might be successful.

The labor associated with roguing fertile plants from female rows (even using linked genetic markers, which are hard to find and rarely tightly linked) has greatly restricted the use of genetic male-sterility in producing hybrid seed. Its future use probably will continue to be restricted chiefly to ornamentals for which less expensive methods of hybrid seed production are not available.

Cytoplasmic Male Sterility

Cytoplasmic male sterility is a form of male-sterility that results from the interaction of sterility genes and a specific cytoplasmic factor. It permits reproduction of the female through the seed, which eliminates the roguing required when genetic male-sterility is used to produce F_1 hybrid seed. Although modifying factors exist, most A-type male-sterile systems involve a single recessive gene *ms* which, when homozygous, makes plants with sterile (S) cytoplasm male-sterile but has no effect on plants with normal (N) cytoplasm. Thus pollen-fertile N *msms* plants can be used to maintain male sterility in S *msms* stocks. The dominant allele *Ms* introduced into the system makes heterozygous S *Msms* plants that are male-fertile (Fig. 5.1). If the commercial hybrid is grown for seed, the male parent must carry the dominant *Ms* fertility-restoring gene. If the hybrid is grown for its vegetative parts only, such as onions, sugarbeets, or pearl millet for forage, the male parent need not carry the *Ms* gene. In such cases, the sterile hybrids will prevent growers from suffering the loss of yield that results from planting seed harvested from the F_1 hybrid. A sterile sorghum × sudangrass F_1 forage hybrid will produce no seed capable of becoming a weed in a succeeding cultivated crop unless *Ms* pollen blows in from neighboring fields.

The first commercial use of cytoplasmic male-sterility to produce hybrid seed occurred in 1944, when an onion hybrid between strain 13-53 and 'Lord Howe Island' was distributed. Strain 13-53 was cytoplasmic male-

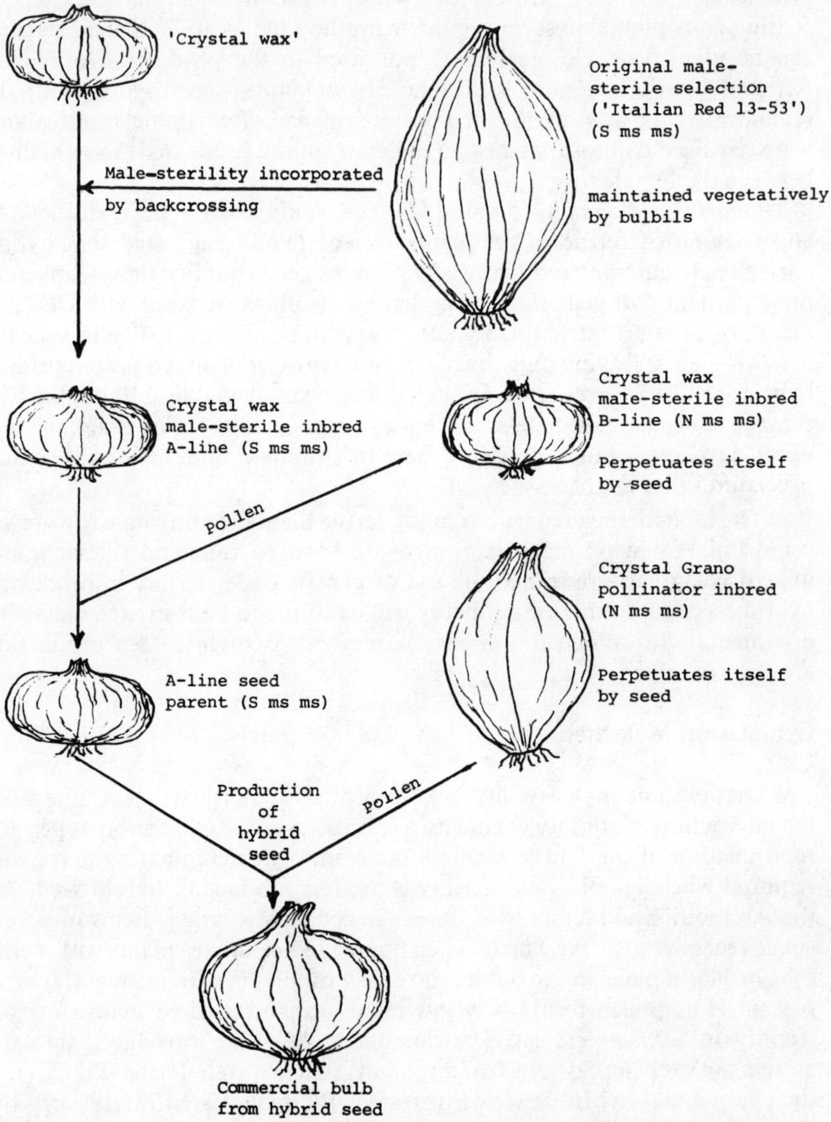

Fig. 5.1. 'Crystal wax' × 'Crystal Grano' hybrid onion seed production. Male-sterility factors S *ms ms* discovered in 'Italian Red 13-53' were incorporated into a Crystal wax inbred by backcrossing. Its sterility is maintained with pollen from a male fertile line of Crystal wax having genetic factors N *ms ms*. Commercial hybrid seed is produced by harvesting seed from the Crystal wax A-line (S *ms ms*) interplanted and pollinated with the Crystal Grano inbred (N *ms ms*).

sterile, the product of research by Jones and Clarke (1943). Seeds for commercial hybrid onion cultivars grown today are produced on a number of *msms* lines from strain 13-53 S cytoplasm. Because the commercial product is a bulb, male fertility need not be restored (Fig. 5.1).

Cytoplasmic male-sterility has been reported in a number of crops, including onions, beets, corn, sorghum, pearl millet, castor beans, and wheat. The trait is found frequently in naturally cross-pollinated species; in self-pollinating species such as wheat, it has been found in wide crosses. The expression of cytoplasmic male-sterility is influenced by environment to the extent that some male-steriles cannot be used for commercial seed production. Burton (1972) found that the environmentally stable Tift 23 A_1 pearl millet male-sterile (A line) produces occasional pollen-shedding heads. The discovery of pollen-shedding sectors on male-sterile heads proved to be caused by cytoplasmic mutations. Seed from these sectors gave rise to pollen-shedding off-spring that, with very few exceptions, were sterility maintainers (B lines). Thus the sterile S cytoplasm mutated to a stable N cytoplasm, a B (maintainer) line. The one or more fertility-restoring exceptions proved to be gene mutations from the sterile *ms* to the dominant fertility restorer *Ms.* Similar cytoplasmic mutation has occurred in male-sterile A_2 pearl millet cytoplasm. The frequency of such mutations is low enough that limited-generation increase would permit continued use of these sterile cytoplasms for commercial hybrid seed production.

If male-sterile S cytoplasm can mutate to male-sterile N cytoplasm, mutation of N to S cytoplasm should also be possible. Burton and Hanna's (1982) success in creating stable S cytoplasm pearl millet mutants by treating seeds of N cytoplasm Tift 23 DB with streptomycin and mitomycin suggests a method for producing S cytoplasm mutants in other species. The S cytoplasm mutants would only express male-sterility if the material being treated were homozygous recessive for the dominant *Ms* fertility restorer gene. Thus, one might need to screen a number of treated genotypes to find one which is homozygous for the *ms* gene.

The successful use of cytoplasmic male-sterility for commercial hybrid seed production requires a stable S cytoplasm, an adequate pollen source, and an effective system of getting the pollen from the male parent to the male-sterile female. Wind, the usual pollinating agent, must be dependable and strong enough to ensure good pollen movement. At Tifton, Ga., where early morning winds are usually low, seed production on male-sterile pearl millet has been improved by using a tractor-powered blower to blow pollen from male rows to male-sterile rows. If grain or fruit is the marketable product, a dependable fertility-restoration system must also be found.

The southern corn blight caused by *Helminthosporium maydis,* Race T, which severely attacked all maize hybrids with cytoplasmic male-sterile T cytoplasm, demonstrated the vulnerability of a hybrid seed production industry based on a single source of male-sterile cytoplasm. For hybrid maize, most seed producers have returned to hand or mechanical emasculation and wind pollination.

Environmentally unstable S cytoplasm, inadequate pollen and pollination systems, and poor fertility restorers are major difficulties in the use of cytoplasmic male sterility in commercial hybrid seed production. But when these problems can be solved, cytoplasmic male sterility offers one of the most effective methods for exploiting hybrid vigor.

Self-Incompatibility

Self-incompatible plants have been found in 66 plant families (Brewbaker, 1957). It may be defined as the inability of normally cross-fertile plants to set seed when self-pollinated. Self-incompatibility is controlled by several genetic systems that usually involve a large number of *S* alleles at a single locus. This results from the interaction of a haploid pollen tube and the diploid style and ovary when they have an *S* allele in common. Pollen-tube growth is arrested, and the male gamete never reaches the egg. Self-fertility (*Sf*) genes exist or arise in many self-incompatible species by mutation or polyploidy. Because there are many *S* alleles, most self-incompatible plants are cross-compatible.

The main advantage of self-incompatibility over cytoplasmic male sterility for hybrid seed production is the fact that all plants in the seed field produce hybrid seed. This increases seed yields and permits bulk harvesting. But the main disadvantage lies in maintaining and increasing parents, particularly for seed-propagated annuals. As a result, only "a few hybrids in the Brassicas are being produced commercially via self-incompatibility mechanisms" (Duvick, 1966). Numerous partial solutions to this problem have been proposed. Self-incompatible plants often set seed when subjected to unusually high or low temperatures. Senescent plants tend to be self-compatible. Self-fertility genes could be carried in lines until they are adequately inbred, at which time such *Sf* genes could be evaluated and removed for hybrid seed production. Producing lines with low self-fertility would assist development, maintenance, and seed increase of the parents, but it would also leave some inbred seed among the commercial hybrid seed.

Exploiting hybrid vigor with self-incompatibility seems most likely to succeed in perennials that can be propagated vegetatively. A good example from our work at Tifton, Ga. is 'Tifhi No. 1' Pensacola bahiagrass. Over a 4-year period, cattle eating this perennial pasture grass, a seed-propagated F_1 hybrid, produced 17% more beef than when they ate the open-pollinated Pensacola bahiagrass check. When selfed, florets of 20% of the plants of Pensacola bahiagrass set less than 2% seed. When grouped in pairs to permit cross-pollination, most of these plants set about 90% seed (Burton, 1955). These self-incompatible clones (in effect, single crosses) could be propagated vegetatively and so were evaluated as clones as well as in diallel hybrid combinations. Clones 14 and 108, the parents of Tifhi 1, gave some of the highest seed yields when evaluated as clones, and gave the highest forage yields when intermated in isolation. Commercial F_1 hybrid seed was produced by harvesting all seed from isolated fields which had been planted vegetatively to alternate strips of the two parents. Combine-harvesting all seed from these fields produced hybrid seed for a number of years. The high cost of labor required to plant the seed fields vegetatively kept this hybrid (based on self-incompatibility) from becoming important economically. Recent research at Tifton, Ga. suggests that mechanical planting can establish seed fields of selected Pensacola bahiagrass clones at a cost low enough to make commercial hybrid seed production by this method economically feasible.

Self-incompatible but cross-compatible clones of bermudagrass were developed that gave high-yielding hybrids when intermated (Burton and Hart, 1967). Seed fields could have been established economically by vegetatively planting alternate strips with these rapidly spreading clones. But the seed yield of the selected bermudagrass clones was too low to make seed production competitive with other crops grown in the seed-producing area.

Amphidiploids

Amphidiploids result when the chromosomes are doubled in hybrids containing genomes with non-homologous chromosomes. Without doubling the chromosomes, such hybrids are usually highly sterile, because there is little if any synapsis between the chromosomes. Doubling the chromosomes in such hybrids should theoretically make them fertile, and might be expected to fix any heterosis carried in the original hybrid. Wheat, tobacco, and other important economic plants originated in nature this way. The extent to which breeders may use this method to fix heterosis needs further investigation.

Chance Hybridization

A seed field, planted to a mixture of two self-fertile lines of a cross-pollinated crop which flower at the same time, should produce 50% single-cross seed and 50% selfed and sibbed seed as a result of random pollen distribution. A seed field planted to a mixture of four self-fertile lines of the same crop should produce 75% of hybrid seed (made up of the six possible single crosses) and 25% of selfed and sibbed seed of the four parents. If the lines have been selected to give high-yield hybrids in all possible combinations, the hybrids will be more vigorous than the selfs and sibs and can suppress or eliminate them soon after emergence. A 6-year study showed that up to 50% of selfed pearl millet seed could be mixed with hybrid seed without significantly reducing forage yields, provided the seed mixture was planted at a rate of 10 kg/ha in 75 cm rows (Burton, 1948b). When seeded at one-fourth this rate, the mixture yielded in proportion to the selfed and hybrid components in it.

'Gahi-1' pearl millet, capable of yielding up to 50% more forage than the common control, is a product of chance hybridization. It is produced by harvesting all of the seed from a field planted to a mixture of equal numbers of pure-line seeds of four inbred lines (13, 18, 23, and 26) that made high-yielding hybrids in all combinations. Plant populations from this seed were 68 to 75% hybrids. The National Foundation Seed Project increases seed of the four Gahi-1 inbreds in isolation and prepares the foundation seed mixture of equal numbers of pure-line seeds of these inbreds. Commercial growers of Gahi-1 purchase foundation seed from the National Foundation

Seed Project and grow it under certification. Only the seed harvested from fields planted with foundation seed can be called Gahi-1. Much of the vigor exhibited by Gahi-1 is lost when it is advanced to the second generation. Gahi-1 has been the most widely grown pearl millet in the USA for many years, and it serves as a check in tests of new hybrids.

The chance-hybridization method has also been used to produce hybrid Sudangrass which yielded 27% more than the controls (Burton et al., 1954) and promises a means of exploiting heterosis in many cross-pollinated crops. For the chance-hybridization method to succeed, the parents must be highly cross-pollinated, flower at the same time, give reasonably good seed yields, and produce hybrids in all combinations that are vigorous enough to crowd out most of the parents in the seedling stage. Because the percentage of chance hybrids increases as the number of parents increases, it is desirable to use at least three or four parents. Parent lines or inbreds that flower at the same time and set seed reasonably well must be evaluated in diallel single-cross yield trials. Yields of a four-line chance hybrid can be predicted quite accurately by averaging the yields of the six single crosses possible from the four lines. Thus, yield data from a ten-line diallel will enable the breeder to predict the yield for many more four-line chance hybrids. The search for good lines can be facilitated by using two or three lines that give the highest-yielding single crosses as testers for new lines.

Synthetic Cultivars

Synthetic cultivars are developed by intercrossing a number of selected lines or plants, harvesting the bulk seed, and replanting it in successive generations. Frequently lines selected for a specific purpose such as "stiff stalk" are pooled together to make a "stiff stalk synthetic." Many forage cultivars are synthetics produced by pooling the seed of several superior clones or by intermating the clones in a polycross nursery and pooling all seed produced. A degree of stability is insured in synthetics such as 'Regal' and 'Tillman' white clover by reconstituting breeders' seed by pooling seed harvested from the component clones.

It may be possible to exploit some heterosis in a highly cross-pollinated species by proper development of synthetic cultivars. Ideally, they should be made up of high-yielding lines that have high-yielding F_1's (considerably more than control cultivars) in most, if not all, hybrid combinations. Thus, lines which flower at the same time and possess other desirable traits must be tested in hybrid combinations for yield alone. If the synthetic is to be made from clones, the yields when selfed should be obtained to predict inbreeding depression.

In most schemes for increasing seed supplies of synthetic cultivars, several generations elapse before the cultivar reaches the public. Usually synthetic cultivar yields level off by the Syn-4 generation, with yields between those of the parent lines and their F_1 hybrids. If the components were highly cross-pollinated, the stabilized synthetic will reach a higher level than if they were only moderately cross-pollinated (Tysdal et al., 1942). In a crop with no self-incompatibility, the amount of selfing can be substantial. In protogynous pearl millet where 100% outcrossing could be expected, 18 to 31% of selfing was observed at Tifton, Ga. (Burton, 1974). Producing seed of a synthetic cultivar under conditions that maximize outcrossing will also maximize the heterosis that can be retained in advanced generations.

REDUCING GENETIC VULNERABILITY

In 1970, an estimated 15% of the maize crop in the USA was destroyed by Southern leaf blight cause by a new virulent Race T of *Helminthosporium maydis.* Where the environment favored the epidemic, entire fields of maize were destroyed.

Race T was unique in that it was virulent only on hybrids in T male-sterile cytoplasm. Before Race T appeared, numerous experiments had revealed no consistent differences between hybrids in T cytoplasm and normal cytoplasm in stalk-breaking, disease susceptibility, moisture percent in the grain, yield, or other important economic characteristics.

By using the T cytoplasmic-male-sterility system in lieu of the more costly detasseling by hand, the hybrid maize industry had been able to save millions of dollars for farmers in the cost of hybrid seed. The industry had used the T male-sterile-cytoplasm exclusively because it was stable, with minimal differences between reciprocal crosses (presumably in different cytoplasms). Cytoplasm had seemed to have little or no effect on important economic traits of the crop.

Although most of the maize crop was in T cytoplasm in 1970, few people thought the crop was vulnerable. The lesson learned from the 1970 epidemic was that genetic uniformity, be it cytoplasmic or nuclear, may increase vulnerability to epidemics.

Most major crops in the USA are remarkably uniform and vulnerable. Most, if not all, crops using a male-sterile cytoplasm system for hybrid seed production have had only one system in use commercially. For many crops, uniformity facilitates harvest and pleases the consumer. The grower seeks the best, most productive hybrid or cultivar. One way to reduce genetic vulnerability without the loss of uniformity for the desired traits would be to use a mixture of genotypes. For example, the highly productive cultivar 'Florunner' peanut is a mixture of genotypes which are uniform for maturity and seed characteristics. Two years after release in 1969, it occupied over 95% of the peanut acreage in Georgia, Florida, and Alabama. Because it is a mixture of genotypes. Florunner is less vulnerable than earlier cultivars which were single pure lines.

Technology Assessment

There must be continual assessment of the technology for breeding better cultivars and getting them into production. The technology used to produce hybrid vigor—not hybrid vigor itself—was responsible for the 1970 corn-blight epiphytotic. Fortunately, the hybrid-seed industry was able to return to hand detasseling to produce a winter crop of maize seed in the southern hemisphere, as an alternate technology for producing hybrid seed. Alternate technologies should be developed for the production and distribution of seed of new hybrids for all major crops. Plant breeders should continually search for new and better technologies in breeding hybrids and ways to put them within reach of all.

Hybrid seed, for example, need not depend on the costly conventional production of seed produced on female strips in seed fields. It can also be produced by planting an isolated field to a mixture of 90% of the cytoplasmic male-sterile female parent and 10% of the fertility-restoring male parent and harvesting all seed in the field. This technique works with pearl millet and should work with sorghum and other similar crops. The 10 to 15% of inbred male parent in the mixture would have little effect on yield, particularly in arid regions of developing nations where several seeds are planted by hand in hills. In those countries, a government agency could produce the parent seed and prepare the foundation seed mixtures to distribute to good farmers who could produce seed for themselves and their neighbors. These hybrids should enable farmers in arid regions to achieve more consistent performance and greater yields.

Germplasm Collection, Evaluation and Use

If genetic uniformity results in genetic vulnerability, more work on genetic diversity would seem to be a logical solution to the problem. The diverse germplasm (landraces) of every major crop should be collected, increased, and put in long-time storage. Seeds of most crops dried to 8% moisture, stored in moisture-proof containers at $-18°$ C, should remain viable for over 100 years. Germplasm collections should be assessed and catalogued for resistance to major pests. Equally important, germplasm should be indexed so that it can be easily and quickly screened for resistance to new pests. Mixtures which have equal quantities of seed of landraces with a common maturity or height could be quickly available to breeders anywhere in the world. If the crop is cross-pollinated, these mixtures should be made of S_1 seed in order to uncover recessive sources of resistance. Such mixtures would be superior to the process of growing germplasm pools which may lose genes, cover recessive genes, and cost more to maintain. The mixtures, like the original landraces, should be labeled, catalogued and put into long-time storage in containers large enough to allow an adequate sample to screen for pest resistance.

Except for genes for resistance, most landraces in a germplasm collection add little, if anything, directly to a hybrid breeding program. At Tifton, inbred lines of pearl millet, developed from F_1 hybrids between a good combining inbred and a landrace, tended to give better inbreds with slightly better combining ability than inbreds developed directly from the landrace. Grouping landraces with one or more important traits in common, such as maturity and height, into pools that can be subjected to population improvement may be the most effective way to develop good material for a hybrid program.

Active Breeding Programs

Well-designed active breeding programs should reduce the genetic vulnerability of a crop and endeavor to develop outstanding hybrids unrelated to those in general use. In order to preserve their cytoplasm, landraces should be used both as females and males in backcrossing programs. Effects of promising new landrace characters for yield or quality should be assessed by an appropriate method, such as the near-isogenic population technique (Burton et al., 1968), before attempting to incorporate them in all hybrids. If cytoplasmic male sterility is used to produce hybrid seed, the breeder should find and develop more than one source. At Tifton, Ga., a number of inbreds with high combining ability are being tested as maintainers on three different male-sterile cytoplasms of pearl millet. One line that is a maintainer for all three cytoplasms, used as the female, will permit the development of the same hybrid in three different cytoplasms. This will reduce the vulnerability of that hybrid to cytoplasm-specific pests.

An active crop breeding program should produce a continuing stream of new hybrids. The old concept, that only the hybrids superior to those in current use should be released, may no longer hold. Some comparable but unrelated hybrids should be named and released as insurance against loss due to epidemics. Others should be held in reserve for the same purpose.

Population improvement to develop new lines should be a part of every well-designed hybrid breeding program. This should consist of repeated cycles of intermating superior selections in each newly formed population to build up the frequency of those genes responsible for the superiority of the selections.

The use of hybrid vigor to improve plants of economic importance has increased tremendously in the past 40 years. There is good reason to believe that its use will continue to increase. Hybrids can maximize the yields required to feed a hungry world. Mechanisms in other species that will permit economical hybrid seed production must be found. Much testing will be required to create better parents, be they inbred lines or clones. Certainly the ultimate efficient utilization of hybrid vigor requires a better understanding of the phenomenon of heterosis.

LITERATURE CITED

Atsmon, Don. 1973. Sex inheritance and hybrid seed production in castor bean. p. 153–156. *In* R. Moar (ed.) Agric. Genet. Sel. Tropics. John Wiley and Sons, New York.

Bashaw, E. C. 1962. Apomixis and sexuality in Buffelgrass. Crop Sci. 2:412–415.

Bauman, L. F. 1959. Evidence of non-allelic gene action in determining yield, ear height, and kernel row number in corn. Agron. J. 57:531–534.

Brewbaker, J. L. 1957. Pollen cytology and self-incompatibility systems in plants. J. Hered. 48:271–277.

————. 1964. Agricultural genetics. Prentice-Hall, Englewood Cliffs, New Jersey. p. 81–85.

Burton, G. W. 1948a. The method of reproduction in common bahiagrass, *Paspalum notatum.* J. Am. Soc. Agron. 40:443–452.

———. 1948b. The performance of various mixtures of hybrid and parent inbred pearl millet, *Pennisetum glauvum* (L.) R. Br. J. Am. Soc. Agron. 40:908–915.

———. 1955. Breeding Pensacola bahiagrass, *Paspalum notatum.* I. Method of reproduction. Agron. J. 47:311-314.

———. 1968a. Heterosis and heterozygosis in pearl millet forage production. Crop Sci. 8:229–230.

———. 1968b. Epistasis in pearl millet forage yields. Crop Sci. 8:365–368.

———. 1972. Natural sterility maintainer and fertility restorer mutants in Tift 23A cytoplasmic male-sterile pearl millet, *Pennisetum typhoides* (Burm.) Stapf. and Hubb. Crop Sci. 12:280–282.

———. 1974. Factors affecting pollen movement and natural crossing in pearl millet. Crop Sci. 14:802–805.

———. 1982. Effect of environment on apomixis in bahiagrass. Crop Sci. 22: 109–111.

———, E. H. DeVane, and J. P. Trimble. 1954. Polycross performance in Sudangrass and its possible significance. Agron. J. 46:223–226.

———, and I. Forbes, Jr. 1960. The genetics and manipulation of obligate apomixis in common bahiagrass (*Paspalum notatum* Flugge). p. 66–71. *In* C. L. Skidmore (ed.) Proc. XIII International Grassland Congress. Univ. of Reading, England. 11–21 July 1960, Alden Press, Oxford, Great Britain.

———, J. B. Gunnels, and R. S. Lowrey. 1968. Yield and quality of early- and late-maturing, near-isogenic populations of pearl millet. Crop Sci. 8:431–434.

———, and W. W. Hanna. 1982. Stable-cytoplasmic male-sterile mutants induced in Tift 23B, pearl millet with mitomycin and streptomycin. Crop Sci. 22:651–652.

———, and R. H. Hart. 1967. Use of self-incompatibility to produce commercial seed-propagated F_1 bermudagrass hybrids. Crop Sci. 7:524–526.

Castro, G. M., C. O. Gardner, and J. H. Lonquist. 1968. Cumulative gene effects and the nature of heterosis in maize crosses involving genetically diverse races. Crop Sci. 8:97–101.

Darwin, Charles. 1877. The effects of cross and self fertilization in the vegetable kingdom. Dappleton and Company, New York.

Davenport, C. G. 1908. Degeneration, albinism and inbreeding. Sci. 28:454–455.

Duvick, D. N. 1966. Influence of morphology and sterility on breeding methodology. p. 85–138. *In* K. J. Frey (ed.) Plant breeding. Iowa State Univ. Press, Ames.

Eberhart, S. A., and A. R. Hallauer. 1968. Genetic effects for yield in single, three-way and double-cross maize hybrids. Crop Sci. 8:377–379.

Galun, Esra. 1973. The use of genetic sex types for hybrid seed production in *Cucumis.* p. 23–56. *In* R. Moav (ed.) Agric. Genet. Sel. Topics. John Wiley and Sons, New York.

Hanson, A. A. 1972. Mesa. p. 39. *In* Grass Varieties in the U.S., Agric. Handbook No. 170, USDA U.S. Governemtn Printing Office, Washington, D.C.

Hull, F. H. 1952. Recurrent selection and overdominance. p. 451–473. *In* J. W. Gowen (ed.) Heterosis. Iowa State Univ. Press, Ames.

Jones, D. F. 1918. The effects of inbreeding and cross-breeding upon development. Conn. Agr. Exp. Stn. Bull. 207.

Jones, H. A., and A. E. Clarke. 1943. Inheritance of male sterility in the onion and the production of hybrid seed. Proc. Am. Soc. Hort. Sci. 43:189–194.

Koelreuter, J. G. 1766. Vorlaufigan Nachricht von einigen das Geschlecht der Pflanzen betreffenden Versuchen und Beobachtungen. Leipzig.

Mendel, Gregor. 1865. Versuche uber pflanzen-hybriden. Naturf. Ver. in Brunn. Verh. 4:3–47.

Rumbaugh, M. D., and J. H. Lonnquist. 1959. Inbreeding depression of diallel crosses of selected lines of corn. Agron. J. 51:407–412.

Shull, G. H. 1908. The composition of a field of maize. Rept. Am. Breeders' Assoc. 4:296–301.

Sprague, G. F., W. A. Russell, L. H. Penny, T. W. Horner, and W. D. Hanson. 1962. Effect of epistasis on grain yield in maize. Crop Sci. 2:205–208.

Taliaferro, C. M., and E. C. Bashaw. 1966. Inheritance and control of obligate apomixis in breeding buffelgrass, *Pennisetum ciliare.* Crop Sci. 6:473–476.

Tysdal, H. M., T. A. Kiesselbach, and H. L. Westover. 1942. Alfalfa breeding. Nebraska Agric. Exp. Stn. Res. Bull. 124.

Wiebe, G. A. 1960. A proposal for hybrid barley. Agron. J. 52:181–182.

SUGGESTED READING

Fehr, W. R., and H. H. Hadley. 1980. Hybridization of crop plants. Crop Sci. Soc. of Am., and Am. Soc. of Agron., Madison, Wis.

Shull, G. H. 1952. Beginnings of the heterosis concept. p. 14–18. *In* J. W. Gowen (ed.) Heterosis. Iowa State Univ. Press, Ames.

PART II.

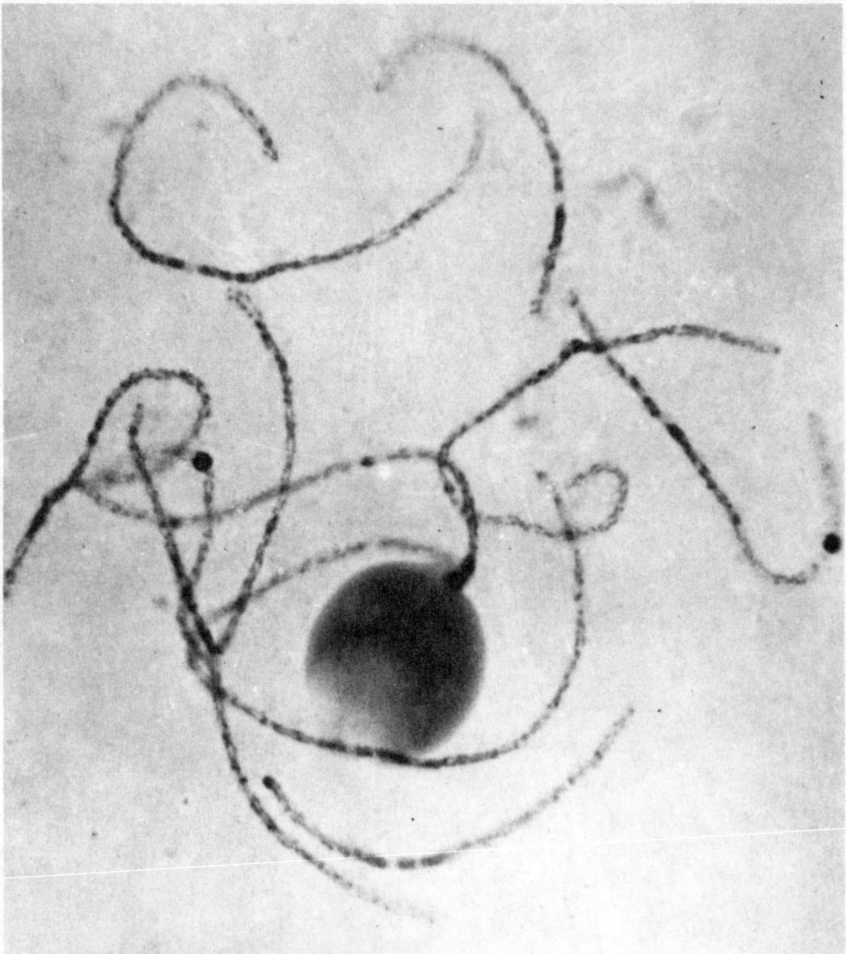

Remodeling the DNA

Chapter 6

Remodeling Crop Chromosomes

ROSALIND MORRIS

University of Nebraska
Lincoln, Nebraska

Do you know that plant scientists have developed potatoes with half the normal chromosome number? Clover with twice the normal chromosome number? A new crop that combines the chromosomes of wheat and rye? Wheat with a piece of a chromosome from a wild grass attached to one of its chromosomes? These are all examples of chromosome remodeling. The methods used to get these conditions and the use of these new or modified crops will be discussed in this chapter. Let us first try to find a general definition that includes all these crop alterations.

WHAT IS CHROMOSOME REMODELING?

Chromosome remodeling as used in this chapter is the manipulation of plant chromosomes by various methods and techniques to create desirable genetic effects. These effects may be the result of increasing or decreasing the doses of genes, combining genes from two different crops, or transferring one or more genes from a wild relative to a crop plant. The definition implies that we have prior knowledge about the chromosomes and their gene content. Before discussing some successes as well as problems in chromosome remodeling, we need to know some established facts about chromosomes of crop plants.

CHROMOSOMES OF CROP PLANTS

The distinguishing characteristics of each crop are due in large part to the genetic information in the form of deoxyribonucleic acid (DNA) molecules that traverse the chromosomes. These DNA molecules contain the chemical code for the pattern of inheritance. The transfer of this genetic information from one generation to the next depends on the consistent behavior of the chromosomes during cell divisions. Since each chromosome

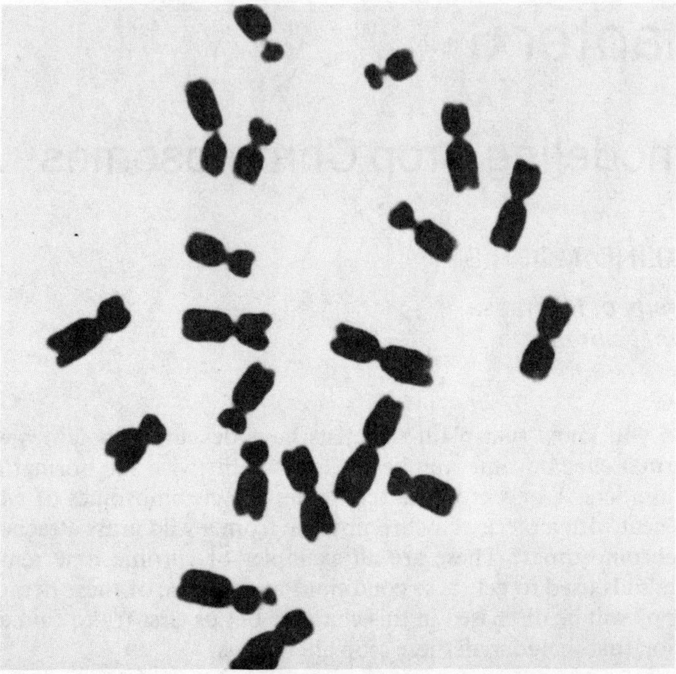

Fig. 6.1. Maize chromosomes at the pairing stage of meiosis. (Courtesy of M. M. Rhoades, Dep. of Biology, Indiana Univ., Bloomington. Slide prepared by D. T. Morgan, Jr.)

can duplicate itself exactly at each cell division, a continuing supply of a crop's unique package of genes, i.e. DNA segments, is assured.

In general, chromosomes are remarkably stable and predictable in their behavior. Thus, one way to characterize the chromosomes of a crop is by number. Take maize, for example. If we germinate a maize seed in an incubator, cut off the tip of a root, stain it, and make a microscope-slide preparation of cells in the meristematic region, we expect to find 20 chromosomes in each cell. This group of chromosomes is called a chromosome complement (Fig. 6.1). The same number of chromosomes would be found in cells from a leaf or any other vegetative part of the plant.

If we dissect an anther from an unopened flower (spikelet) in the maize tassel and squash the meiotic cells in a drop of stain, we can see, through the microscope, cells with 10 pairs of chromosomes (Fig. 6.2). The 20 chromosomes that we observed in the root cells have come together in 10 pairs because there are basically 10 different chromosomes in maize. The vegetative or somatic cells have two doses of each kind of chromosome and identical (homologous) chromosomes are attracted to each other during meiosis. In the early phase of the pairing period, the maize chromosomes can be differentiated by length, the position of the centromere (a specialized region that gives mobility to the chromosome), and a pattern of dark- and light-staining regions (Fig. 6.3). These morphological differences are useful when we want to identify a specific chromosome that has been modified structurally.

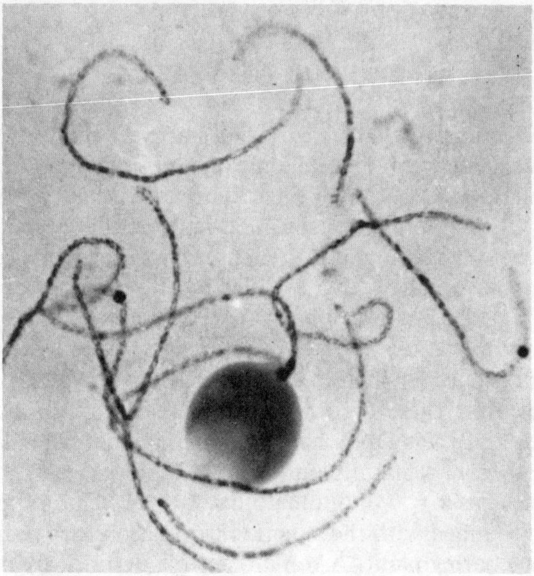

Fig. 6.2. Maize somatic chromosome complement. (Courtesy of J. D. Horn and D. B. Walden, Dep. of Plant Sciences, Univ. of Western Ontario, London.)

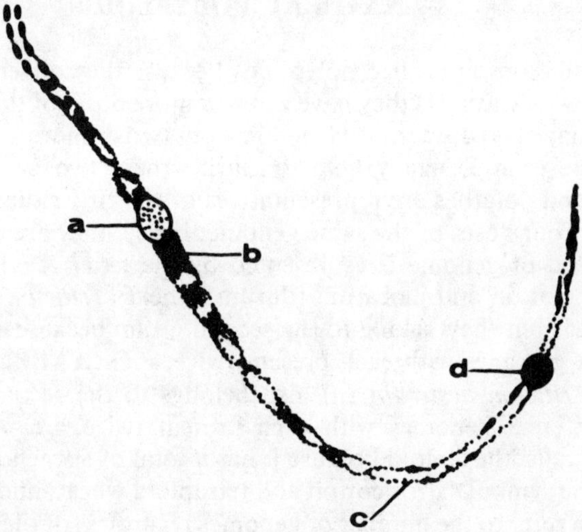

Fig. 6.3. Maize chromosome showing centromere (a), dark-staining area (b), light-staining area (c), and knob (d).

Maize pollen or ovules are the ultimate products of meiosis and have the reduced chromosome number (10) in their nuclei. This gametic chromosome number is symbolized by n. Therefore, the 20 chromosomes in the somatic maize cells would be 2n.

CROP GENOMES

We have indicated that maize has a basic set of 10 different chromosomes. The basic set is called a genome, symbolized by x. Maize is a diploid (di = two, and ploid = number of sets) because it has two doses of the one genome, i.e. 2x, in its somatic cells. You may wonder at this point what the difference is between 2x and 2n. Remember that 2n refers to the number of chromosomes in somatic cells regardless of the number of genomes. In a diploid, 2n = 2x but, as we shall see later, there are polyploid conditions where 2n = 4x or 6x, etc.

Sorghum is another diploid crop with 10 different chromosomes in its genome. However, these chromosomes differ in appearance from maize chromosomes and their DNA has the genetic code that makes a sorghum plant different from a maize plant. Not all diploid crops have x = 10. For example, barley has x = 7 and tomato has x = 12. Chromosomes in a barley seed are programmed with the genetic information that results in the characteristics of a barley plant. A tomato seed is destined by its genes to produce what we recognize as a tomato plant.

NATURAL POLYPLOIDS

Some crops are called polyploids because they differ from diploids in the following ways: (1) they have three or more doses of the same genome or (2) they have two doses of each genome but two or more kinds of genomes.

Some examples may help to clarify these two types of polyploids. Alfalfa and potatoes are representatives of the first situation because each crop has four doses of the same genome; hence, they are called tetraploids. If the kind of genome is represented by the letter A, the tetraploids are AAAA. Cotton and macaroni (durum) wheat (*Triticum turgidum* L.) are tetraploids but they belong to the second group because each crop has two different genomes with each present twice, e.g., AABB. Bread (aestivum) wheat (*Triticum aestivum* L.) also belongs to the second group and has three different genomes with each present twice, i.e. AABBDD. Bread wheat is called hexaploid because it has a total of six genomes. By the way, we use the symbol 4x for cotton and tetraploid wheat, and 6x for hexaploid wheat, to refer to the number of genomes regardless of their composition.

CHROMOSOME REMODELING—PRECEDENTS IN NATURE

Increases in Chromosome Number

The polyploid crops that we have just mentioned are examples of chromosome remodeling in nature because they have evolved under natural conditions. The primitive ancestors of present-day crops were diploids.

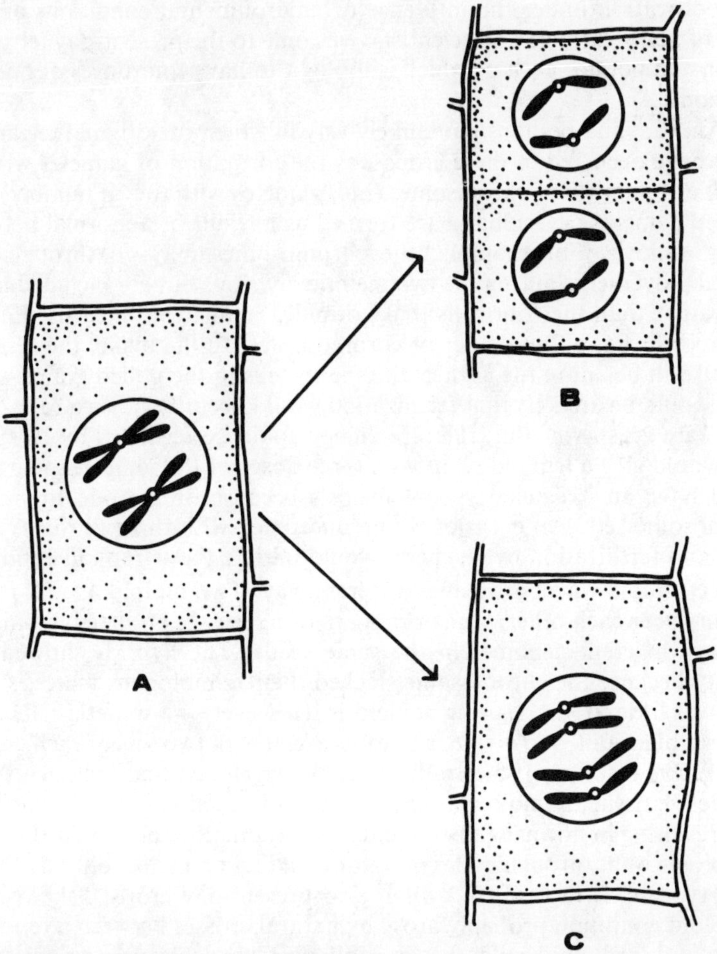

Fig. 6.4. Somatic diploid cell, with two homologous chromosomes and their sister chromatids (A), which gives rise to two diploid daughter cells by normal cell division (B), or one tetraploid cell when cell division fails (C).

Nature, according to and like humans, makes occasional errors. In alfalfa or potatoes, perhaps cell division failed to occur in a somatic cell of a diploid ancestral plant. The chromosomes divided, but the two linear halves (chromatids) of each chromosome stayed in the same cell instead of separating into two cells (Fig. 6.4). Thus, a tetraploid condition arose. Assuming that the tetraploid cell divided normally at the next cycle of cell division, a sector of tetraploid cells developed. The size of the sector depended on when the event occurred in the development of the affected plant. The earlier the occurrence, the larger the sector because of more intervening cell divisions. Alfalfa and potatoes reproduce sexually through the fertilization process as well as asexually through vegetative parts of the plants. If the tetraploid sector included flowers or some of the vegetatively reproducing tissue, the tetraploid condition could be perpetuated. Thus, over centuries of modification under the influence of environmental conditions and the work of many farmers and scientists, we come to the present-day tetraploid alfalfa and potato, each of which is thought to have four doses of one kind of genome.

Another and perhaps more likely way in which the original tetraploids might have occurred in these crops was the formation of gametes with the unreduced number of chromsomes (i.e., gametes with the 2n number). Unreduced gametes occasionally are formed as a result of abnormal behavior during meiosis. For example, the chromosomes may go through their normal movements during the two meiotic divisions but are included in two cells rather than the four cells that normally result from meiosis. Each of the two cells has a 2n number of chromosomes. In the ovule, the resultant egg cell will be 2n; in the anther, the sperm cells in the pollen grains will be 2n. It would be unlikely that fertilization would occur between a rare 2n egg and a rare 2n sperm. But, the rare 2n egg could be fertilized by 1n sperm. This would give a triploid plant with three doses of the same genome and it would have an irregular type of meiosis because of the odd number of chromosome sets. An occasional unreduced egg with three genomes could occur and fertilization by 1n sperm would initiate the tetraploid condition.

Tetraploid wheat arose in a different way. Two diploid ancestors were growing near each other. One or more natural crosses occurred, combining the two different genomes in the same seeds. The hybrids showed high sterility because the chromosomes lacked their homologous mates, so that there was little, if any, pairing at meiosis. However, if a somatic cell in such a hybrid plant failed to divide, a tetraploid cell with two sets of each genome would have resulted. If a tetraploid sector developed that included the reproductive tissue, homologous chromosomes in each kind of genome could pair, resulting in normal gametes and seed production. No doubt these new tetraploids went through cycles of natural and human selection, just like the other type of tetraploid, in evolving to present-day crops. Likewise, the hexaploid condition probably arose by natural crosses between a tetraploid wheat and a diploid wild species, followed by a chromosome-doubling event. Tetraploid cotton arose in a manner similar to that of tetraploid wheat.

Decreases in Chromosome Number

A reduction in chromosome number from 2n in the somatic cells to 1n in the gametes is a natural process in plants as in other organisms and is necessary to maintain a constant 2n condition after two gametes unite in fertilization. However, an occasional plant occurs with the reduced number of chromosomes in its somatic cells. This plant is called a haploid. How does this condition arise? In another of nature's capricious errors, fertilization does not take place but the 1n egg cell divides and a 1n embryo develops.

What happens to a crop plant when the chromosome number is reduced by half in its somatic cells? It depends on whether the crop is a diploid or a polyploid species. If halving occurs in a diploid crop, the somatic cells have only one genome present. This monoploid situation can have disruptive effects on vigor and fertility. Another effect is that recessive mutations, which are concealed by their dominant alleles in the diploid, will be expressed in the monoploid. Since recessive mutations often have harmful effects, monoploid plants with this type of mutation rarely survive.

If the chromosome number halves in a polyploid, the chromosome makeup of the resultant haploid (called polyhaploid when it comes from a polyploid) depends on whether the polyploid state is attributable to the multiplication of one genome or the diploid state of two or more different genomes. In the latter case, the polyhaploid would have one dose of each genome, a situation similar to that in a haploid from a diploid. In the former case, if the polyploid, say a tetraploid, has four doses of the same genome, the polyhaploid would still have two sets of chromosomes and would be essentially like a diploid. The allelic states of the genes, i.e. homozygous or heterozygous, would be influenced by whether the crop is self-fertilizing or cross-fertilizing, respectively.

Exchanges of Segments Between Chromosomes

Another type of chromosome change that is useful in chromosome remodeling is breakage and exchange of pieces. Events of this type affect the structure of chromosomes. Their ends normally have properties that prevent them from fusing with any other chromosome ends. However, occasionally a chromosome thread may break into two or more pieces. We do not know how these breaks occur under natural conditions because we cannot detect them until later. We can speculate that there may be stress on the thread at the time the chromosome is replicating. Or, the plants may be exposed to some environmental condition, such as high temperature or natural ionizing radiation, which can break chromosomes. Whatever the cause, a broken end is able to fuse with another broken end for a certain period of time after the break has occurred. The type of fusion that interests us from a chromosome remodeling standpoint is an exchange of broken pieces between different chromosomes (Fig. 6.5). This is called a reciprocal translocation. It is obvious that such a mutual exchange of segments between chromosomes is accompanied by an exchange of genes. We shall see how this kind of chromosome abnormality can be used to transfer desirable genes from a wild plant to a cultivated crop.

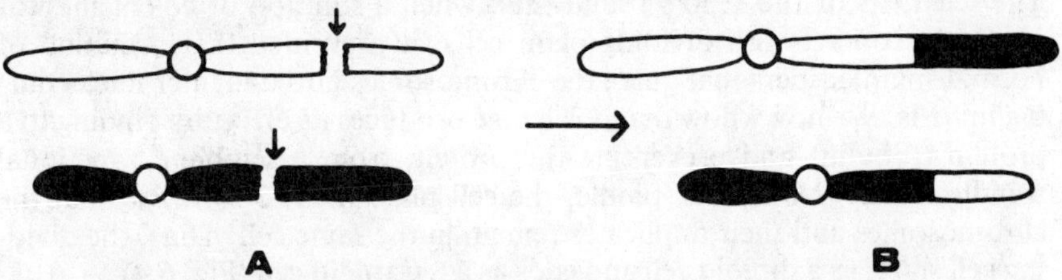

Fig. 6.5. Reciprocal translocation resulting from breaks in two different (nonhomologous) chromosomes (A), and exchange of segments (B).

EXPERIMENTS IN CHANGING CHROMOSOME NUMBERS

Over the years, scientists have sought to mimic nature in remodeling crop chromosomes but with some very important differences. Whereas naturally occurring variations in chromosome number or structure are fortuitous events and may or may not improve a crop, scientists have some clear objectives in mind when they try to bring about alterations. Furthermore, a number of methods or techniques have been developed that increase the chances of producing desired changes.

Before attempting to restructure a crop by chromosome remodeling, we need to consider the crop's present status. Many improvements in crop cultivars come about by breeding methods described in other parts of this book. These methods add, subtract, combine, or rearrange genes without any changes in chromosome number or structure. However, some unique situations can be achieved only by chromosome manipulation. For example, a change from a natural diploid to a tetraploid state can result in large fruits, seeds, and other plant parts. A reduction in chromosome number from a natural tetraploid to a diploid state simplifies genetic and breeding studies. The transfer of a gene for disease resistance from a wild species to a crop cultivar enhances the value of that cultivar but is usually not achieved by hybridization and selection methods alone. Use of chromosome remodeling methods requires laboratory and greenhouse facilities as well as persons trained in chromosome techniques and use of a microscope.

We are going to discuss some examples of chromosome remodeling in crop plants where scientists have developed or used methods that increase the chances of making a desired change in a crop. As you become familiar with these examples, try to evaluate the objectives sought, the method or technique used, and the degree of success attained.

Induced Chromosome Doubling

One aspect of plant chromosome remodeling began in the 1930's, when several scientists discovered that a chemical could bring about chromosome doubling during mitosis. This chemical was a natural product of a plant called autumn crocus, *Colchicum autumnale* L., hence its name colchicine. The scientists of the 1930's found that, when a solution of colchicine was placed in contact with dividing plant cells, it prevented the formation of normal spindle fibers that guide the chromosomes into daughter nuclei during mitosis. We now know that colchicine produces its effect by binding to a protein (tubulin) and preventing the protein from assembling functional spindle fibers. Without a spindle, the cell plate fails to form, so that the chromosomes and their duplicates remain in the same cell. Thus, the dividing cell starts as a diploid cell and ends as a tetraploid cell (Fig. 6.4).

Two additional significant findings were made. One was that colchicine had the same effect on the cells of all kinds of plants, including various crop plants. The other important observation was that colchicine produced its drug-like effect on the spindle only as long as it was in contact with a cell. When it was removed, the cell could resume a normal pattern of behavior, forming a spindle and dividing at appropriate times. If, from previous studies, the approximate time that it took cells to go through one cycle of division was known, colchicine could be applied just long enough to produce one round of chromosome doubling.

Colchicine is usually applied to germinating seeds or seedlings, so that it can be taken up by shoots or roots and make its way to the actively dividing regions of the plants. Only a portion of the total cells in a plant undergo chromosome doubling because not all of the cells are in an active state of division at one time. Hence, a diploid plant after treatment will have a mixture of diploid and tetraploid cells. The tetraploid state has to include the male and female reproductive organs if it is to be transmitted to the next generation. This requirement can be waived if the fertilized egg (zygote) is doubled, but it is difficult to time the colchicine treatment with the division of the zygote.

In the early years of the so-called "colchicine era", crop scientists were optimistic about the potential of the "instant polyploids" because, when compared with diploids, they produced larger flowers, fruits, seeds, leaves, and tubers. However, the polyploids had some disadvantages. They were later in maturing and were more sensitive to low temperatures because of a higher water content in the cells. They produced fewer seeds because of an effect of the extra chromosome sets on fertility. We have to remember that the polyploid state involves not only the added bulk of chromosome material but also an increase in the dose of each gene. If a gene is present in four doses instead of two doses, it may behave differently in its individual action and its interaction with other genes, including genes that are located in certain organelles of the cytoplasm, namely, chloroplasts and mitochondria. The added doses of genes in the nuclei may cause differences in the reaction of diploid plants and the derived polyploids to the same environmental conditions.

It was soon realized that there is a better chance of improving a crop species by chromosome doubling if it has certain characteristics. One of the most important is a low chromosome number, preferably in a diploid. This is based on the assumption that crop plants, like other plant species, have evolved with a chromosome number that is at or close to the optimum level for harmonious functions. If the chromosome number is raised much above this level, there are disruptive effects on vigor and fertility. Plants that are already polyploids would be more vulnerable than diploids.

Two other desired characteristics are that the crop can reproduce not only sexually but also vegetatively, such as by cuttings or tubers, and that the plant parts of commercial value are vegetative, i.e. clover leaves for hay, sugar beet roots for sugar, etc. These characteristics sidestep the problem of low seed yields in newly induced polyploids.

It is also important that the crop reproduce sexually by cross-pollination rather than self-pollination. Why? A self-pollinating crop would have a high proportion of its genes in homozygous condition. This greatly restricts the number of tetraploid genotypes from which to select those that are better adapted to certain growing conditions. On the other hand, cross-pollination brings together the different allelic states of genes that exist in different plants of a crop. The benefit is a high frequency of heterozygous genotypes, which provide a much wider genetic base for selection of adapted types.

A group of crop species that has most or all of the desired characteristics for successful polyploidy is the perennial forages, i.e. grasses and legumes that constitute pasture and hay crops (Dewey, 1980). Let us consider red clover as an example of a perennial forage crop that has been quite successful as a tetraploid but needs additional improvements. Red clover has only 14 chromosomes in its somatic cells, a comparatively low number. It is grown for its vegetative above-ground parts and is cross-fertilizing, with pollination by bees.

Swedish scientists were among the early pioneers in induced-polyploidy studies who applied colchicine solution to germinating clover seeds or to the plumule, i.e. the potential shoot, of very young seedlings. The resulting plants had a mixture of diploid and tetraploid tissues. The leaves and flowers were larger in the tetraploid sectors than in the diploid sectors. Crosses between the large flowers were made by hand. The resulting seeds were germinated and chromosome numbers were checked in the root tips. Only those with a doubled chromosome number, $2n = 28$, were saved. Tetraploids were produced from many commercial cultivars of red clover and also from different plants within cultivars. The cross-fertilizing trait kept many genes in the heterozygous state, so that different genotypes could be obtained at the tetraploid level within a cultivar as well as between cultivars.

The newly induced tetraploids of red clover, like those of other crop types, showed a general lack of adaptation, especially under adverse conditions. However, it was possible to select genotypes with earlier and more rapid growth in the 2nd year after planting, and with resistance to drought and to certain diseases and insects. The larger leaves of the tetraploids contributed to higher forage yields than did the smaller leaves of the diploids. The major disadvantage for the commercial use of tetraploid red clover has been the low yield of seed. In addition to some irregularities in chromosome behavior at meiosis leading to nonfunctional gametes, there are few flowers on the tetraploid plants. Honeybees (*Apis mellifera* L.) find it difficult to get nectar from the larger flowers. Bumblebees (*Bombyx mori* L.) spend more time at each flower on a tetraploid plant than on a diploid plant, so they do not transfer as much pollen to other flowers or plants. Dewey (1980) thinks that it would be possible to select for fertility by doubling additional sources of red clover and crossing them with the existing tetraploid populations.

Suppose you want to induce chromosome doubling in a certain diploid crop. Put on your thinking cap and ask yourself these questions. What do I hope to gain by making a tetraploid crop? What are the advantages and/or limitations of this crop for tetraploidy? How much time (in terms of plant generations) and labor am I prepared to give in order to develop an adapted tetraploid crop? A search of the literature on what has already been done on your crop and a liberal amount of thought will enable you to make an intelligent decision.

Induced Chromosome Halving

We have compared haploid plants derived from diploids and from two kinds of polyploids on page 114. All three types of haploids have uses in genetic and breeding studies, but we shall choose for illustration the type of haploid that is derived from a tetraploid with four doses of the same genome.

A group of scientists at the University of Wisconsin led by S. J. Peloquin foresaw some advantages of using polyhaploids in the potato. They reasoned that the polyhaploids would be easier to use than the tetraploid in crosses to introduce valuable genes from related *Solanum* diploids or to determine the inheritance patterns of many characters. The polyhaploids have 24 chromosomes in somatic cells whereas the tetraploids have 48 chromosomes. It is obvious that chromosome maneuvers and gene segregations would be more complicated in the tetraploid than in the polyhaploid.

The first polyhaploids obtained in the potato were from crosses between the tetraploid common potato used as the female parent and a diploid group, Phureja ($2n = 24$), used as the male parent (Peloquin et al., 1966). The male was homozygous for a dominant gene that gave purple pigment in the seedlings, whereas the female was homozygous for the recessive allele (no pigment). Seedlings with no pigment were selected as tentative polyhaploids, then checked by appearance and chromosome number. Crosses between distant plant relatives, as in this case, increase the chances that the egg will be stimulated by the pollen to develop into an embryo without fertilization. However, the frequency of polyhaploids from this source was too low to make extensive use of them in genetic and breeding studies, so additional methods were developed.

A boost in the occurrence of polyhaploids resulted from cutting off the upper parts of the potato plants when they started to flower and by placing them in jars of water in an air-conditioned greenhouse. These conditions gave more fruit from the crosses and consequently more seeds per pollination, so that the chances for haploid events were increased. The scientists obtained a further increase in polyhaploid frequency by using certain tetraploid cultivars as the female or seed parents and selections of Phureja or other diploid sources as male parents or pollinators.

The polyhaploids ($2n = 24$ chromosomes) could be crossed with many diploid species ($2n = 24$ chromosomes). The hybrid progeny (F_1), which also had the diploid number ($2n = 24$), varied widely in tuber characteristics, such as shape, color, number of eyes, and quality (Fig. 6.6). These hybrids

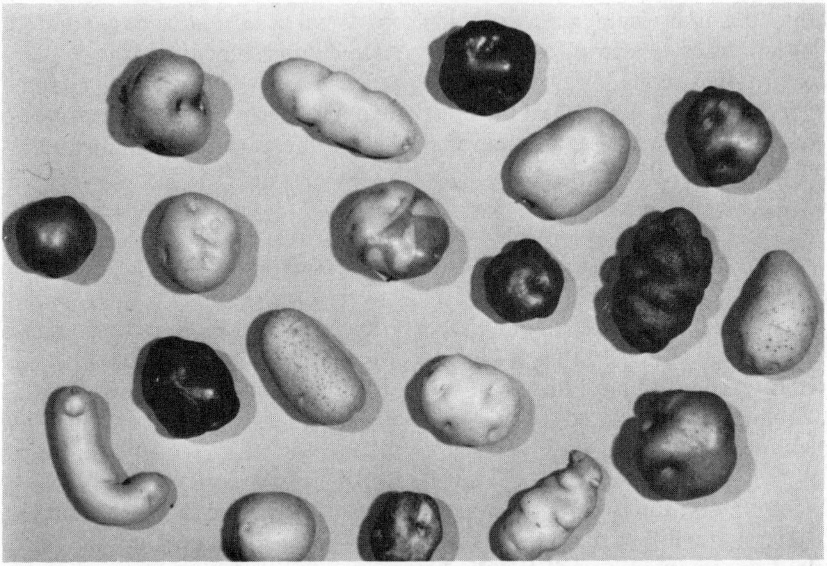

Fig. 6.6 Variations in potato tubers of F_1 hybrids from crosses between haploid *Solanum tuberosum* L. Group Tuberosum and diploid *S. tuberosum* L. Group Phureja. (Courtesy of S. J. Peloquin, Dep. of Horticulture, Univ. of Wisconsin, Madison.)

and the progenies derived from them proved to be excellent material to study the inheritance patterns of a number of traits at the diploid level. They also provided a simple way of increasing genetic diversity in the potato crop. In general, however, the diploid hybrids were not as vigorous or productive as tetraploids, so the best hybrids were crossed as male parents with tetraploid potato cultivars. Normally, crosses of this type would give a high frequency of triploids, combining a 2n egg from the tetraploid with 1n pollen from the diploid. Instead, these potato crosses gave mostly tetraploid offspring, which had an unusual amount of vigor in vegetative growth and tuber yields. Why?

Potato scientists (Mok and Peloquin, 1975) looked at meiosis and pollen in the diploid hybrids and found an explanation for the tetraploid offspring and their vigor. The hybrids produced mostly 2n pollen because of some unusual events during meiosis. In the potato, normally the two cell plates develop simultaneously at the end of the second meiotic division, forming four microspores. A recessive gene, present in the diploid hybrids, caused the two spindles of the second meiotic division to lie parallel to each other instead of being at a 60° angle (Fig. 6.7). This unusual alignment caused only one cell plate to be formed across the two spindles and resulted in two 2n microspores. The genetic consequence of these events was that the 2n gametes had a high proportion of genes in the heterozygous state. Since heterozygosity of numerous genes is thought to be responsible for the added vigor in hybrids compared with their parents, the gene for parallel spindles

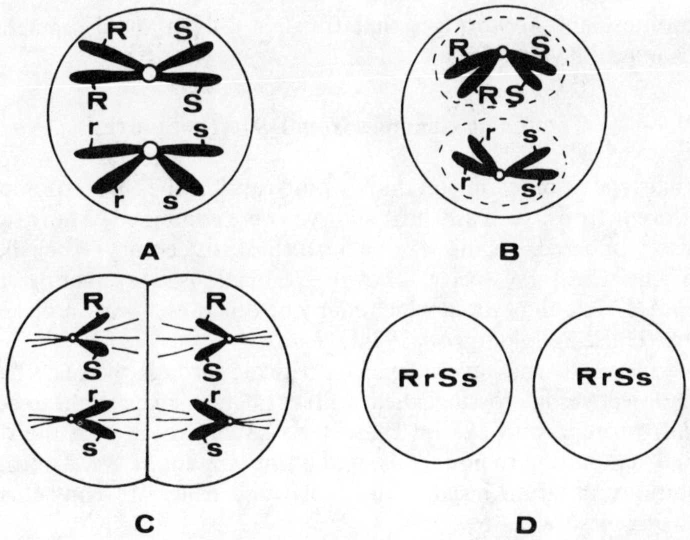

Fig. 6.7. Gene for parallel spindles may lead to 2n microspores by the following process: A. chromosome pair heterozygous for two genes, *Rr Ss*. B. end of first meiotic division. C. parallel spindles and one cell plate at second meiotic division. D. resultant 2n spores and gametes.

provided a mechanism whereby heterozygosity and its effect on vigor could be transferred almost intact from the diploid hybrids to the tetraploid level. Regardless of the genetic constitution of the 2n gametes from the tetraploid, i.e. homozygous or heterozygous for individual genes, the tetraploid progeny would have a high proportion of heterozygous genotypes.

The next phase in the potato story would be the development of improved potatoes by conventional breeding and selection methods, but that is beyond the scope of this chapter. Review the various conditions and techniques that contributed to chromosome remodeling in the potato. Perhaps, some day, you can apply them to some other polyploid crop.

CROP GENETIC BRIDGES OF DIFFERENT SIZES

A genetic bridge is a method whereby genes are exchanged between plant sources or transferred in one direction, such as from a wild species to a crop. Plant breeders are continuously building bridges between old and new crop cultivars in their crossing and selection programs. They may try to transfer genes from related wild plants to crops by breeding methods. However, some of these objectives cannot be accomplished without some ingenious chromosome manipulations.

We shall follow the development of various types of genetic bridges, the largest involving transfers of genomes, smaller bridges involving single chromosomes, and footbridges that transfer only a small segment of a chromosome with valuable genes.

Combining Genomes from Different Sources

What type of plant do you think would result if the chromosome sets of two different crops, such as wheat and rye, were combined? Many scientists have also wondered, all the way back to the 19th century. Their development of the wheat–rye species known as triticale is a fascinating story. It telescopes in time the way in which many of our present-day crops evolved in nature (Hulse and Spurgeon, 1974).

The scientists who have worked on triticale through the years have had definite objectives in crossing wheat with rye. They have sought to combine the yield, protein, and baking characteristics of wheat with the drought tolerance, adaptation to poor soils, and lysine content of rye. This has been an ambitious program because most of these traits are controlled by a number of genes.

Triticale development has involved crosses between diploid rye (one kind of genome) and both tetraploid wheat (two kinds of genomes) and hexaploid wheat (three kinds of genomes). The more successful triticales have come from using tetraploid wheat, so we shall consider this type from now on. It was called hexaploid triticale because its 2n chromosome number consisted of six sets (Fig. 6.8). When rye pollen was applied to wheat pistils, the hybrid seeds were shrivelled because of poor endosperm development

Fig. 6.8. Hexaploid triticale (center) and its two parents, durum wheat (left) and rye (right). (Courtesy of R. Dodds, Dep. of Agricultural Communications, Univ. of Nebraska-Lincoln.)

and usually could not produce viable plants. However, by cutting the embryos out of the seeds and placing them in glass tubes containing the needed nutrients, the F_1 seedlings developed and were transferred to pots of soil. At maturity, these hybrids would usually not be expected to produce any seeds because they had only one dose of the wheat and rye genomes, the haploid state for each crop. Doubling the chromosome number by colchicine (page 116) gave two doses of each genome in sectors of the hybrids, so that seeds were produced. The plants grown from these seeds would have the doubled chromosome number in all their cells and tissues. These plants were called amphiploids because they consisted of double doses of the genomes of wheat and rye. They were also called primary triticales.

After primary triticale lines involving various types of wheat and rye had been developed in several countries, crosses were made between them to get the best combinations of wheat and rye genes for adaptation to different regions. A cooperative research program between the University of Manitoba in Canada and the International Wheat and Maize Research Center in Mexico gave a surge of effort to hexaploid triticale development. As a result of a chance natural cross between a triticale line and wheat in a Mexican experimental plot, some highly desirable genes for reduced height, daylength insensitivity, and increased fertility became part of the triticale heritage.

A major and continuing problem in triticale has been seed wrinkling because of abnormal endosperm development, which has been postulated to be associated with the amount and kind of DNA in the rye chromosomes (Bennett, 1981). Cytological studies have shown that the DNA at the ends of rye chromosomes replicates later than DNA in other parts of rye chromosomes or in wheat chromosomes. As a result, the sister chromatids of rye may not separate completely during mitotic divisions in the endosperm nuclei, which are not separated by cell walls in the early phases of endosperm development. The persistent contacts between rye chromatids lead to abnormalities in the nuclei, including high polyploid numbers, and result in shriveled seeds. The solution would be to get rid of the late-replicating DNA, which does not seem to be needed for the transfer of genetic information. By using a new staining method called Giemsa, we can detect its presence as dark bands on the ends of the rye chromosomes (Gill and Kimber, 1974; Fig. 6.9). Already in some triticale lines nature has obligingly eliminated some of these chromosome tips, with marked improvement in kernel development. Therefore, attempts are being made to remove about 18% of the nonessential DNA in rye from the triticale lines with the hope of overcoming the problems that keep triticale from becoming a major crop.

Wheat and rye are by no means the only combinations that scientists have attempted. Crosses have been made between wheat and barley, radish and cabbage, potato and tomato, black currant and gooseberry, to mention only a few. Success has been elusive. The desired parts of each crop were not always recovered in the amphiploid, especially for the radish–cabbage and the potato–tomato crosses. Would you like to take up the challenge of overcoming such obstacles?

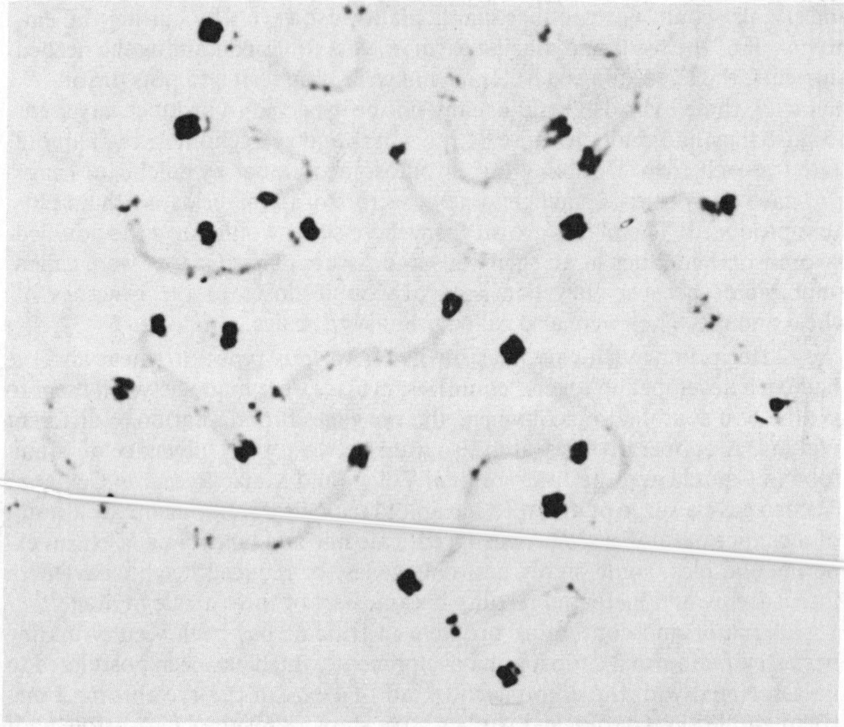

Fig. 6.9. Rye chromosome complement in a root-tip cell showing the banding pattern revealed by the Giemsa staining method. (Courtesy of T. Endo and B. S. Gill. Dep. of Plant Pathology, Kansas State Univ., Manhattan.)

Substituting Chromosomes

When genomes of crop species are combined with whole genomes of related wild species, genes for undesirable traits usually accompany the genes for desirable traits from the wild species. For example, a gene for resistance to a disease occurs in a wild genome along with genes for late maturity and small seeds. The logical next step would be to eliminate all the chromosomes of the wild genome except the one carrying the desired gene. How can we do this?

We can find a good model in tobacco, which is a tetraploid (2n = 48) with two different genomes, S and T (Gerstel and Burk, 1960). The objective was to transfer a dominant gene for resistance to tobacco mosaic virus (TMV) from a diploid species, *Nicotiana glutinosa* L. with genome G (2n = 24), to *N. tabacum.* An amphiploid (SSTTGG) that combined the two genomes of *N. tabacum* with the one genome of *N. glutinosa* was obtained from a cross between the two species followed by chromosome doubling. The amphiploid was backcrossed to a *N. tabacum* cultivar in order to eliminate all the *N. glutinosa* chromosomes except the one carrying the gene for resistance to TMV. The progeny of this first backcross had the genome constitution SSTTG. The G chromosomes were unpaired at meiosis and lagged

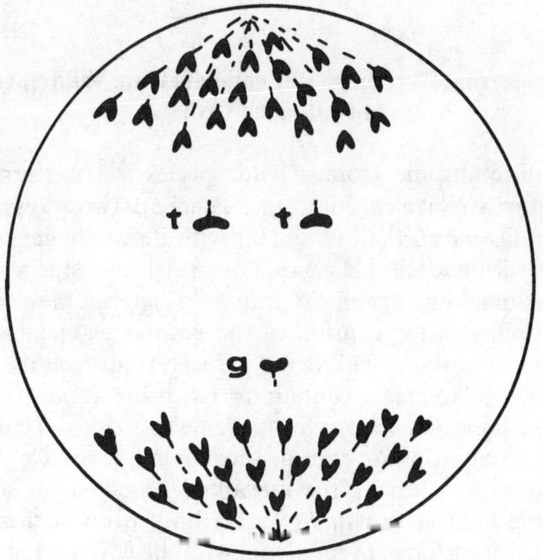

Fig. 6.10. Tobacco meiotic cell showing an unpaired *Nicotiana glutinosa* chromosome (g) going to the lower pole with 23 *N. tabacum* chromosomes and two homologous unpaired *N. tabacum* chromosomes (t) going to the upper pole.

in their movements, so that many of them were not included in the daughter nuclei. After another backcross followed by self-pollination, a tobacco line was obtained that bred true for resistance to TMV. However, it had the same chromosome number as tetraploid tobacco (2n = 48). The explanation was that the *N. glutinosa* chromosome with the gene for resistance had substituted for a *N. tabacum* chromosome and had been transmitted through male and female gametes during self-pollination.

How did this chromosome substitution come about? The most likely way was through an irregularity in chromosome behavior during meiosis, say in a resistant plant from the second backcross. This plant could have 24 pairs of *N. tabacum* chromosomes and one unpaired *N. glutinosa* chromosome (with the gene for resistance) at meiosis. Suppose that two homologous *N. tabacum* chromosomes failed to pair. By chance, the *N. glutinosa* chromosome was included, along with 23 *N. tabacum* chromosomes, in a nucleus at the end of the first meiotic division and eventually in a gamete (Fig. 6.10, lower half of cell). If, during self-pollination, a male and a female gamete of this type came together, the resultant plant would have a pair of the *N. glutinosa* chromosomes and 23 pairs of the *N. tabacum* chromosomes. This plant would be the start of the substitution line. The *N. glutinosa* chromosome must have had some of the same genes as the missing *N. tabacum* chromosome because the plants with the substitution had normal vigor and fertility. However, the *N. glutinosa* chromosome also contained genes for slow growth, low yield, and some other unfavorable traits.

In this situation, is there any way that we can eliminate the unwanted genes in the introduced chromosome? We shall seek answers in the next section.

Transferring Chromosome Segments from Wild Species to Cultivated Crops

When a chromosome from a wild species is transferred to a crop species, it is called an alien chromosome. There are two ways of remodeling the alien chromosome so that a segment with desirable genes is separated from segments with undesirable genes. One way is by using x-ray radiation, a chromosome-breaking agent, to induce a lateral break in the alien chromosome, followed by a fusion of the desired segment with a broken end of a crop chromosome. This is the induced translocation method (Fig. 6.5). Another way is to create conditions that bring about pairing between the alien chromosome and a crop chromosome at meiosis. Then, if crossing-over occurs in an appropriate region, the wanted genes are transferred to the crop chromosome, leaving the unwanted genes on the alien chromosome. This is the induced crossing-over method. Both of these procedures have been used in transferring genes from wild species to wheat.

E. R. Sears (1956), at the University of Missouri, first demonstrated the induced-translocation method. He wanted to transfer a dominant gene for leaf rust (caused by *Puccinia recondita* Rob. ex Desm. f. sp. *tritici*) resistance (symbolized by *Lr*) from a wild diploid grass called aegilops to a susceptible hexaploid wheat cultivar, 'Chinese Spring'. He could not get viable seeds from crossing this aegilops species with Chinese Spring, but he did get seeds from crossing it with a tetraploid wheat. The amphiploid, derived by doubling the chromosome number of the hybrid, could be crossed successfully with Chinese Spring. Thus, the tetraploid species provided a bridge over which the gene *Lr* could pass from the diploid to the hexaploid. However, the progeny of the cross between the amphiploid and Chinese Spring still had all the aegilops chromosomes in a single dose. In order to reduce their number, two additional backcrosses were made to Chinese Spring, with tests for leaf rust between each cross to be sure that the gene *Lr* was retained.

After these backcrosses, Sears found some resistant plants with a full complement of wheat chromosomes and a modified aegilops chromosome, which had lost one arm by a division of its centromere across the chromosome instead of in the linear plane of the chromosome. This sometimes happens when a chromosome is unpaired at meiosis, as was the aegilops chromosome. The other arm was present in a double dose because the aegilops chromosome was divided into its two chromatids at the time the centromere misdivided and the sister arms remained attached to their centromere. This modified chromosome is called an isochromosome. Sears treated these plants with x-rays to induce reciprocal translocations between the aegilops isochromosome and a wheat chromosome. The chances of transferring the *Lr* gene to a wheat chromosome were increased by using plants with the isochromosome, which had two doses of *Lr,* rather than an unmodified chromosome, which would have only one dose of *Lr*.

Sears tested with leaf rust over 6,000 progeny from crosses using irradiated plants and made extensive observations on the chromosomes at meiosis in the resistant plants. He found some plants in which a segment of the aegilops chromosome containing the *Lr* gene had been broken by the x-ray treatment and had fused with a broken piece of a wheat chromosome. This hybrid chromosome could pair at meiosis with its wheat homologue and could be transmitted via pollen and eggs. Eventually a line was developed that had the same chromosome number as hexaploid wheat but had two doses of the hybrid wheat-aegilops chromosome. The plants were like Chinese Spring except for leaf-rust resistance and a slightly delayed maturity. The line was called 'Transfer' and has been used in breeding programs to transfer the *Lr* gene to wheat cultivars.

Before discussing the induced crossing-over method, we must be aware of some facts concerning chromosome pairing in wheat (Sears, 1976). Tetraploid and hexaploid wheats have a gene, *Ph*, which allows only homologous chromosomes to pair at meiosis. An alien chromosome from a species related to wheat may have one or more segments in common with those of a wheat chromosome as well as some differing segments. In the presence of the *Ph* gene, pairing cannot occur between these two chromosomes. If *Ph* is changed to an inactive form through mutation or suppression, some pairing can occur between the alien chromosome and a related wheat chromosome.

Mutations to an inactive form, *ph*, have been induced at the *Ph* locus by treatment with x-rays or a chemical, ethyl methane sulfonate (EMS) (Wall et al., 1971; Sears, 1977). By a series of detailed chromosome manipulations, we can get wheat plants with the mutant *ph* gene and also with single doses of an alien chromosome and a related wheat chromosome. These conditions promote pairing between the two chromosomes. With pairing comes the chance for crossing-over and the transfer of desirable alien genes to wheat. The *ph* mutants have been valuable in making such transfers from wild relatives, such as species of aegilops and *Agropyron,* to wheat.

The activity of the *Ph* gene is suppressed when it is in the presence of genomes from certain aegilops species, such as *Triticum speltoides* (Tausch) Gren. ex Richter. Riley and his colleagues (1968) at the Plant Breeding Institute in Cambridge, England, used this approach to transfer a dominant gene (*Yr*) for stripe or yellow rust (caused by *Puccinia striiformis* West) resistance from another aegilops species [*T. comosum* (Sibth. & Sm.) Richter] to wheat. By developing an amphiploid and back-crossing it to wheat, they derived a plant with all the wheat chromosomes and the single *T. comosum* chromosome with the *Yr* gene. When this plant was crossed with *T. speltoides,* the rust-resistant progeny had single doses of the wheat and *T. speltoides* genomes as well as the *Yr*-bearing aegilops chromosome. The wheat gene *Ph* was present but it was suppressed by the *T. speltoides* genome, so that pairing could occur between the *Yr*-bearing aegilops chromosome and a related wheat chromosome. After further crosses with wheat and self-pollinations using rust-resistant plants, the scientists developed a line called 'Compair'. It had the same chromosome number as wheat but one of the chromosome pairs was remodeled as a result of crossing-over between the aegilops chromosome and a wheat chromosome. This pair had segments of both chromosomes and was homozygous for the *Yr* gene carried on the aegilops segment.

In the preceding section on substituting chromosomes, we mentioned that a *N. glutinosa* chromosome donated a gene for resistance to tobacco-mosaic virus to commercial tobacco along with genes for some unfavorable traits. Perhaps you are wondering if any further research was done to eliminate these traits. Yes, the line with the substituted *N. glutinosa* chromosome was backcrossed to commercial tobacco a number of times until the undesirable traits disappeared. A cytological study of the substituted chromosome showed that it consisted of both *N. glutinosa* and *N. tabacum* segments. Apparently, during the backcrosses, pairing had occurred between the *N. glutinosa* chromosome and a related *N. tabacum* chromosome without the special manipulations needed in wheat. Through crossing-over, remodeling of the *N. glutinosa* chromosome had occurred. The segment with the gene for resistance to TMV remained but the segment with the undesirable genes had apparently been eliminated.

SUMMARY

You have been introduced to some of the intriguing ways in which crop chromosomes can be remodeled in number or structure.

The natural state of crops may be diploid with two doses of a genome, or polyploid with three or more genomes of one or more types. The chromosome-doubling chemical, colchicine, has been used to obtain tetraploids from many diploid crops. Some crops have more suitable characteristics for tetraploidy than others, but all newly induced tetraploids require adaptation and selection. A halving of the somatic chromosome number (haploidy) can result from failure of a sperm to fertilize an egg. Haploids derived from the natural tetraploid potato are actually diploids and have facilitated genetic and breeding studies.

Genetic bridges for the exchange or transfer of genes involve crosses between crop species or between a crop species and a wild species. The combination of wheat and rye genomes has produced a new crop called triticale, with traits of both crops but with problems to be overcome, such as seed wrinkling. Alien chromosomes with one or more desirable genes have been substituted singly for related crop chromosomes, but they often introduce undesirable genes as well. The desirable segments can be separated from the undesirable segments by means of a chromosome-breaking agent, such as x-radiation, or by genetic conditions that promote pairing and crossing-over between the alien chromosome and a related crop chromosome.

Chromosome remodeling is similar to a game of chess. We make certain moves, i.e. crosses, treatments, etc., anticipating that our partner (nature) will respond with predictable events. Sometimes, like a chess partner, nature surprises us with an unexpected move, such as the production of a haploid or the spontaneous pairing between an alien chromosome and a tobacco chromosome. These unexpected incidents accentuate the challenges and excitement of this kind of research.

LITERATURE CITED

Bennett, M. D. 1981. Nuclear instability and its manipulation in plant breeding. Philos. Trans. R. Soc. London Ser. B 292:475–485.

Dewey, D. R. 1980. Some applications and misapplications of induced polyploidy to plant breeding. p. 445–470. *In* W. H. Lewis (ed.) Polyploidy: biological relevance. Plenum Publ. Corp., New York.

Gerstel, D. U., and L. G. Burk. 1960. Controlled introgression in *Nicotiana*: a cytological study. Tob. Sci. 4:147–150.

Gill, B. S., and G. Kimber. 1974. The Giemsa C-banded karyotype of rye. Proc. Natl. Acad. Sci. USA 71:1247–1249.

Hulse, J. H., and D. Spurgeon. 1974. Triticale. Sci. Am. 231:72–80.

Mok, D. W. S., and S. J. Peloquin. 1975. Three mechanisms of 2n pollen formation in diploid potatoes. Can. J. Genet. Cytol. 17:217–225.

Peloquin, S. J., R. W. Hougas, and A. C. Gabert. 1966. Haploidy as a new approach to the cytogenetics and breeding of *Solanum tuberosum*. p. 21–28. *In* R. Riley and K. R. Lewis (ed.) Chromosome manipulations and plant genetics. Plenum Publ. Corp., New York.

Riley, R., V. Chapman, and R. Johnson. 1968. The incorporation of alien disease resistance in wheat by genetic interference with the regulation of meiotic chromosome synapsis. Genet. Res. 12:199–219.

Sears, E. R. 1956. The transfer of leaf-rust resistance from *Aegilops umbellulata* to wheat. p. 1–22. *In* Symp. on Genetics in Plant Breeding. Brookhaven Natl. Lab., Upton, N.Y.

———. 1976. Genetic control of chromosome pairing in wheat. Annu. Rev. Genet. 10:31–51.

———. 1977. An induced mutant with homoeologous pairing in common wheat. Can. J. Genet. Cytol. 19:585–593.

Wall, A. M., R. Riley, and V. Chapman. 1971. Wheat mutants permitting homoeologous meiotic chromosome pairing. Genet. Res. 18:311–328.

SUGGESTED READING

Eigsti, O. J., and P. Dustin, Jr. 1955. Colchicine. Iowa State Univ. Press, Ames.

Kasha, K. J. (ed.) 1974. Haploids in higher plants. Univ. of Guelph, Guelph, Ontario, Canada.

Lewis, W. H. 1980. Polyploidy: biological relevance. Plenum Publ. Corp., New York.

Simmonds, N. W. 1979. Principles of crop improvement. Longman, Inc., New York, p. 279–314.

Chapter 7

In Vitro Crop Breeding

S. L. LADD AND M. R. PAULE

Colorado State University
Fort Collins, Colorado

In vitro crop breeding encompasses essentially all cell and tissue culture techniques that assist in propagating, studying, and manipulating the genetics of plants without use of the sexual cycle. Basically, in vitro culture is the process of propagating, by sterile technique, cells, tissues, organs, or protoplasts.

In vitro methodologies are important techniques with newly recognized potential to assist the crop breeder in attaining crop improvement goals, developing improved cultivars, and learning more about the species. Success with in vitro technology has been limited thus far to a few crops. However, crop breeders now recognize that recombinant deoxyribonucleic acid (DNA) techniques can be applied to breeding programs through cell culture. In vitro culture also provides a means to study the physiology, biochemistry, genetics, growth, and development of crop species at the molecular level.

Contributions of conventional crop breeding to agriculture have been outstanding for several decades. Such contributions will continue, aided by in vitro techniques. It is important for crop breeders to realize the value of these tools and utilize them. It is equally important for scientists working with in vitro technology to recognize limitations to their methods.

In vitro plant culture had its beginnings in 1902 when Haberlandt theorized that cells could be cultured. However, it was not until the late 1930's that other workers achieved success with cell culture (Nickell, 1973; Street, 1977).

Although most crop species have been cultured in vitro, few species of economic importance can reliably be regenerated to plants. However, investigations into in vitro techniques for most species are still in their infancy.

The discovery of genetic recombination in bacteria using in vitro DNA molecules enhanced interest in culturing plant cells. Techniques used in bacterial culture provided models of genetic manipulation readily transferable to plant cell cultures. Prospects are good for applications of recombinant DNA methods for solving crop improvement problems.

CULTURES

Callus

Callus cultures are masses of cells derived from plant tissue growing on a nutrient medium. The medium is made up of inorganic salts, a carbon source (usually sucrose), auxins, and cytokinins. The composition of the medium is critical for the species to be cultured. To produce a good callus, nutrients must promote rapid cell growth. Tobacco and carrot, are cultured easily. Other species such as oats, wheat, and sugarcane can now be consistently regenerated from cell culture. All of these crops have provided valuable data that will make successful culture and regeneration of other species easier.

Several plant tissues may be used to initiate callus cultures, including roots, stems, leaves, meristems, and anthers. However, in a particular species one type of tissue may produce callus more readily than another.

To initiate callus, tissue is removed from a plant and surface sterilized to kill bacteria and fungi that would contaminate the culture. Some workers prefer to generate callus from plants grown in aseptic conditions with surface-sterilized seed to reduce problems from contaminating microorganisms.

The isolated tissue is grown on sterile medium at 20 to 25°C and exposed to light. If the auxin/cytokinin balance of the medium is appropriate, a mass of callus will be produced. Subculturing the callus to fresh medium usually occurs at regular intervals of 2 to 8 weeks. Regular subculturing sometimes helps to maintain totipotency, the ability to regenerate into plants. The need for regular subculturing is not fully understood but may relate to the regular, cyclic organization and deterioration of pseudoembryos within the culture (Nadar et al., 1978).

The development of callus is governed by hormones added to the medium, primarily auxins and cytokinins. Altering the hormone levels may induce the callus to produce shoots and/or roots. Required hormone balances are critical for each species and often will vary from cultivar to cultivar. One of the useful auxins for the initiation and promotion of callus production is 2,4-dichlorophenoxyacetic acid (2,4-D), a synthetic growth regulator.

Tobacco is propagated through true callus culture, so more is known about in vitro culture of tobacco than any other species. However, tobacco is not propagated commercially through in vitro culture, partly because many of the mutants selected are not of commercial importance and partly because of the ease with which tobacco is propagated through seed.

Sugarcane also is easily propagated in vitro but shows chromosomal instability. Regenerated plants are stable and are propagated by vegetative cuttings. Thus, the crop breeder can use in vitro cultures to develop new genetic entities and recover stable clones.

Regeneration techniques have not been developed for many species;

Table 7.1. Some crop species for which plant regeneration has been demonstrated.

Alfalfa	Cauliflower	Lily	Red clover
Almond	Celery	Maize	Rice
Artichoke	Citrus	Millet	Rose
Asparagus	Coffee	Oat	Ryegrass
Aspen	Cotton	Oil palm	Sorghum
Barley	Date palm	Onion	Soybean
Begonia	*Datura*	Orange	Sugar beet
Broccoli	Eggplant	Pea	Sugarcane
Brussels sprout	Flax	Pepper	Taro
Red cabbage	Grape	Petunia	Tobacco
Cacao	Kale	Pineapple	Tomato
Carrot	Ladino clover	Potato	Wheat
Cassava	Lettuce	Pumpkin	

this limits the successful use of tissue culture. However, research is continually overcoming barriers and answering questions and should lead to wider use of tissue culture in crop improvement. Table 7.1 lists species for which regeneration has been demonstrated. A comprehensive list was provided by Vasil et al. (1979).

If variant characters selected in callus are to be useful to the plant breeder, they must be heritable in the regenerated plant. Genetic transmission of characters selected in culture has been reported in a number of species. However, the characters may not be expressed in the regenerated plant or transmitted to subsequent generations. That is, they are apparently not of genetic origin (epigenetic) or are chimeras that do not affect the reproductive tissues. In addition, factors that are selectable in culture may not be related to characters desired in the intact plant. Conversely, characters for which a plant breeder may wish to select may not be selectable in culture. Yield may be one of the difficult factors to select for in vitro. Yield is a multigenic characteristic and a function of the whole plant and its interaction with the environment. Shepard et al. (1980) selected protoclones (clones derived from protoplasts) of 'Russet Burbank' through protoplast culture and field-tested 65 derived potato plant lines. These lines differed in a number of traits such as growth habit, days to maturity, disease reaction, photoperiod requirement, and tuber characteristics. None of the selected traits was directly related to increased tuber yield.

More extensive investigation can be expected toward improvements in yield and tolerance to disease, salt, and drought. The first report of a potential new commercial cultivar developed through in vitro culture was a disease-resistant sugarcane clone (Nickell, 1973). After field testing revealed weak plant tops, the new clone was not released for commercial use but was used as parental germplasm in the breeding program.

Suspension

Cell suspension cultures consist of single cells or small aggregates of cells in a liquid medium. The culture container usually is rotated or shaken to induce and maintain cell dispersal and enhance exchange of gases between the cells and the gaseous component of their environment. The liquid medium commonly is quite similar to the solid medium used for callus except agar is not used. The liquid medium components may be altered to promote rapid growth and cell separation.

A common sequence for initiating suspension cell cultures begins with obtaining cells from callus culture, transferring them to liquid medium, and incubating the cultures on a shaker. To enhance cell separation, it is most desirable to use friable callus, i.e. callus in which the cells easily separate from one another.

Culture Systems

Cell suspension cultures can be closed or open systems in either continuous or batch culture. In batch culture, cells are placed in a small culture container, such as a 250-ml Erlenmeyer flask or a larger vessel such as a carboy. The cultures, occupying about 25% of the volume of the container, are shaken continuously, either under supplemental light or in complete darkness, until they reach maximum cell concentration. Maximum density is reached when the medium components, space, or other factors become limiting, and subcultures must be made if cell multiplication is to continue. Cultures also have minimum growth densities that are dependent upon culture conditions. If subcultures are below these minimum densities, cell division and culture growth will cease.

Genetic studies usually are conducted in small, closed systems. This permits large numbers of cell lines to be exposed to many treatments and selection systems in the laboratory. Nabors (1976) points out that 100 ml of suspension culture may contain 10^7 cells, providing an extremely large population for exposure to selection.

For the crop breeder, cell suspension and callus cultures hold both promise and frustration. Cultures often are cytologically unstable, as in oat cultures, where cells may be missing all or parts of a chromosome, making it difficult to propagate specific genotypes. On the other hand, this genetic instability could be a valuable source of new mutants. For example, one may be able to use the unstable cultures to produce monosomics or trisomics for cytogenetic studies.

Cell cultures have been used extensively for selection of desired genotypes. Chemicals or metabolic products may be added to the medium as selective agents to eliminate susceptible cells in favor of tolerant cells. Examples of plants regenerated after treatment with various selective agents in cell cultures are listed in Table 7.2. Chaleff (1981) and Handro (1981) provide detailed references to such experiments.

Anther and Pollen

Anther and pollen culture are used to produce monoploids or haploids. Although mutations are easily induced in cultured cells, many mutations are recessive and, therefore, not readily detectable because the cells are diploid

Table 7.2. Variant types of crop plants selected through in vitro culture.

Plant	Selective agent	Plant	Selective agent
Alfalfa	NaCl	Sugarcane	Disease
Carrot	Amino acid analog	Tobacco	Disease
Datura	Aminopterin		Herbicide
Maize	Disease		NaCl
Pepper	NaCl		Cold
Petunia	Streptomycin		Amino acid analog
Potato	Disease		Antibiotic
	Amino acid analog		Fungicide
Rice	Amino acid analog	Tomato	Aluminum
	NaCl	White clover	Herbicide

or polyploid. Monoploids may be produced by anther culture of diploids to permit direct expression of recessive mutations. Plants produced through anther culture of polyploids are not true monoploids and may not permit direct expression of mutations, but anther culture helps simplify their detection.

A second use of anther and pollen culture is the production of totally homozygous diploids by doubling the chromosome complement of monoploids produced in the culture system.

Haploids occur with low frequency outside the culture system, and the ability to produce them at higher frequencies would be a major advance for the crop breeder. In vitro haploid production has been accomplished in at least 79 species according to a recent report by Schaeffer et al. (1979).

One must not assume that all cells or plantlets produced in an anther culture system are haploid or monoploid. Somatic (2n) cells surrounding the microspores may also produce callus. The callus from the somatic cells may completely surround the haploid cells, making it extremely difficult to identify and propagate the haploid callus. Therefore, it is necessary to confirm the chromosome number of each culture or plant.

Anther culture is a relatively simple technique in some species, such as tobacco, but extremely difficult in others. Success in inducing haploid production in vitro is often cultivar dependent, e.g., some wheat cultivars readily produce haploids while others produce none. Experience in this and other areas of anther culture suggests that the "reluctant" cultivars may be induced to produce haploids given the right set of environmental conditions. For best haploid production, optimum conditions for the following should be determined for each cultivar or species: microspore development stage; medium composition; anther pretreatment; and source, condition, and age of the plant from which the anthers were taken (Collins, 1977).

Homozygous lines have been produced in a single generation by doubling the chromosome number of recovered haploids. With standard breeding techniques, a plant breeder must self-pollinate for several generations before reaching a satisfactory level of homozygosity. Also, severe inbreeding depression encountered in some out-crossing species may make the recovery of homozygous lines difficult by conventional breeding methods. Although this problem may be overcome by doubling recovered haploids, one should not rely solely on haploids to produce homozygous lines until experience has shown that the technique is fruitful. For example, the method was used in the 1950's to produce homozygous lines of maize, but useful lines were difficult to recover because of the problems associated with testing the recovered lines.

Genetic instability in anther culture is common. Generally, the longer a cell line is cultured the greater the instability. In vitro environments that stabilize cultures will need to be developed to maintain the selected variant cultures.

The potential for anther culture has been demonstrated by its use in the selection of superior tobacco breeding lines (Collins, 1977) and in the production and testing of dihaploid lines (Schaeffer et al., 1979).

Embryo

Embryo culture was one of the early applications of in vitro technology to plant breeding and has been used in numerous instances to obtain intergeneric or interspecific hybrids. With embryo culture, the embryo is excised from the developing seed a few days after fertilization and is cultured on a solid or liquid medium under controlled environmental conditions to produce a seedling plant that can be transferred to soil and produce a mature plant.

Embryo culture is useful in rescuing embryos of sexual crosses between distantly related species or genera that fail because the hybrid embryo aborts. The presence of two incompatible genomes often results in a breakdown of the embryonic developmental cycle. Critical factors may be the loss of endosperm or inadequate nutrition of the embryo. If the hybrid embryo is excised before it aborts and is placed on an optimum culture medium, the required nutrients may permit continued embryonic development.

Culture conditions usually include a 2- to 4-week incubation in total darkness at 20°C or less, followed by a short incubation above 20°C while exposed to light, after which the developing embryo is transferred to a solid medium for enhancement of root and shoot growth. After roots and shoots are sufficiently developed, the plant may be transferred to soil and kept at high humidity and moderate temperatures until established.

Embryo culture media may include inorganic salts, sucrose, vitamins, amino acids, hormones, and nutritionally undefined substances such as coconut milk. Younger embryos require more complex media than older ones.

Uses of Embryo Culture

Embryo culture has been used to produce hybrids in several species. For example, Keim (1953) demonstrated the feasibility of using embryo culture to produce viable interspecific hybrids from crosses between tetraploid and diploid *Trifolium* species. Hybrid wheat embryos have been successfully cultured (Schaeffer et al., 1979), and embryo culture has been used to speed seedless grape improvement and to improve stone fruits, including peaches, nectarines, and plums.

Interspecific hybrids have been obtained through embryo culture of species within the genera *Gossypium* (Stewart and Hsu, 1978), *Lilium* (Emsweller and Uhring, 1962), and *Phaseolus* (Honma, 1956; Mok et al., 1978). Intergeneric crosses using embryo culture have included *Triticum* × *Hordeum* hybrids.

In addition to permitting production of some hybrids that would be impossible to obtain otherwise, embryo culture affords a means to study and overcome seed dormancy. Removal of the embryo from its normal environment within the seed will overcome inhibition created by an impenetrable seed coat. Also, hormones, enzymes, or other chemicals may be added to the medium to promote chemical changes within the embryo, thus counteracting metabolically induced dormancy.

Protoplast

Microbiologists have demonstrated fusion of single cell microorganisms, and animal scientists have induced fusion of animal cells. However, the rigid plant cell wall makes cell fusion in plants difficult. As long as the cell wall is intact, cell fusion is virtually impossible. The cell wall inhibits the uptake of DNA, plasmids, cell organelles, and bacteria, a distinct problem for the researcher attempting molecular recombination. A solution is removal of the cell wall to expose the naked cell—the protoplast.

Although protoplasts have been isolated from nearly all parts of the plant, leaf mesophyll cells and cells from suspension cultures have been most useful. To remove the cell wall, cells usually are exposed to hydrolytic enzymes that degrade cell wall materials. Once the protective cell wall is removed, protoplasts must be maintained under carefully controlled conditions.

Protoplasts require culture media and other environmental conditions similar to those required for cell suspensions. To prevent rupture of the naked cells, careful attention must be given to the osmotic potential of the medium. Specific requirements for protoplast harvest and growth vary among species, and among cultivars of the same species. The yield, function, and viability of protoplasts may depend also on the conditions under which the source plants were grown.

Protoplast culture is still in the developmental stage but shows potential success in several areas. For example, bacteria can be introduced into protoplasts for study of the interaction between bacteria and plant cells. Such studies may ultimately lead to the development of symbiotic relationships and nitrogen fixation in non-legumes. Protoplasts are also convenient tools for genetic and/or organelle manipulations; for example, foreign DNA, plasmids, chloroplasts, or mitochondria may be introduced into a protoplast for genetic manipulation and study. The resulting cells may be exposed to toxins or toxin analogues to select genetic variants (Carlson, 1973; Matern et al., 1978).

Table 7.3. Intergeneric and interspecific hybrids produced through protoplast fusion.

Intergeneric hybrids	
Carrot × *Aegopodium*	Soybean × Alfalfa
Carrot × Barley	Soybean × Barley
Carrot × Petunia	Soybean × Maize
Carrot × Tobacco	Soybean × Meadow saffron
Maize × Oat	Soybean × Pea
Maize × Sorghum	Soybean × Rape
Petunia × Fava bean	Soybean × Tobacco
Petunia × *Parthenocissus*	Tomato × Potato
Interspecific hybrids	
Dacus carota × *Dacus capillifolius*	
Datura innoxia × *Datura discolor*	
Datura innoxia × *Datura stramonium*	
Nicotiana glauca × *Nicotiana langsdorffii*	
Nicotiana sylvestris × *Nicotiana knightiana*	
Nicotiana tabacum × *Nicotiana debneyi*	
Nicotiana tabacum × *Nicotiana sylvestris*	
Nicotiana tabacum × *Nicotiana glauca*	
Petunia hybrida × *Petunia parodii*	

Protoplast Fusion

Somatic, or parasexual, hybridization is an important use for protoplasts, particularly when sexual interspecific and intergeneric crosses fail. For example, the crossing of sorghum and sugarcane may make the production of sugar possible in arid areas if somatic hybridization can occur between the two crops.

Fusion of protoplasts has been demonstrated in a number of species (Table 7.3), see Chaleff (1981) for details. Carlson (1973) produced somatic hybrid plants through fusion of protoplasts of *Nicotiana glauca* Graham and *N. langsdorffii* Weinm. Morphologically and biochemically the hybrid produced via protoplast fusion was identical to a sexually produced *N. glauca* × *N. langsdorffii* amphidiploid. *Petunia* and *Parthenocissus* protoplasts were fused by Power et al. (1975). The resulting cells were not true hybrids since all *Petunia* chromosomes were lost, but some *Petunia* characteristics did persist for more than a year.

The identification of somatic hybrids produced by protoplast fusion is difficult. In a cultured population of protoplasts formed by mixing two species and inducing fusion, protoplasts will result from the union of the two species as well as from the union of protoplasts of the same species. In addition some protoplasts will not have fused with any other protoplasts. Fusion experiments must be carefully designed knowing the growth requirements so that all but the desired protoplasts will be eliminated. Carlson et al. (1972) used differential growth requirements of the hybrid to eliminate protoplasts of the two donor species. Power et al. (1975) removed one of the donor species by use of differential growth requirements. Others have demonstrated the use of drugs or drug analogues to select hybrid protoplasts or eliminate the parental protoplasts (Cocking et al., 1974; White and Vasil, 1979). The simplest methods of somatic hybrid selection use visually detectable genetic markers (e.g. chlorophyll deficiency) to permit direct identification of the hybrid cells or plants.

Table 7.4. Crop species from which protoplasts have been isolated and cultured.

Alfalfa	Cotton	*Parthenocissus*	Sorghum
Asparagus	Cucumber	Pea	Soybean
Barley	*Datura*	Pearl millet	Sugar beet
Bean	Flax	Petunia	Sugarcane
Belladonna	Grape	Potato	Tobacco
Buttercup	Kale	Rape	Tomato
Carrot	Maize	Rice	Wheat
Cassava	Orange	Rose	

Cell wall formation and plant regeneration for hybrid protoplasts requires modification of the culture medium and growth conditions. Regeneration conditions are often quite similar to those required for regeneration from callus. Sustained cell division is a prerequisite to plant regeneration from protoplasts. This initial step in regeneration from protoplasts has been demonstrated in several species (Table 7.4). More details on particular instances may be found in Gamborg et al. (1981).

MOLECULAR APPROACHES TO ALTERING PLANT GENOTYPES

In most of the techniques discussed so far, the production of plant cells or whole plants with the desired phenotypes has relied upon random alteration of the genotype of the cultured cell and a time-consuming selection process. What if the pertinent gene could be individually isolated, modified appropriately by well controlled steps in the laboratory, and then replaced in the plant where it would elicit the necessary change in phenotype? Though a visionary dream a decade ago, most of the necessary techniques to accomplish these results are now available. Dubbed "genetic engineering" in the press, molecular genetic engineering is a more accurate term for this activity. It includes the techniques of gene isolation by recombinant DNA methods, subsequent alteration of the recombinant gene by in vitro mutagenesis techniques if necessary, introduction of the gene into the plant, and recognition of the altered genotype.

Recombinant DNA Methods

Recombinant DNA technology is based upon the ability to fractionate and then join fragments of DNA from widely different sources (Fig. 7.1). One of the fragments, the vector, must carry the appropriate information to allow the DNA to be replicated in a host cell. The recombinant DNA is introduced into the host cell where it becomes part of the genetic makeup of the cell and of its progeny. In many cases, if a complete gene is contained within the recombinant DNA, it will be expressed and its expression regulated as though it were always part of the cell's genotype. In this way, the phenotype of the recipient cell can be suitably altered to meet crop improvement goals.

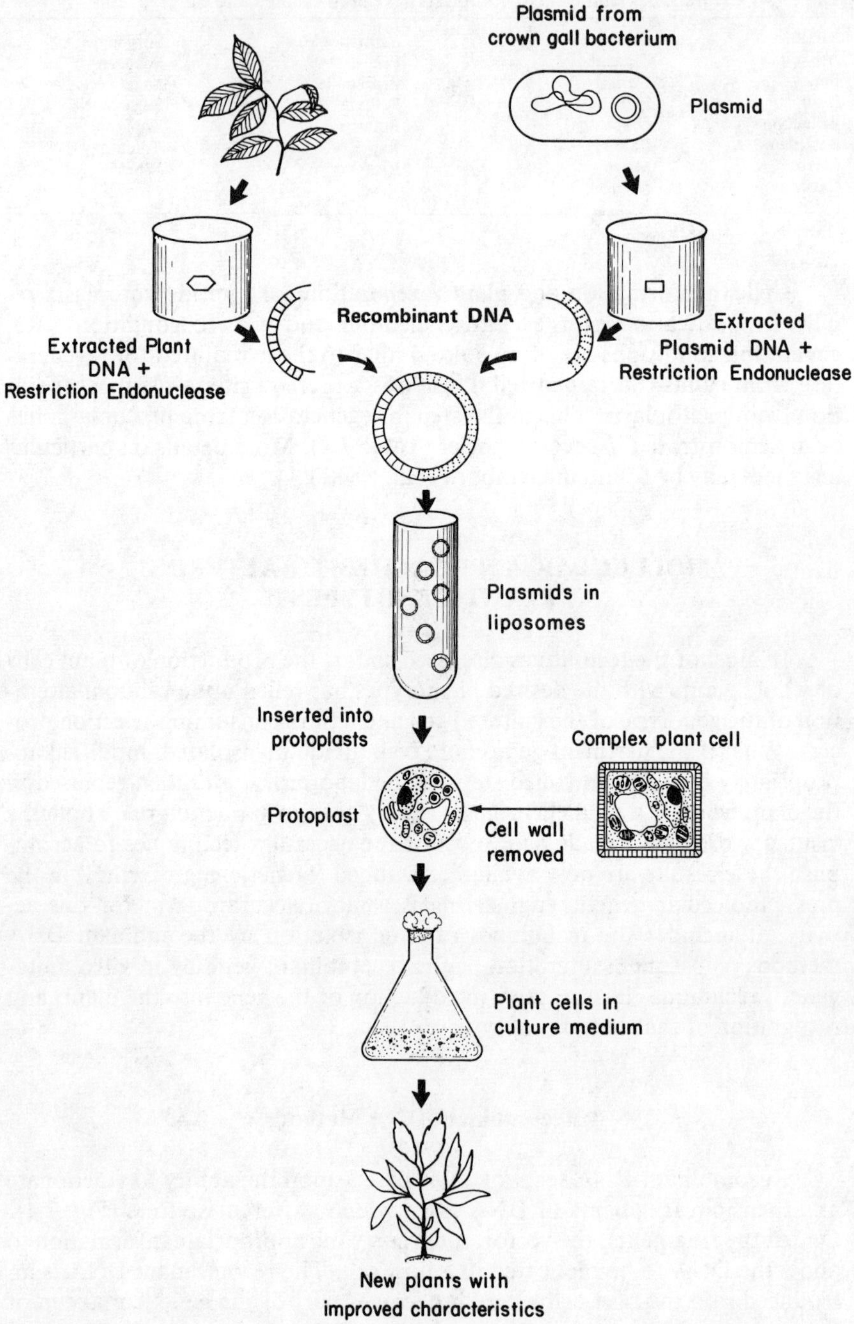

Fig. 7.1. Steps in the molecular genetic engineering of improved crop cultivars.

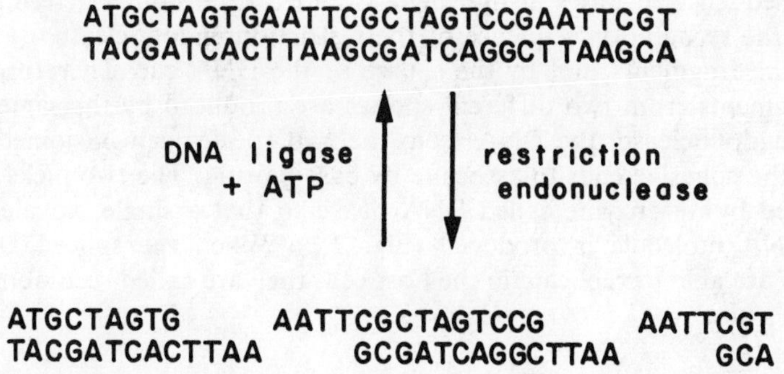

Fig. 7.2. Action of restriction endonuclease and ligase on DNA. The unpaired strands are the sticky ends.

Splicing

The crux of the recombinant DNA technique is the splicing together of DNA fragments. Several methods are utilized: ligation of fragments with cohesive ends produced by restriction endonuclease cleavage, blunt-end ligation, synthetic linkers, and synthetic tailing.

The scientists must first identify and isolate a DNA molecule containing the gene that codes for the appropriate genetic characteristic to be transferred. Then the first step in carrying out the recombinant DNA experiment is the production of a DNA fragment containing the gene of interest. This is accomplished by using an enzyme called restriction endonuclease. Restriction endonucleases are a large group of enzymes isolated from bacterial cells. They hydrolyze the phosphodiester backbone of DNA at specific sites determined by a sequence of bases in the DNA. A four- to six-base pair sequence is sufficient to allow recognition and cutting by a restriction enzyme. More than 200 restriction endonucleases are known with more than 100 different recognition sites. These sites occur at random in any given DNA molecule. Often it is possible to find a restriction enzyme that produces a DNA fragment containing the gene of interest. Sometimes combinations of different restriction endonucleases are needed to get the desired DNA fragment.

Cohesive End Ligation

When restriction endonucleases cut double-stranded DNA molecules, many of them make staggered cuts in the two strands, producing short single-stranded extensions on the resulting DNA fragments. Since the single-stranded extensions from two DNA fragments complement one another (in the Watson-Crick base-pairing sense), they can "stick" together to re-form a long double-stranded DNA molecule containing two phosphodiester backbone breaks (called nicks). These types of ends are called cohesive or sticky ends (Fig. 7.2).

The sequence of bases in the single-stranded extension is determined solely by the recognition sequence of the restriction endonuclease used to produce the fragments, not by the source of the DNA cut. Therefore, if DNA fragments from two different species are produced by the same restriction endonuclease, the DNA from the two species can be joined by allowing the cohesive ends to associate by base-pairing. The two nicks can be repaired by an enzyme called DNA ligase so that a single, covalently joined DNA molecule is produced (Fig. 7.2). When the spliced DNA molecules are able to replicate in the host cell, they are called recombinant DNA molecules.

Blunt End Ligation

A second method of splicing relies upon the finding that the DNA ligase can link two DNA fragments even if their ends are blunt (i.e., have no single-stranded extensions). However, the efficiency is at least an order of magnitude (10 ×) higher when the sticky ends are present.

Synthetic Linkers

A variation of the above procedure involves the blunt end ligation of chemically prepared linker DNA molecules to the blunt ends of the DNA fragments. The ligation reaction efficiency can be increased by using an excess of the linkers. The base sequence of the linkers is chosen so that after they have been attached to the DNA fragments, they can be cut with an appropriate restriction endonuclease to produce sticky ends—ends which will work more efficiently in subsequent ligation reactions.

Synthetic Tailing

Finally, specific complementary homopolymer (a polymer made of a single nucleotide) can be added to the ends of the two DNA's to be joined using the enzyme terminal transferase. This enzyme uses nucleoside triphosphates as substrates to extend the length of one of the DNA strands at a blunt end. For example, the homopolymer for guanine (G) can be added to one fragment and the homopolymer for cytosine (C) to the other. Since C and G are a Watson–Crick pair, these ends are sticky toward each other and can be efficiently ligated to form a recombinant.

Vectors

The first recombinant DNA experiments utilized bacterial plasmid DNA as vectors and bacterial cells as hosts. Plasmids are small, circular, DNA molecules that replicate autonomously in the bacterial cell. Plasmids carry one or more genes that confer resistance to antibiotics on their host cells. In recombinant DNA experiments, recombinant plasmids are capable of replicating in the bacterial cell and they express the antibiotic-resistance genes carried in their DNA. The expression of antibiotic resistance allows selection of those cells containing the plasmid when an appropriate antibiotic is used to kill all the cells not containing plasmids. Selection is important because only one out of $1{:}10^3$ to $1{:}10^6$ cells is capable of taking up the recombinant DNA (being transformed). In plant systems, two types of DNA appear promising for use as vectors: those of plant viral origin and the plasmid from the plant bacterial pathogen *Agrobacterium.*

Plant Viruses

Although most plant viruses have ribonucleic acid (RNA) rather than DNA as their genetic material and are not useful as vectors in recombinant DNA experiments, some do have double-stranded DNA genomes and can serve as vectors. For example, cauliflower mosaic virus DNA will infect and cause production of virus particles when rubbed on susceptible leaves. Small additional DNA fragments, up to about 250 base-pairs in length, can be spliced to the viral DNA and propagated in plant leaves. This limitation on the additional DNA that can be tolerated by the virus severely limits its use as a vector. Restructuring of the viral DNA by recombinant techniques may lead to a more suitable form capable of carrying entire genes. Similar problems were overcome in using bacterial viruses as vectors. A more severe problem with the plant viral DNA vector is the possibility that its DNA remains in the cell's cytoplasm to replicate and function. If it is not transported to the nucleus and recombined with the chromosomal DNA of the plant cell, it is unlikely that heritable traits can be introduced into the plant.

Crown Gall Plasmid

Agrobacterium tumefaciens and its relatives cause plant disease (crown gall by *A. tumefaciens*) by transferring a plasmid, the tumor-inducing or Ti plasmid, to the plant cell. A specific piece of the Ti plasmid, transferred or T-DNA, is incorporated into the host plant's nuclear DNA where it is inherited as part of the transformed plant cell's genotype. Genes that are normally in T-DNA are expressed. Among other things, these genes stimulate plant cells to grow rapidly (a form of plant cancer). Cells from callus cultures or root slices transformed in this way can be selected by their ability to grow without the addition of hormones. Selectability of transformed cells is one of the critical properties needed in a vector DNA. Presumably, foreign DNA inserted into regions of the T-DNA would also be expressed, thus altering the plant cell's genotype in desirable ways. Research on the regions of the T-DNA that may be replaced by foreign DNA is currently in progress. There appears to be no practical limit to the size of the DNA fragments that may be inserted into the Ti-plasmid. Up to 40,000 base pairs have been incorporated, more than enough to code even the most complex genes. A limitation in the use of *A. tumefaciens* is its inability to infect monocotyledons. It is possible that this trait could be engineered out of the organism and its T-DNA by appropriate genetic manipulation.

TRANSFORMATION—INTRODUCTION OF DNA INTO THE PLANT CELL

Introduction of the recombinant DNA into plant cells represents the next step in molecular genetic engineering. Freshly prepared protoplasts of plant cells effectively take up DNA from the surrounding milieu. The addition of polycations (e.g. diethylaminoethyl-dextran, poly-L-lysine, poly-L-ornithine) stimulates this uptake. Thus, protoplasts can be mixed with preparations of recombinant DNA, including DNA constructed using vectors suited only for propagation and selection in bacterial cells, and some plant cells will acquire the DNA.

One of the problems with the above approach is that DNA-degrading enzymes occur in the medium surrounding the protoplasts. With some plant species, free DNA introduced into the medium is degraded rapidly enough to reduce significantly the efficiency of DNA uptake. To prevent degradation, the DNA can be packaged before its introduction into the medium. For example, the DNA can be coated with specific proteins whose normal function is to protect viral DNA's (such as the capsid protein of tobacco mosaic virus). Alternatively, the recombinant DNA can be left inside of a host bacterial cell that is converted to a protoplast by lysozyme treatment. This bacterial protoplast can be fused with the plant protoplast so that the cytoplasms of the two cells mix and the recombinant DNA is introduced into the plant cell without ever contacting the extracellular medium.

Thus, despite some technical difficulties with introducing DNA into plant cells, protoplasts can be effectively used to produce plant cells with altered genetic makeup. These cells can then be used to regenerate whole plants with altered genotypes. Early work and the complexities of introducing DNA into the plant cell were reviewed by Kado and Kleinhofs (1980).

PROBLEMS AND APPROACHES TO SOLUTIONS

So far, we have dealt with the mechanics of introducing new genes into plant cells. A complex set of additional steps is involved in the correct expression and functioning of the gene once in the cell. Problems can arise at any one of these steps.

Replication

The DNA, once incorporated into the host plant cell genome, must be replicated so that all daughter cells will receive copies of the new gene. In eukaryotic cells, each round of DNA replication starts at a large number of sites within each chromosome. These sites, called replication origins, are specific sequences within the DNA. These sequences are slightly different in each species so that origins from some species (e.g. present in genes originally isolated from bacteria) are not likely to be functional in plant DNA. This may cause anomalous replication of the inserted DNA. It is possible that normal replication could arise from plant origins next to the inserted DNA.

Gene Expression

Similarly, the conversion of genetic information resident in the DNA of a cell into a functional protein requires the presence of a number of specific base sequences. These sequences are variously involved in recognition by the enzymes that produce an RNA copy of the DNA sequences, by enzymes that process the primary RNA transcript into a mature, functional RNA, and by the proteins involved in translating the information coded in the RNA into a protein sequence. Each of these processes requires that specific recognition sequence be coded in the DNA and, as with replication origins, it is not yet clear whether these signals are identical for all species.

Gene Regulation

An understanding of how gene expression is regulated is important for applying molecular genetic recombination. The regulation of gene expression in eukaryotic plants and animals is poorly understood. The introduced gene must be expressed in the proper tissues. Genes to produce pigments involved in photosynthesis, for example, are of little use in root cells, which seldom see light!

Phenotypic Expression

One other problem may be difficult to solve. Proteins function in a particular environment and, when transplanted to a new one, may not function as effectively. For example, the nitrogen-fixing enzyme complex of the symbiotic bacterium (*Rhizobium*) of legumes converts atmospheric nitrogen gas to ammonia, a form usable by plants. One goal of genetic engineers is to reduce the need for nitrogen fertilizers by introducing this gene complex into non-leguminous plants. Unfortunately, one of the enzymes in the complex is extremely sensitive to inhibition by oxygen and only functions effectively in the anaerobic environment of the root nodule of the legumes. Merely introducing the genes for this enzyme complex into a plant will not yield the desired results of nitrogen fixation unless this enzyme is protected. Similar problems of function and regulation of enzymes for specific duties in a defined environment may arise. However, the potential rewards of solving such problems and gaining the basic information about the functioning of the plant cell are well worth pursuing.

Since the basis for problems of gene function and expression resides in the nucleotide sequence, they may be solved by altering the sequence of the gene before introducing it into the plant cell. Powerful new techniques have been developed which allow such changes in DNA sequence with relative ease and precision.

Promoter Sequences

The first step in expressing a gene is the synthesis of an RNA copy of the DNA. The beginning and end of the region of the DNA that is copied are signaled by specific DNA sequences, the promoter and the terminator, respectively. Also, in the region of the promoter are often other short base sequences (probably less than 100 bases long) that serve as sites for regulation of the expression of the gene. Without these sequences, the gene either cannot be expressed at all, or is expressed in an anomalous and uncontrolled fashion. Genes introduced into a cell by recombinant DNA techniques do not necessarily carry with them proper promoter and regulatory sequences. But these sequences can be borrowed from genes that are known to function properly in the host plant. The introduced gene, lacking a functional promoter, may be inserted downstream of the host promoter. Downstream is the direction in which transcription of DNA to RNA occurs. The inserted gene would be transcribed when the host promoter initiates transcription of the associated host genes.

Regulatory Sequences

As more is learned about how the regulatory sequences function, appropriate regulatory sequences can be appended to the gene being inserted. An example of this approach was the appending of the lactose (lac) operon control region to genes cloned in *Escherichia coli.* The inserted gene was expressed only under conditions in which the lac operon would be expressed—that is, only when lactose is added to the growth medium for the cells.

Chemically Synthesized Genes

A second example of approaches to such problems is the use of chemically synthesized, predetermined sequences of DNA. Simple methods are now available to synthesize any short DNA sequence either in a test tube by classical chemical manipulations or automatically using a machine called a DNA synthesizer. Molecules containing promoters, regulatory sequences, modified regions of gene sequences, or even entire genes can be made. These chemically synthesized DNA fragments can be used to replace undesired fragments of isolated genes to produce designer genes that function in unique or even bizarre ways. A well publicized example of this approach was the attachment of the synthetic DNA sequence for human somatostatin (an important hormone, found in hypothalmic extracts, which inhibits the secretion of growth hormone) to the end of the first gene in the lac operon of *E. coli*. This gene is expressed in *E. coli* whenever lactose is present in the medium and the protein coded by the first gene is produced with the somatostatin polypeptide attached. The enzyme-hormone conjugate was readily isolated in a pure state and the biologically active somatostatin was chemically detached from the enzyme. This experiment showed that functional DNA molecules can be designed using even chemically synthesized DNA fragments. Further, it demonstrated the tremendous potential for creating genetic diversity in a highly controlled manner in crop plants.

FUTURE PROSPECTS AND POTENTIAL

Carlson (1975) has suggested, ". . .cell and tissue culture will probably be very effective with some plants in attacking some problems. They certainly will not be a panacea for solving all plant improvement problems." However, in vitro techniques show promise for use by virtually all plant geneticists and breeders in some aspect of their work. Let us briefly summarize and reflect on the potential of in vitro technology and consider the influence that it may have on crop improvement.

Screening for Variants

The culture system easily lends itself to screening and selection for many characteristics. If one wishes to select for salt tolerance, as a number of workers, including Nabors et al. (1980), Dix and Street (1975), and Croughan et al. (1978) have done, one needs only to add salt to the cell culture medium and propagate the surviving cells. If the goal is disease resistance, it may be possible to add to the culture system a pathogen product, such as a toxin, as has been accomplished in maize by Gengenbach and his co-workers (Gengenbach and Green, 1975; Gengenbach et al., 1977) and in sugarcane by Heinz and his co-workers (Heinz, 1972; Heinz et al., 1969). Unfortunately many pathogens do not produce toxins, thus limiting in vitro selection. Other chemicals such as herbicides or pollutants might also be added to the medium for purposes of selection.

The reaction of a cell to selection in culture, however, may not be the same as that of a regenerated plant growing in the field; or a variant character selected in vitro may prove not to be inherited. In addition, it is highly probable that undesirable mutations will occur in the same cell(s) as desirable mutations; thus, regenerated plants with desired mutations may also have unwanted characteristics. This happened with one high-yielding, disease-resistant sugarcane clone developed and selected through the culture system. Unfortunately, it had weak and brittle tops, which made the plant susceptible to breakage. So screening methods must be developed to identify and eliminate undesirable mutations. Screening under conditions of commercial crop production ultimately should be used to determine the effectiveness of selection at the single cell level.

Germplasm Preservation

In vitro techniques can provide assistance in germplasm conservation, thus helping to prevent the loss of genetic variability in crop plants.

It may be feasible to preserve genetic materials through cryogenic systems that maintain temperatures of −196°C for storage of callus and/or cell cultures. In vitro instability seems to be more pronounced in actively dividing cells or in cell cultures that have reached their maximum density; therefore, drastically reducing or halting activity by cryogenic storage may help avoid genetic changes. For further discussion of cryogenic preservation of germplasm refer to Bajaj (1979).

Propagation

In vitro systems can be used for mass propagation of a selected genotype—indefinitely if desired. If a desirable genotype is selected, either within or outside the culture environment, it may be cultured, multiplied, and regenerated into a plant. In theory, all regenerated plants should be exact genetic duplicates of the original genotype. Thus a highly advantageous genotype may be propagated at will without concern for the dilution or random loss of individual alleles normally experienced in sexual crossing systems.

Propagation is not always simple, however, because genetic and/or chromosomal instability of cell and tissue cultures is common in many species. Thus reproduction of a genotype often is all but impossible. Many genetic changes may occur in the culture system; however, instability within culture apparently is genetically controlled, and it may be possible to find genetic or chemical means to induce stability.

Culture systems are also quite effective for propagation of disease-free plants through meristem culture. Meristems usually are free of pathogens or may be pretreated to eliminate them, and meristem cultures generally are relatively stable genetically. Consequently, meristem culture provides a simple method for mass propagation of disease-free plants of stable genotype. Pathogen-free propagules of such crops as orchids, potatoes, carnations, grapes, and strawberries have been cultured and shipped around the word for several years. Recently Phillips and Collins (1979) developed virus-free red clover plants through meristem culture.

Molecular Genetic Engineering

As more is learned about how the presence or absence of particular proteins affects plant phenotypes, we will be able to predict how adding or taking away the genes for those proteins will improve the crop plant. When more is known about how gene expression is regulated, we will be more able to design regulatory sequence–promoter gene recombinants that will be expressed in the proper plant tissues to be most useful in altering phenotype. Soon, it may be possible to alter the amino acid sequence of the protein (by changing its gene) so that undesirable properties of the protein can be eliminated. We may, in fact, be on the threshold of being able to design our crops in the same way that we currently design our farm machinery.

Cell culture technology has adopted many methods used in the study of microorganisms. There are differences, such as nuclear and chromosomal organization, between microorganisms and plant cells that necessitate modification of techniques.

In vitro systems—callus, cell, protoplast, and organ culture and recombinant DNA techniques—provide an excellent opportunity for study of the plant cell, genetic manipulation, creation of variability, selection, and propagation. These are tools we must fully utilize. The short-term benefits are promising—the long-term development essential.

LITERATURE CITED

Bajaj, Y. P. S. 1979. Technology and prospects of cryopreservation of germplasm. Euphytica 28:267–285.

Carlson, P. S. 1973. Methionine sulfoximine-resistant mutants of tobacco. Science 180:1366–1368.

———. 1975. Crop improvement through techniques of plant cell and tissue culture. BioScience 25:747–749.

———, H. H. Smith, and R. D. Dearing. 1972. Parasexual interspecific plant hybridization. Proc. Natl. Acad. Sci. USA 69:2292–2294.

Chaleff, R. S. 1981. Genetics of higher plants. Cambridge Univ. Press, New York.

Cocking, E. C., J. B. Power, P. K. Evans, F. Safwat, E. M. Frearson, C. Hayward, S. F. Berry, and D. George. 1974. Naturally occurring differential drug sensitivities of cultured plant protoplasts. Plant Sci. Lett. 3:341–350.

Collins, G. B. 1977. Production and utilization of anther-derived haploids in crop plants. Crop Sci. 17:583–586.

Croughan, T. P., S. J. Stavarek, and D. W. Rains. 1978. Selection of a NaCl tolerant line of cultured alfalfa cells. Crop Sci. 18:959–963.

Day, P. R., P. S. Carlson, O. L. Gamborg, E. G. Jaworski, A. Maretzki, O. E. Nelson, I. M. Sussex, and J. G. Torrey. 1977. Somatic cell genetic manipulation in plants. BioScience 27:116–118.

Dix, P. J., and H. E. Street. 1975. Sodium chloride-resistant cultured cell lines from *Nicotiana sylvestris* and *Capsicum annuum.* Plant Sci. Lett. 5:231–237.

Emsweller, S. L., and J. Uhring. 1962. Endosperm-embryo incompatibility in *Lilium* species hybrids. p. 360–367. *In* J. C. Garnaud (ed.) Advances in horticulture science and their applications, Vol. 2. Pergamon Press, Inc., Elmsford, N.Y.

Gamborg, O. L., J. P. Shyluk, and E. A. Shahin. 1981. Isolation, fusion, and culture of plant protoplasts. p. 115–153. *In* T. A. Thorpe (ed.) Plant tissue culture. Academic Press, Inc., New York.

Gengenbach, B. G., and C. E. Green. 1975. Selection of T-cytoplasm maize callus cultures resistant to *Helminthosporium maydis* race T pathotoxin. Crop Sci. 15:645–649.

———, C. E. Green, and C. M. Donovan. 1977. Inheritance of selected pathotoxin resistance in maize plants regenerated from cell cultures. Proc. Natl. Acad. Sci. USA 74:5113–5117.

Handro, W. 1981. Mutagenesis and in vitro selection. p. 155–180. *In* T. A. Thorpe (ed.) Plant tissue culture. Academic Press, Inc., New York.

Heinz, D. J. 1972. New procedures for sugarcane breeders. Proc. Int. Soc. Sugarcane Technol. 14:372–380.

———, G. P. Mee, and L. G. Nickell. 1969. Chromosome numbers of some *Saccharum* species hybrids and their cell suspension cultures. Am. J. Bot. 56:450–456.

Honma, S. 1956. A bean interspecific hybrid. J. Hered. 47:217–220.

Kado, C. I., and A. Kleinhofs. 1980. Genetic modification of plant cells through uptake of foreign DNA. p. 47–80. *In* I. K. Vasil (ed.) Perspectives in plant cell and tissue culture. Int. Rev. Cytol. Suppl. 11B.

Keim, W. F. 1953. Interspecific hybridization in *Trifolium* utilizing embryo culture techniques. Agron. J. 45:601–606.

Matern, U., G. Strobel, and J. Shepard. 1978. Reaction to phytotoxins in a potato population derived from mesophyll protoplasts. Proc. Natl. Acad. Sci. USA 75: 4935–4939.

Mok, D. W., M. C. Mok, and A. Rabakoarihanta. 1978. Interspecific hybridization of *Phaseolus vulgaris* with *P. lunatus* and *P. acutifolius.* Theor. Appl. Genet. 52:209–215.

Nabors, M. W. 1976. Using spontaneously occurring and induced mutations to obtain agriculturally useful plants. BioScience 26:761–767.

———, S. E. Gibbs, C. S. Bernstein, and M. E. Meis. 1980. NaCl-tolerant tobacco plants from cultured cells. Z. Pflanzenphysiol. 97:13–17.

Nadar, H. M., S. Soepraptopo, D. J. Heinz, and S. L. Ladd. 1978. Fine structure of sugarcane (*Saccharum* sp.) callus and the role of auxin in embryogenesis. Crop Sci. 18:210–216.

Nickell, L. G. 1973. Test-tube approaches to bypass sex. Hawaii. Plant. Rec. 58: 293–314.

Phillips, G. C., and G. B. Collins. 1979. Virus symptom-free plants of red clover using meristem culture. Crop Sci. 19:213–216.

Power, J. B., E. M. Frearson, C. Hayward, and E. C. Cocking. 1975. Some consequences of the fusion and selective culture of *Petunia* and *Parthenocissus* protoplasts. Plant Sci. Lett. 5:197–207.

Schaeffer, G. W., P. S. Baenziger, and J. Worley. 1979. Haploid plant development from anthers and in vitro embryo culture of wheat. Crop Sci. 19:697–702.

Shepard, J. F., D. Bidney, and E. Shahin. 1980. Potato protoplasts in crop improvement. Science 208:17–24.

Stewart, J. M., and C. L. Hsu. 1978. Hybridization of diploid and tetraploid cottons through in-ovulo embryo culture. J. Hered. 69:404–408.

Street, H. E. (ed.) 1977. Plant tissue and cell culture. Univ. of California Press, Berkeley.

Vasil, I. K., M. R. Ahuja, and V. Vasil. 1979. Plant tissue cultures in genetics and plant breeding. Adv. Genet. 20:127–215.

White, D. W. R., and I. K. Vasil. 1979. Use of amino acid analogue-resistant cell lines for selection of *Nicotiana sylvestris* somatic cell hybrids. Theor. Appl. Genet. 55:107–112.

SUGGESTED READING

Barz, W., E. Reinhard, and M. H. Zenk (ed.) 1977. Plant tissue culture and its biotechnological application. Springer-Verlag New York, Inc., New York.

Freifelder, D. 1983. Molecular biology. A comprehensive introduction to prokaryotes and eukaryotes. Science Books International, Boston.

Old, R. W., and S. B. Primrose. 1980. Principles of gene manipulation. An introduction to genetic engineering. Studies in microbiology, Vol. 2. Univ. of California Press, Berkeley.

Reinert, J., and Y. P. S. Bajaj (ed.) 1977. Plant cell, tissue, and organ culture. Springer-Verlag New York, Inc., New York.

Rubenstein, I., R. L. Phillips, C. E. Green, and B. G. Gengenbach (ed.) 1979. Molecular biology of plants. Academic Press, Inc., New York.

Sharp, W. R., P. O. Larsen, E. F. Paddock, and V. Raghaven (ed.) 1979. Plant cell and tissue culture: principles and applications. Ohio State Univ. Press, Columbus.

Chapter 8

Induced Mutations

B. SIGURBJÖRNSSON

Agricultural Research Institute
Reykjavik, Iceland

Mutations, defined as changes in genetic material, are a part of the basic phenomena of life. If mutations never occurred, living material would not have developed and adapted to different ecological conditions. Throughout history mutations have occurred spontaneously, caused by a number of natural phenomena such as cosmic and ultraviolet radiation. Recently people have discovered the means to induce mutations artificially (Fig. 8.1). An example of the integral part mutations play in our lives is the fact that as you are reading these lines, mutations are occurring in the genetic material of some of your body cells. Such mutations may lead to a gradual weakening of the body, and accompanying symptoms of advancing age.

Since all living things have been selected for adaptability to their environments, most changes in their genetic makeup are normally detrimental. Mutations, spontaneous or induced, are usually harmful, and cells bearing new mutations tend to lose out in competition with unaffected cells. Since the type of genetic change involved in a mutation is more or less random, it does happen, though rarely, that such a change improves the ability of the organism to survive, grow, and reproduce. In that case, the cell containing the improved genetic material will have an advantage over the old form and may replace the old type of tissue or organism. Through recombination between this new genetic material and the existing genetic material, a range of new types of organisms is formed. These types, better adapted to the environment, may survive and increase, and the poorly adapted organisms will disappear.

All variability in living organisms is ultimately caused by mutation. The patterns seen among plant and animals result from the interaction between the environment, mutations, and their recombinations. That is exactly what evolution is all about. The three components of evolution are: (1) mutation, (2) recombination through hybridization, and (3) selection. Plant breeding, which could be called guided evolution, depends on the same principles. Breeders grow different genotypes of a crop in an environment

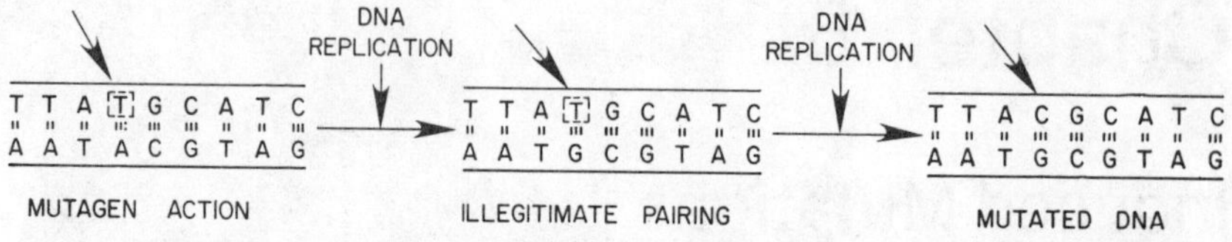

Fig. 8.1. Genetic changes that alter the DNA code may be induced by chemical treatment or radiation. In this illustration, a change in thymine (T) permits it to pair illegitimately with guanine (G). The resultant daughter strand carries the mutated genetic code.

to see which genotype performs the best. They hybridize these genotypes to obtain genetic recombination of desired traits and test them in the appropriate environments.

Breeders then select those types which will suit the needs of the grower, and ultimately they release improved commercial cultivars. Thus, the breeders normally exercise control over two of the cornerstones of natural evolution. They choose the parents for crossing to obtain recombinations, and they choose the environment in which they test and select the progeny. With the ability to induce mutations, breeders also gain control over the third cornerstone. They can increase or create new genetic variability in the plant material.

Mutation, in its broadest sense, covers any change in the genetic material at the gene and chromosome level. The Dutch botanist, Hugo de Vries, used the term for the first time at the turn of this century, although earlier Darwin had felt that "species were mutable productions". Similarly, Linnaeus had noticed a number of "mutations" which complicated his classification and nomenclature system but at that time the nature of these phenomena was not understood. De Vries noticed morphological changes in *Oenothera lamarckiana* Ser. and called them "mutations." However, these changes were not of the type we now normally refer to as "mutations." Some of them had two sets of chromosomes (diploid). Others either lacked or had extra chromosomes (aneuploid). Wilhelm Konrad von Roentgen discovered x-rays in 1895, and in 1896 Henri Becquerel discovered that uranium gives off radiation. In 1897, Pierre and Marie Curie isolated radium. In 1904, De Vries had the foresight to suggest the use of the newly discovered radiation phenomena to induce mutations.

Muller (1927) successfully used x-rays to induce mutations in the fruit fly. A year later, Stadler (1928a, b) induced mutations in barley and maize. At the same time, many workers tried to induce mutations with various chemicals.

MUTAGENS

A number of agents are capable of inducing mutations in plants. Such agents are called mutagens, some of which occur in nature and can cause "spontaneous mutations." Relatively few of the numerous known mutagens are used routinely for mutation induction in higher plants.

Mutagens are generally classified as physical and chemical.

Physical Mutagens

Typical physical mutagens are different types of radiations. Breeders most frequently use x-rays, gamma rays, ultraviolet rays, and neutrons as mutagens. All of these radiations (with the exception of ultraviolet rays) ionize atoms in a tissue by detaching electrons from the atoms. The ionizations from irradiation with x-rays, gamma rays, and beta rays, occur sparsely along the track of the ionizing particle. When the ionizing agent consists of an atomic nucleus, such as alpha particles or recoil nuclei, knocked out by fast neutron radiation, the ionizations are more densely concentrated. The ion density is expressed in linear energy transfer (LET), which is the energy dissipated per unit length along the tracks of the ionizing particles. The ionizations may result in a regrouping of the molecules along the track of ions left by the radiation. Such regroupings cause chemical change that may lead to gene mutations or to chromosomal breaks and rearrangements.

In the ionization process, positive radical ions and free electrons are produced. In a biological system the electrons are trapped and the radical ion, which is both unstable and reactive, can react with other molecules. The free electrons in water solutions can polarize water molecules and become hydrated electrons. The free radicals generated in solution will eventually recombine with each other to form stable products. Molecular oxygen reacts readily with radiation-induced free radicals to form peroxyradicals. The presence of oxygen is therefore important in changing and enhancing the effect of irradiation in biological systems. In tissues with low water content such as dry seeds, the most commonly irradiated plant part, the radiation-induced radicals decay slowly. If the water content of the seeds is high, the radicals decay rapidly. So in irradiating seeds, the breeder must ascertain the water content and oxygen status in order to obtain repeatable results. Since the ratio of gene mutations to gross chromosomal alterations is lower in materials exposed to oxygen, breeders will generally wish to reduce or prevent the oxygen enhancement.

Another important chemical reaction caused by irradiation in water solutions is the recombination of free radicals to form hydrogen peroxide. The free radicals and hydrogen peroxide can react with molecules to form chemical changes similar to those taking place after ionization of a molecule, the effect depending on the use of low or high radiation. The mutation induced by irradiation could be attributed to the radiation striking the molecules in the biological material or indirectly through the chemicals produced by the ionizing radiation.

Within plant cells, certain regions are more sensitive to radiation than others. These regions, called radiosensitive targets, correspond generally with the location of DNA in the cells. Irradiation can affect either one or both of the DNA strands of the double helix. If the breaks involve a single strand, the linear integrity of the DNA molecule is still intact and the repair of the break restores normal DNA. When the breaks involve both of the strands, they are not as readily repaired and this is the most important radiation-induced damage of DNA resulting in detectable mutations.

Ionizing radiations differ in their effectiveness in causing mutations. Higher doses (in amount of energy absorbed by the tissue) of gamma and x-rays than of fast neutrons are needed to induce mutations. Another way to explain this idea is to say that thé relative biological effectiveness (RBE) of fast neutrons is higher than that of gamma and x-rays. The RBE is affected by several factors, but is primarily a function of LET up to a certain maximum, because inactivation of a radiosensitive target requires a minimum number of ionizations within a small critical volume. Fast neutrons, with their higher ion density (LET), would increase the probability of sufficient ionizations occurring within the critical volume. A maximum exists, because more ionizations than required for target inactivation would have no additional effect. Irradiation of plants also causes physiological damage, e.g. stunted growth and reduced viability, which are proportional to radiation dose. The breeder is interested in radiation treatments with high RBE to obtain the maximum number of mutations. This can be accomplished by selecting radiation treatments with high ionization density or by adjusting the conditions of radiation treatment.

X-rays are produced in special machines by bombarding tungsten or molybdenum with electrons. During the early years of the use of induced mutations in plant breeding, x-ray was the only readily available mutagen.

Gamma irradiations are normally performed using the radioactive isotopes cobalt-60 or cesium-137. Both x-rays and gamma rays are electromagnetic radiations that are primarily used to irradiate seeds or vegetative tissue. The gamma rays have shorter wavelength with more energy per proton than x-rays. Gamma sources are frequently used in greenhouses, growth chambers, and in outside fields (called gamma fields) for the irradiation of whole plants.

Ultraviolet light, although not an ionizing radiation at the wavelengths commonly employed, is used as a mutagen to treat pollen grains. Ultraviolet wavelengths, in the range of 2,500 to 2,900 nm from germicidal mercury vapor lamps are biologically effective since this is the region of maximum absorption by nucleic acids.

In recent years, neutrons have gained popularity in use by breeders because treatment techniques have been improved and standardized. Neutrons are normally obtained from a nuclear reactor where uranium-235 fuel is undergoing nuclear fission. Since a variety of radiations is produced by the reactor (fast and thermal neutrons, gamma rays, etc.), uniformity in dosimetry and in procedures was difficult to attain. With the development of facilities for neutron irradiation in nuclear reactor pools and the adoption of standardized procedures of treatment and dosimetry, the use of fast neutrons has become an important and effective method of mutation induction.

Beta radiation emitted by internal radioisotopes, such as radioactive phosphorus, ^{32}P, radioactive sulfur, ^{35}S, and radioactive carbon, ^{14}C, has been extensively used in radiobiology. Two crop cultivars have been released as a result of multiplication of mutants that originated from feeding ^{32}P to the plant.

Further information on physical mutagens can be obtained from the paper by Constantin (1975) and from the *Manual on Mutation Breeding* (International Atomic Energy Agency, 1977).

Chemical Mutagens

Most plant breeders do not have easy access to efficient radiation sources. Chemical mutagens are more readily available, and the ratio of mutational to undesirable modifications is somewhat better for chemical mutagens than for irradiation. Hence, chemical mutagens are becoming increasingly popular. The application of chemical mutagenesis in practical plant breeding has been rather slow. By 1976 there were just over 20 crop cultivars developed by chemically induced mutations.

For practical plant breeding, the most interesting group of chemical mutagens are alkylating agents. These compounds bear one or more reactive alkyl groups capable of being transferred to other molecules at positions where the electron density is sufficiently high. The alkylating agents react with DNA by alkylating the phosphate groups as well as the purine and pyrimidine bases. Among the 30 to 40 chemical mutagens in the category, some of the most powerful and useful mutagens are ethyl methanesulfonate (EMS), diethylsulfate (DES), ethylenimine (EI), N-nitroso-N-methyl urethane (NMUT), and N-nitro-N-methyl urea (NMU), N-nitrose-N-ethyl urethane (NEUT) and N-nitrose-N-ethyl urea (NEU). Extreme care should be taken in using these compounds because they are known to be carcinogens.

Another group of useful mutagens are base analogues. They are closely related to the DNA bases, and can be incorporated into DNA in place of normal bases without hindering its replication. Because they differ from the normal bases in certain ways, these mutagens can give rise to pairing errors in DNA which result in mutations (Fig. 8.1). An example of a frequently used analogue is 5-bromouracil (BU). A related compound, maleic hydrazide (MH), causes chromosome aberrations. These compounds have not proved effective as plant mutagens, but may prove more useful when applied to cell cultures.

Nilan et al. (1975) reported another highly efficient mutagen, sodium azide (NaN_3), a respiratory inhibitor for plants and animals which induces gene mutation in barley without chromosome breakage. It is effective on barley, peas, and diploid wheat but not on polyploid wheat or oats.

METHODS OF APPLICATION

Physical Mutagens

Whole plants or any part of the plant can be treated by radiation. However, because of the physical bulk, special installations are needed to irradiate whole plants or large plant parts. Gamma fields are expensive to build and maintain. They are normally circular fields with pie-shaped sectors and an earthen wall to protect the surrounding area from the radiation. In the center of the field a powerful gamma source, normally cobalt-60, is located either underground or in a shielded tower. The source is either lowered or raised to expose the plants growing in the field to radiation. Since the radiation dose decreases with the square of the distance, treatment intensity can be varied by planting or placing the material at various distances from the source.

Although gamma fields have not been very useful for practical breeding purposes, they have contributed to the study of radiation effects on whole plants and have added to our knowledge of radiobiology as well as to radioecology.

For treating vegetative parts, such as branches of fruit trees or entire plants, smaller gamma chambers with controlled environmental conditions are useful. Such chambers are similar in design to a gamma field except that they have a completely enclosed treatment area.

Gamma fields and chambers are used primarily for treatment of vegetatively propagated crops and for chronic irradiation of meristems. Pollen grains also are often irradiated with ionizing radiations and ultraviolet rays. The main advantage of pollen irradiation, as compared to irradiation of seeds or growing plants, is that it rarely produces chimeras or mutated sectors. The chief disadvantage is the short time pollen grains are viable.

Seeds are most commonly used for irradiation. They offer a number of advantages (Nilan et al., 1961). They are easy to handle and store. They can be subjected to a variety of physical and chemical environments. They can be desiccated, soaked, heated, or frozen. They can be maintained for extended periods of time in a vacuum almost free of oxygen as well as under high pressure of oxygen or other gases. When dry, seeds are almost inert biologically and severe environments cause no significant biological damage. This is fortunate since rigid conditions are necessary for proper control of the factors that modify radiation damage induced in the seed. Also, dry dormant seeds are easy to handle and can be shipped over great distances. Seeds are most commonly treated with gamma rays in a gamma cell and with fast neutrons in the standard facilities available in many pool-type reactors.

A great number of studies have been done on the different biological, environmental, and chemical factors which can modify the effect of radiation treatment. These factors modify mutagenic effectiveness (the number of mutations per unit of dose and mutagenic efficiency (ratio of the number of mutations to the damage such as sterility, stunted growth, etc.) of both physical and chemical mutagens in cells of higher plants.

Oxygen is a major modifying factor of biological (genetic) damage caused by x- and gamma rays in dry dormant seeds and its effect is, in turn, influenced by temperature, seed water content, radiation energy, and hydrogen ion concentration. Oxygen and most other external factors have little direct influence on the biological effects caused by densely ionizing radiations such as neutrons or by chemical mutagens. Oxygen only affects chemical mutagens indirectly through its influence on metabolic and cellular activity. The effect of low LET radiations can be modified most effectively by adjusting the amount of oxygen and best results are obtained when plant parts are irradiated under oxygen-free conditions.

The effects of x-rays and gamma-rays are also influenced by the water content of seeds, effects in part due to the relationship between water and oxygen content of the seed. In the "dry" range (1 to 16% moisture) barley seeds with a content of 3 to 6% moisture show increased radiosensitivity (Singh and Singh, 1975). Soaked or wet seeds also show increased radiosensitivity (Conger et al., 1968).

Specific information on treatment and radiation dose for individual crops, including vegetatively propagated crops, is provided in the *Manual on Mutation Breeding* (International Atomic Energy Agency, 1977).

Chemical Mutagens

Chemical mutagens are most commonly used to treat seeds and, rarely, buds and cuttings, which can be either dormant or actively growing when soaked in the mutagenic solution. The chemical can also be introduced into a plant tissue such as the stem, leaf, or inflorescence. Because the roots are sensitive to chemical mutagens, the mutagen should not be introduced into the growing medium. Pollen grains may be exposed to a vapor of the mutagen.

Some of the chemical mutagens can cause a higher number of mutations than radiation, but good results may depend on careful attention to the chemical concentration, duration of treatment, temperature, and pH of the mutagenic solution, and water content of the seeds. Other factors that influence the effects of chemical mutagens are often species- and mutagen-specific. Usually it is important to rinse seeds in water after treatment in order to prevent damage. If the seeds are not planted immediately, they must be dried. With certain mutagens, especially EMS, drying of seeds after treatment can markedly enhance injury, even resulting in total loss.

MUTATION BREEDING

Genetic variability resulting from induced mutations appears to be equivalent to that occurring naturally. Therefore the basic principles for the use of induced variability are similar to those for natural variability. A clearly defined objective, knowledge of the crop, careful choice of parents, skillful application of plant breeding principles, effective use of selection criteria, and thorough testing of the resulting lines are fundamental to any plant breeding program.

In addition, the breeder who uses induced mutations should be familiar with mutagens and their application, the nature of induced mutations, and have the means and ability to identify mutants.

Choosing Parents

After having established the objectives of the breeding program and the necessity for using induced mutations to solve particular problems, the breeder must consider the selection of parent stock for mutagen treatment.

Normally a high-yielding, well-adapted cultivar is chosen, especially when the cultivar lacks a single important attribute, such as strength of stem, resistance to a particular disease or pest, or early maturity.

However, because mutations are induced randomly and the mutagen rarely changes only one particular gene, the effect of mutagenic treatments on quantitatively inherited characters should be considered. All the known mutagenic agents, applied at levels intended to yield an appreciable amount of visible mutations, also induce a considerable variation in quantitatively inherited characters (Brock, 1965; Gaul, 1965; Scossiroli, 1970). Thus, although the goal of the selection is a certain desirable mutated trait, the background genotype would very likely be altered in several other ways. These secondary effects of the mutagen may or may not be important. If no selection is applied, the random effect of mutagen treatments and the preponderance of deleterious mutations may decrease the average performance of the offspring. Furthermore, the mean for a desired character may shift away from the direction of the previous selection (Brock, 1976). However, when the selection pressure was applied to pick out vigorous and well-adapted types which contained the desired alteration in a single character, it was possible to breed superior cultivars without going through a crossbreeding program. A majority (117 out of 207) of the seed-propagated crop cultivars that have resulted to date from induced mutations have been direct multiplications of selected mutants (Sigurbjörnsson and Micke, 1974).

However, it is not always necessary to choose a well-adapted, high-yielding genotype for a starting point. If the objective is to provide new sources of variability for use in a crossbreeding program, it may be sensible to start with a genotype in which the desired mutation can be identified most readily. Identifying mutants for early maturity is easier when a late-maturing cultivar is treated with mutagens. Similarly, dwarf mutants are more easily identified in a tall cultivar.

In choosing parent seeds of self-fertilizing species for mutagen treatment, genetic purity is important because mixtures or outcrosses with different cultivars or genotypes will complicate the identification of mutants. The best insurance is to use elite or breeder's seedstock for mutagen treatment, since it is produced under isolated conditions, and is guaranteed true to type.

M1 Generation

The successive generations of descendants from mutagen-treated seeds or plants are designated M1, M2, M3, etc. to distinguish them from generations following hybridization which are designated F1, F2, F3, etc.

The M1 generation starts with the germination of mutagen-treated seeds. Only dominant mutations will be expressed in the M1 generation. The M1 plants will be heterozygous for newly induced mutant genes and will segregate into mutant and non-mutant phenotypes in the M2 generation. Newly induced recessive mutations will be expressed and observed in the M2 generation following self-fertilization of the M1 plants.

Identification and isolation of mutants in cross-fertilizing species requires controlled pollination when producing the M2 generation. In self-incompatible species, homozygous mutants can only be expressed in the M3 generation following mating of sister M2 plants.

The mutagenic treatment causes physiological damage or injury to the seed or plant part, often resulting in significant reduction in germination, seedling growth, and fertility of the M1 plants. Hence, the main concern is to produce as many fertile M1 plants as possible to obtain a sufficiently large M2 population so that selection of desired mutants can be effective. Some of these injuries are discussed below.

Indicators of M1 Injury

The degree of physiological injury caused by the mutagen is usually correlated with the frequency of mutations induced, especially with regard to radiation (Gaul. 1959a). Therefore, a quantitative determination of M1 injury is commonly used as a check on the effectiveness of the mutagen treatment (Constantin, 1975).

The most common indicators of M1 injury are: (1) effects on plant growth, (2) cytological effects, and (3) induced sterility.

Plant Growth

Reduction in seedling height is most commonly used as an indicator of mutagen effect. Seed germination is not a good indicator for radiation treatments; seeds germinate even after being exposed to radiation doses which will kill the seedling or the adult plant. However, chemical mutagens can drastically affect seed germination. The effects of fast neutron treatments on growth in the M1 generation are normally much more uniform than the effects of x-ray or gamma-ray treatments. This difference is due to the greater sensitivity of x-ray and gamma-ray treatments to various modifying factors, especially oxygen. Emergence under field conditions or survival under laboratory or field conditions are also useful guidelines to assess mutagen damage to M1 plant growth.

Cytological Effects

The effects of mutagen treatments can be observed under the microscope in various ways. These range from stickiness and clumping of the chromosomes to induced chromosome mutations which can be observed in cells during mitosis and meiosis. The frequency and types of chromosomal aberrations are often used to determine the effectiveness of mutagenic treatments. Most commonly, the first mitotic cycle in root or shoot tip preparations of germinating treated seeds is examined. The frequency of chromosome or chromatid bridges at anaphase is proportional to the radiation dose (Gaul, 1963). Extensive studies have also been done on the chromosomal aberrations appearing in meiosis. Such studies have given us a much clearer understanding of the induction and nature of mutations. There are distinct differences between the cytological effect caused by ionizing radiations and the effects of chemical mutagens. The latter generally cause fewer structural effects (Sparrow, 1960; Gaul, 1963).

Sterility

One of the main disadvantages of using high doses or concentrations of a mutagen is the sterility it may cause in M1 plants resulting in reduction in M2 population size. Mutagen-induced sterility may be caused by chromosome aberrations, gene mutations, cytoplasmic mutations, and physiological effects. Chromosome aberrations are probably the major cause. There does not seem to be a relationship between the degree of sterility in the M1 generation and the frequency and spectrum of mutations found in the M2 generation (Gaul, 1965). Sterility is normally indexed by counting the number of inflorescences per plant, number of florets per inflorescence, number of seeds per inflorescence, or number of seeds or fruits per plant.

Selecting Dose Level

In selecting appropriate dose levels for physical and chemical mutagens, the various physiological and cytological effects on the M1 generation caused by the mutagen must be considered. The maximum number of induced mutations results when the number of fertile M1 plants is maximized to produce a sufficiently large M2 generation. In general, when using sparsely ionizing radiations (x- and gamma-rays) in greenhouse tests, the optimum dose for cereals should cause 30 to 50% reduction in seedling height. When densely ionizing radiations (neutrons) are used, the height reduction should be 15 to 30%. With chemical mutagens, the reduction should be 10 to 30%. In practice, the optimum dose is often achieved by using three separate doses (with an untreated control). One dose or level should be chosen based on reduction in seedling height in laboratory and field tests. The other two doses should be about 10% higher and lower, respectively.

There are differences between the efficiency of various mutagens; some will induce a high mutation frequency and a low level of damage and sterility in the treated generation. On the basis of the number of mutations relative to sterility, EMS appears to be four times more efficient than x-rays (Gaul et al., 1972). Constantin (1975) has compiled data to show that for each 10% increase in sterility caused by EMS, the number of mutants per 1,000 M2 seedlings increases by nearly 22 times. This is 14 times the number obtained with gamma rays and 4½ times that obtained with neutrons.

Size of the M1 Population

The size of the M1 population, i.e. the number of seeds to be treated, is partly governed by the effectiveness (number of mutations per unit dose) and efficiency (number of mutations relative to sterility or other adverse effects) of the mutagen. As can be expected these results will vary for different plant species. Even within a species, cultivars respond differently to mutagen treatment. For example, when two cultivars of barley were exposed for the same length of time with the same concentration of EMS, one cultivar produced 68 and the other 11.5 mutants per 1,000 M2 seedlings.

Brock (1976) calculated M2 progeny size and M1 population requirements according to the various mutant segregation ratios and the probability of the occurrence of the homozygous mutant. Assuming a 50% lethality in the M1 generation of 5,000 plants, 2,500 fertile survivors would be tested in the M2 generation. The M2 generation with 20 individuals per progeny would then be 20 × 2,500 = 50,000 plants. With lower mutation frequencies, higher populations would be required. However, the breeder may wish to increase the number of M2's examined to increase the number of mutants available for evaluation.

Chimeras

Mutations are induced in single cells, and mutants are observed in the progeny of the mutated cell. When mutagenic treatment is applied to multicellular organs such as seeds, buds, or growing plants, each cell has a probability of mutation. Progeny of mutant cells and the adjacent normal cells produce tissues of different genotypes. This mixture of genotypes is called a chimera. The embryos of higher plants possess numerous meristematic cells. Each has the potential to contribute to the formation of various plant parts. After treatment the resultant M1 plant may be a chimera of different genotypes.

The mutagenic treatment of a dormant embryo induces mutations at random in individual cells. A mutation induced in a single viable meristematic cell will occur in either all or part of that cell's descendants, depending on the stage in the cell's divisional process when the mutagenic treatment took place. Thus the part of the plant which arose from the mutated cell will be different from the rest of the plant. Plants having tissues or organs consisting of cells of different genetic composition are said to be chimeric; the mutated part is commonly referred to as a sector (Gaul, 1959b).

The M1 plants are normally chimeric. Throughout the growth of the

M1 plant, the opportunity exists for competition between mutant and non-mutant cells. This phenomenon is called diplontic selection and its occurrence influences the frequency of mutants observed in the M2 generation. Dominant mutations can be expressed in the M1 plant but only in that chimerical sector which results from the original mutated cell. Polyploidy causes a further complication for recessive mutations which are only revealed phenotypically when all alleles at a locus are present in the homozygous recessive condition.

Other Considerations of the M1 Generation

Because effects of mutagens are detrimental, the M1 generation must be handled with care. Lack of special consideration for mutagen-treated seeds during treatment and throughout the growing of the M1 generation has often resulted in the failure of mutation breeding programs. Even if the mutagen treatments induce a number of mutations, the breeder must provide favorable conditions for the expression of mutations. It is important to have a well-prepared seed bed with adequate moisture to ensure optimum seedling growth and development. Most of the induced mutations are contained in the first tillers or flowering stems produced.

Outcrossing can be a problem in normally self-fertilizing species. Since mutagen treatment causes considerable male sterility, the M1 generation is subject to fertilization by foreign pollen, causing outcrossing which will mask the expression of mutations. To prevent outcrossing, the nursery can be grown in isolation, planting could be delayed, or inflorescences of M1 plants covered with paper or plastic bags. The M1 generation should receive the best possible care with regard to weed, insect, and disease control since the objective is to transfer as many mutations as possible to the M2 generation where selection for desired phenotypes is done.

M2 and Subsequent Generations

Success in discovering mutants in the M2 and subsequent segregating generations depends not only on success in treating and raising the M1 generation, as explained above, but also in the way the M1 plants are sampled for harvest. Several different methods are used, largely depending on the ontogenetic development pattern of the species and the method of screening used. In practice, however, the selection of breeding methods often depends on the relative availability and cost of field or greenhouse space, machinery, labor and time. One can harvest and plant all surviving seeds of the M1 generation. One can also sample the M1 generation with as good or better results. Normally one then selects the primary tillers, main branches, or first fruit clusters, etc., on the M1 plant, since these will produce the highest number of mutants.

Redei (1974) has reviewed the considerations in choosing the size of the progenies to be examined for the most efficient detection of recessive mutants in the M2 and subsequent generations. He recommended a large M1 population and small M2 families. Usual breeding procedures, such as single seed descent or plant-to-row methods, should be used in the mutation breeding program to identify the rather low frequency of mutant phenotypes.

Success in identifying mutants in the M2 and M3 population may depend to a large extent on ease of detection. Techniques of applying heavy selection pressure, such as subjecting the segregating population to a pathogen or pest, and mechanically sifting out lightweight seeds, are common practices in breeding programs. Success depends on rigorous selection and discarding of useless material. Gustafsson (1975) described this situation: "It is indeed astonishing, when glancing back, what a multitude of experiments have been devoted . . . to the blind induction of mutations in higher plants, creating 'worthless junk' instead of meaningful selections."

Induced mutations are generally regarded as occurring at random throughout the genome (Brock, 1976). Experience with the erectoides mutants (short, compact plants and spikes) and with the eceriferum locus in barley cast doubt on this generalization (Persson and Hagberg, 1969; Lundqvist et al., 1968). Nilan (1972a) also found that chlorophyll-deficient mutations in barley were induced more frequently with one mutagen than with another. There are differences between various types of radiations, between radiations and chemical mutagens, and among various chemical mutagens.

Breeders have long hoped to be able to induce specific mutations selectively by the use of either particular mutagens or treatment methods. They dreamed of chemical mutagens which would recognize and specifically mutate particular genes. The chance of obtaining a chemical mutagen which would recognize the sequence of a structural gene appears very remote. Brock (1976) pointed out several ways to gain control over the mutational process other than by specifically affecting the structural gene. Genes in bacteria appear to be variously sensitive to mutagens depending on whether they are active or dormant. Auerbach (1967) suggested controlling the mutational process by manipulating the physical, cellular, or genetic environments to influence the type and the frequency of mutations that are recovered.

Constantin (1975) expressed doubt that any mutagen induces a specific mutation. However, he believed experience with the erectoides, eceriferum, and chlorophyll-deficient mutants warranted a continued search for evidence of mutagenic specificity.

There seems to be no limit, however, to the kinds of traits that can be improved by mutation. Commercial cultivars of induced mutant origin have been improved in most of the possible attributes (Sigurbjörnsson, 1975). Practical and basic studies on mutation induction in higher plants have shown that no aspect of plant morphology or physiology is immune from mutagenic action; in principle we can induce any gene mutation which has occurred naturally as well as traits that have not yet occurred spontaneously or have been lost from the existing populations. Table 8.1 indicates the range of characters improved with the use of mutagens.

Table 8.1. Improved characters in mutant cultivars.

Improved characters	Cereal†		Other crops‡	
	No.	%	No.	%
Plant morphology				
reduced plant height	88	64.3	7	9.2
stiff straw, lodging resistance	67	48.9	6	7.9
Increased earliness				
5 to 10 days	29	21.2	13	17.1
> 10 days	3	2.2	1	1.3
Increased yield				
3 to 10%	48	35.0	33	43.4
> 10%	16	11.7	8	10.5
Seed characteristics				
morphology (size, color)	11	8.0	2	2.6
quality (protein, oil, malting, baking)	47	34.3	8	10.5
Fruit, fiber and leaf				
morphology	--	--	1	1.3
quality	--	--	10	13.1
Resistance to disease				
fungi	34	24.8	7	9.2
bacteria	--	--	2	2.6
virus	--	--	2	2.6
others (nematode, insect)	--	--	3	3.9
Other characters				
adaptability	11	8.0	4	5.3
threshability	2	1.4	2	2.6
easy harvesting	--	--	7	9.2

† A total of 137 cultivars.
‡ A total of 76 cultivars.

EXAMPLES OF SUCCESSFUL USES OF INDUCED MUTANTS

Muller (1927) foresaw the possibilities of using radiation to induce mutations for plant breeding purposes. Stadler (1928b), influenced by the overwhelming number of deleterious mutations he observed in maize, did not feel that this new method would assist the plant breeder. Up until the 1970's, very few scientists used induced mutations to improve crop plants, and the few cultivars released that carried mutated genotypes were largely the byproducts of theoretical studies.

The total number of cultivars which were developed using induced mutations includes nearly all crop types (Table 8.2), and represents various breeding systems, growth habits, and propagation methods. In 1981 some 213 induced mutant cultivars were recorded in seed propagated crops, some 20 in vegetatively propagated crops, and at least 223 in ornamental plants (Micki and Donini, 1981).

A detailed account of the use of mutations in plant breeding programs was reported by Sigurbjörnsson and Micke (1973). The first commercial success with induced mutations was reported in 1934 with the release of a new tobacco cultivar 'Chlorina,' characterized by pale color and high quality leaf. The mutant was induced by x-ray treatment in 1930. The next mutant cultivar, released in 1950, was the white mustard cultivar 'Primex' which resulted from x-ray treatments done in 1941 by the Swedish Seed Association at Svalof (Sigurbjörnsson and Micke, 1974).

Table 8.2. Number of seed-propagated cultivars developed by induced mutation.

Crops	Total developed	Direct mutants	Mutant used in cross	Percent developed through crosses with mutant
Cereal: barley, bread and durum wheat, rice rye, oat, pearl millet, maize	137	69	68	49.6
Proteoleaginous: soybean, peanut, lupine, sunflower, castor bean	22	15	7	31.8
Fiber: flax, cotton, jute	8	8	--	--
Vegetable: onion, lettuce, spinach, pepper, bean, pea, tomato	26	20	6	23.0
Other crops: mustard, clover, tobacco, sesame, etc.	20	14	6	30.0
	213	126	87	40.8

Table 8.3. Mutagens used for crop breeding and the number of cultivars resulting from each.

	Treatment performed					
Type of mutagen used	Before 1954	1955–1960	1961–1965	1966–1970	1971–1975	Total no.
X-rays	35	23	6	0	0	64
Gamma rays	0	14	11	18	7	50
Thermal neutrons	7	10	1	2	0	20
Fast neutrons	0	3	0	0	0	3
Chemical mutagens	3	2	4	11	1	21
Combined mutagens	0	1	1	0	0	2
	45	53	23	31	8	160

All types of physical mutagens and a number of chemical mutagens have been used to induce mutations. At the beginning, before isotopes became readily available and nuclear reactors accessible, x-rays were the most common radiation used (Table 8.3).

Swedish scientists were pioneers in the practical uses of induced mutations and developed a great deal of the present technology. Foremost among the scientists was Ake Gustafsson, who worked with barley. A brief review of mutation breeding of barley in Sweden will illustrate the practical use of this technology.

In addition to being an important crop in Sweden and many other countries, barley is popular among research workers for other reasons. Its cytogenetic makeup is well known; the diploid chromosome number of 14 makes it a good experimental plant. The genetics of barley has been well studied and this self-fertilized plant is among the most versatile and widely grown crops (Nilan, 1974).

Following x-ray treatments of the barley cultivar 'Bonus' in 1945 and 1949, Gustafsson (1975) isolated two mutants. A direct multiplication of these two mutants gave rise in 1960 to the cultivar 'Pallas' which had a marked increase in resistance to lodging, and in 1962 to the cultivar 'Mari' which matured earlier than Bonus. Pallas became the standard cultivar in Sweden and Denmark for testing. In 1966, almost 40% of the barley acreage in Denmark was sown with Pallas. Mari became popular in northern regions where its remarkable earliness and resistance to adverse weather broke through climatic barriers and extended barley cultivation to the north. These two mutant cultivars have since given rise to 10 other cultivars through hybridization. These cultivars resulting from Bonus through induced mutations are now widely grown throughout Europe and have done remarkably well even under tropical conditions in South America (Sigurbjörnsson, 1975).

In the USA the barley cultivar 'Pennrad' was derived from thermal neutron treatment, and 'Luther' was the first commercial barley cultivar developed using a chemical mutagen. Both were highly successful (Nilan, 1972b), but have since been replaced by derivatives. Induced mutations have been used in the development and release of 58 commercial mutant barley cultivars. Less than half of these cultivars resulted from direct multiplication of induced mutants, the rest from the use of induced mutants in hybridization programs (Micke and Donini, 1981).

In addition to improvement in the agronomic characteristics by mutagen treatments, valuable progress has been made in nutritional quality of barley. Like other cereal grains, barley is deficient in lysine content. A spontaneous mutant, high in lysine, known as 'Hyproly,' was discovered in a low-yielding Ethiopian cultivar by Hagberg and Karlsson (1969). But Hyproly is not adapted to European growing conditions. In spite of extensive backcrossing to modern northern European genotypes, the yield of derivatives has been 80 to 85% of the currently grown cultivars (Hagberg, 1978).

Doll et al. (1974) extensively treated the barley cultivars 'Bomi' and 'Carlsberg II' with mutagens. After screening about 15,000 lines for lysine content, they found 20 mutants which had different amino acid composition, with six characterized by high lysine contents. One of them, 'Bomi 1508,' resulting from treatment with ethylenimine, was a single gene mutant with a 40% increase in lysine content over its parent, Bomi.

Genetic studies of high lysine mutants have confirmed that the high lysine characters of Hyproly and Bomi 1508 are due to two independent recessive genes and that they act in an additive manner to increase the seed lysine content (Muench et al., 1976). However, the progress in improving the yield of high lysine cultivars derived from Bomi 1508 has been limited (Doll and Koie, 1978).

The next example of successful use of mutagens in a breeding program concerns a crop completely different from the examples cited above and illustrates a problem which could hardly have been solved without an induced mutation.

All peppermint oil produced in the USA is obtained from the steam distillation of the herbage of a single triploid clonal strain of *Mentha piperita* L., known as the cultivar 'Mitcham.' This cultivar replaced all previously grown clonal strains in about 1890 and has been cultivated continuously since that time through vegetative propagation by stolons. The annual value of peppermint grown in the USA is about $25 million. It flavors chewing gum, toothpaste, and candy products worth billions of dollars. This industry was threatened by severe attacks on Mitcham by *Verticillum* wilt. A comprehensive program to combat this threat was undertaken. Wild types of spearmint are highly resistant to wilt. Although the resistance could be incorporated in Mitcham, the quality of mint oil was unacceptable. Other attempts to combat wilt through cultural methods, use of soil fumigants, antibiotics, propane flaming, etc., were not successful either.

Murray (1971), after irradiation treatment of a large number of stolons, grew about 6 million plants and, in the second year, subjected this population to severe infestations of wilt. The stand was reduced to approximately 1%. The survivor plants were selected and advanced for further testing. The irradiation and screening program produced seven highly wilt-resistant strains, some having earlier maturity, smaller leaf, or a more erect habit than Mitcham. In 1971, the mutant cultivar 'Todd's Mitcham' was released. This cultivar was more resistant to wilt than spearmint and has the acceptable oil quality of its parent cultivar. Since then, a second mutant cultivar, 'Murray Mitcham,' has been released.

In addition to the use of induced mutation for the direct purpose of breeding improved crop cultivars, there are a number of indirect uses of mutagens for plant breeding and genetics. Transfer of genetic information between species which do not interbreed is an example of such use of mutagens. Sears (1956) pioneered this approach by transferring a gene for leaf rust resistance from the primitive *Aegilops umbellulata* Zhuk. to bread wheat. This effort involved the production of a hybrid (hexaploid amphidiploid) with 42 chromosomes from the cross between *A. umbellulata* (14 chromosomes) and *Triticum dicoccoides* Korn. (28 chromosomes). A segment of the *A. umbellulata* chromosome carrying leaf rust resistance in the hybrid was sliced off by an x-ray treatment of pollen. The segment with the resistance gene became attached to a *T. dicoccoides* chromosome. Of over 6,000 offspring, 132 were resistant to leaf rust and 40 of these carried a translocated chromosome. In one instance the self-pollinated off-spring were distinguishable from normal control plants only by their resistance and slightly later maturity. The transferred segment was evidently rather small with only a few genes other than the gene for leaf rust resistance. Similar transfers of disease-resistant genes, utilizing radiation, have been reported with other crops (Driscoll, 1965).

Mutations can contribute to the conversion of non-domesticated wild species into crop plants. This effect was first demonstrated by von Sengbusch (1942) with lupine. He succeeded in ridding lupine of a poisonous alkaloid compound by identifying mutants of low alkaloid plants, using quick methods of chemical analysis to screen large populations (6 to 7 million plants) for low frequency spontaneous mutations (1:100,000). Several mutant types of practical interest have been subsequently induced in lupine by radiation (Gustafsson and Gadd, 1965).

Evolution under natural conditions results in the accumulation of genes advantageous to the survival and reproductive ability of the wild plant. Some of these genes can lower the value of the plant for agricultural uses. Where they occur as dominant alleles in the wild plant, their activity can be eliminated by induced deletion or mutation to recessive alleles, or they can be suppressed by modification of the genes (Brock, 1976). Thus, induced mutations will have a role in converting the large number of potentially valuable tropical and semi-tropical legume species to crop plants by removing such common barriers as toxins and other anti-nutritional factors, such as seed shattering, long seed dormancy, etc.

VARIOUS USES OF INDUCED MUTATIONS

Induced mutations can be used in a variety of ways.

1. An induced mutant may rectify a specific fault of an otherwise well-adapted cultivar without significantly altering the rest of the genotype. The mutant is then used as a parent instead of the original cultivar. For example, the mutant induced in the red-kerneled Mexican wheat cultivar 'Sonora-64' carried amber-colored grains which were preferred by Indian consumers. The Indian mutant cultivar named 'Sharbati Sonora' is therefore preferred over Sonora-64 as crossing material (Sigurbjörnsson, 1971).

It is common for a plant breeder who wants a desired attribute in an adapted high-yielding cultivar to screen collections of primitive material or obsolete cultivars and use a laborious crossing and backcrossing program to transfer the desired gene to an adapted genotype. Inducing the desired change in an adapted genotype by mutation either directly or in cross breeding could reduce the time and effort. A series of highly successful navybean cultivars have been released in Michigan from crosses to an x-ray-induced bush-like mutant. Many bush-type beans occur naturally but this particular one was induced in 'Michelite,' an adapted bean cultivar (Gustafsson, 1975).

2. The success and popularity of a cultivar may have undesirable side effects. Because of high and stable yield, a cultivar or group of related cultivars may become so widely used that local, lower-yielding cultivars disappear. The successful cultivar may have a narrow genetic base and be vulnerable to attacks and destruction by diseases and pests.

The highly successful Mexican wheat cultivars of the "Green Revolution" owe much of their popularity to short and strong stems originating from the Japanese cultivar, 'Norin 10.' The high-yielding rice cultivars developed in the Philippines owe their short, strong straw to a spontaneous mutant from Taiwan, 'Deo-Geo-Woo-Gen' (Hu et al., 1970). In contrast, the short straw mutants of barley result from a number of different mutations and thus rest on a wider genetic base. Konzak (1976) isolated about 100 short-strawed mutants in winter wheat and about 200 in spring wheat. Many of these mutants are genetically different from Norin 10; some are recessive, others dominant. A number of different short-strawed mutants of rice have been reported. One such mutant with straw 25 cm shorter than the parent cultivar, 'Calrose,' was released to growers in California as 'Calrose 76' (Rutger et al., 1976; Fig. 8.2). This Calrose 76 gene source has been used to develop six additional semidwarf rice cultivars,

Fig. 8.2. The semi-dwarf mutant cultivar 'Calrose 76' is 30 cm shorter than the tall California rice cultivar CS-M3. (Photograph provided and published with the permission of Dr. J. N. Rutger, USDA, ARS.)

which average 15% higher grain yield than the tall cultivars they replaced (Rutger, 1982).

3. Often the breeder is unable to locate a suitable source in existing collections or through exploration. The gene may exist, but simply cannot be found. In these instances the logical approach is to induce desired mutations. Thus, in Indonesia, a search of the world rice collection for early maturing lines identified lines which needed 120 to 125 days for ripening but were low in yield. After treating local cultivars with a mutagen, some lines were isolated which ripened in 110 to 125 days and retained their high yield (Ismachin and Mikaelsen, 1976).

4. Crossing two related mutants often gives an unexpected degree of hybrid vigor. This phenomenon has been found in self-pollinated as well as in cross-pollinated crops (Micke, 1974; Ramirez et al., 1969). Stoilov and Daskaloff (1976) have found that about 10% of their maize mutants used as inbred lines in Bulgaria gave increased yield, with some combinations up to 25% higher.

5. Vegetatively propagated crops are of two types: those that are also easily propagated by sexual means and those that are difficult or impossible to propagate sexually. It is not easy to generalize on the role of induced mutations for this diverse group of crops. Crops that have a functional sexual mechanism can be improved with standard breeding methods up to the time that the improved genotype is propagated for commercial production. Induced mutations are useful in breeding all vegetatively propagated crops and are essential for the improvement of crops that have no mechanism of sexual recombination (Broertjes and van Harten, 1978). An advantage of these crops is that a genotype may be propagated and used

directly. Most vegetatively propagated crops, however, have a longer cycle of sexual reproduction, if any, than seed-propagated species. Breeding these crops can be very time consuming. Furthermore, because of high heterozygousity, sexual breeding of such crops is often very complicated (Nybom, 1970).

Spontaneous mutations or "sports" have been used extensively to improve the apple crop, especially with regard to easily identifiable characteristics such as fruit color and compact growth. In many fruit crops, both deciduous and citrus, there are valuable cultivars with qualities widely recognized by consumers which might be affected by crossbreeding. Therefore mutations, spontaneous or induced, are an effective means of improving specific attributes of such cultivars (Lapins, 1973). Research is needed to improve the methods of application of inducing mutations in fruit crops, especially with regard to chimera formation and tissue homogeneity. Chimera formation could be avoided if adventitious buds were used. Radiation has been most commonly used in mutation breeding of fruit trees. There is a need to increase the penetration of chemical mutagens into bud meristems and otherwise improve application methods.

6. Breeders of vegetatively propagated crops say that induced mutations could become a more important breeding method for these crops if in vitro tissue and single cell culture techniques were further developed. This prediction would be especially true if whole plants could be routinely developed from haploid cells.

FUTURE OF INDUCED MUTATIONS IN CROP BREEDING

The role of induced mutations in plant breeding has long been a source of controversy among breeders and agronomists. Some believe that induced mutations are shortcut solutions to various breeding problems. This belief may stem from the fact that the breeding of most crops requires several segregating generations, each one taking at least a year in the field. Breeding thus requires long-term planning and patience.

The use of induced mutations can be a shortcut in some instances. Brock (1976) has calculated on the basis of records of released mutant cultivars that the average time which elapsed from the end of the mutagenic treatment to the release of the mutant cultivar was 8.9 years for cultivars which were direct multiplications of mutants as compared with 18.0 years for cultivars arising from crossing programs. However, these figures are based on a small number of cultivars and are confounded by the difference in time for cultivar testing required in different countries.

A few years ago there was active opposition to the use of induced mutations by plant breeders in developed countries. These attitudes reflected a backlash following the great enthusiasm and broad claims made by "mutation breeders" during the early years of the atomic age (Sigurbjörnsson, 1972).

The steadily increasing number of crop cultivars of induced mutant origin, the increasing use of mutants in crossbreeding programs, and the growing interest in these techniques seems to indicate that induced mutations are now taking their place as standard breeding methods.

In evolution as well as in plant breeding, mutation is the ultimate source of variability. Methods to increase the rate of mutation at times when appropriate selection pressures can be applied increases the breeder's chances of success.

Since all variability stems from mutation it has been argued that there is no need to preserve existing genetic variability, because variability for breeding purposes can be generated at will. It is indeed possible to induce almost any gene mutation that exists naturally, many types which have never occurred, and those which have occurred and disappeared. However, agricultural practices have changed. The requirements for performance of a crop plant today are quite different from those of earlier agriculture. Yet a useful mutation will only be retained if it is selected and propagated. It is therefore probable that many types of mutations with potential use for modern agriculture have already occurred during evolution; since they did not improve the plant's chances of survival, they were eliminated during nautral selection. Thus, when breeders induce mutations while applying specific screening procedures, they are likely to discover useful mutations that do not exist in naturally evolved populations. The induction of high lysine mutants in barley (Doll et al., 1974) and of wilt resistance in peppermint (Murray, 1971) illustrate this point.

Can induced mutations then replace germplasm exploration and conservation? The answer is definitely no. Evolution has shaped the modern plant over a long time through gradual changes. The performance of plants depends on complex genotypes that include linked gene complexes. Where such sets of genes occur, they can be transferred to other genotypes, but they are unlikely to be produced as a result of random mutation (Brock, 1976). These complexes are the necessary building blocks for further genetic advance by recombination with existing genes. The importance of induced mutations at the single-gene level is the probability that we can induce and retain the full range of genetic variability that exists in natural gene pools. The rapid rate of erosion of our genetic resources, partly caused by the spread of modern high-yielding cultivars, should be a cause for concern to all breeders regardless of the breeding methods they use for improving crop plants.

LITERATURE CITED

Auerbach, C. 1967. The chemical production of mutations. Science 158:1141–1147.

Brock, R. D. 1965. Induced mutations affecting quantitative characters. p. 451–464. *In* The use of induced mutations in plant breeding. (Suppl. to Radiation Bot. Vol. 5) Pergamon Press Ltd, Oxford.

———. 1976. Prospects and perspectives in mutation breeding. p. 117–132. *In* A. Muhammed, R. Aksel, and R. C. von Borstel (ed.) Genetic diversity in plants. Plenum Press, New York.

Broertjes, C., and A. M. van Harten. 1978. Application of mutation breeding methods in the improvement of vegetatively propagated crops. Elsevier, Amsterdam.

Conger, B. V., R. A. Nilan, and C. F. Konzak. 1968. Post-irradiation oxygen sensitivity of barley seeds varying slightly in water content. Radiat. Bot. 8:31–36.

Constantin, M. J. 1975. Mutations for chlorophyll deficiency in barley. Comparative effects of physical and chemical mutagens. p. 96–112. *In* H. Gaul (ed.) Barley genetics III. Proc. 3rd Int. Barley Genet. Symp. Garching, Germany, 7–12 July. Verlag Karl Thiemig, Munich.

Doll, H., and B. Koie. 1978. Influence of the high-lysine gene from barley mutant 1508 on grain, carbohydrate, and protein yield. p. 107–114. *In* Seed protein improvement by nuclear techniques. I.A.E.A., Vienna.

———, ———, and B. O. Eggum. 1974. Induced high lysine mutants in barley. Radiat. Bot. 14:73–80.

Driscoll, C. J. 1965. Induced intergeneric transfers of chromosome segments. p. 727 739. *In* The use of induced mutations in plant breeding. (Suppl. to Radiation Bot. Vol. 5). Pergamon Press Ltd., Oxford.

Gaul, H. 1959a. Determination of suitable radiation dose in mutation experiments. p. 65–69. Proc. 2nd Congr. Eur. Assoc. for Research on Plant Breed. Cologne, West Germany. 6–10 July 1959. M. DuMont, Schaubery, Cologne, West Germany.

———. 1959b. Über die Chimarenbildung in Gerstenpflanzen nach Roentgenbestrahlung von Samen. Flora 147:207–241.

———. 1963. Mutationen in der Pflanzenzuchtung. Z. Pflanzenzuecht. 50:194–307.

———. 1965. The concept of macro- and micro- mutations and results on induced micro-mutations in barley. p. 408–426. *In* The use of induced mutations in plant breeding. (Suppl. to Radiat. Bot. Vol. 5). Pergamon Press Ltd., Oxford.

———, G. Frimmel, T. Gichner, and E. Ulonska. 1972. Efficiency of mutagenesis. p. 121–139. *In* Induced mutations and plant improvement. I.A.E.A., Vienna.

Gustafsson, A. 1975. Mutations in plant breeding—a glance back and a look forward. p. 81–95. *In* O. F. Nygaard, H. I. Aldler, and W. K. Sinclair (ed.) Proc. 5th Int. Congr. Radiat. Res. 14–20 July 1974. Academic Press, Inc., New York.

———, and I. Gadd. 1965. Mutations and crop improvement. II. The genus *Lupinus* (Leguminosae). Hereditas 53:15–39.

Hagberg, A. 1978. The Svalöv cereal protein quality program. p. 91–105. *In* Seed protein improvement by nuclear techniques. I.A.E.A., Vienna.

———, and K. E. Karlsson. 1969. Breeding for high protein and quality in barley. p. 17–21. *In* New approaches to breeding for improved plant protein. I.A.E.A., Vienna.

Hu, C. H., H. B. Wu, and H. W. Li. 1970. Present status of rice breeding by induced mutations in Taiwan, Republic of China. p. 13–19. *In* Rice breeding with induced mutation II. I.A.E.A., Vienna.

International Atomic Energy Agency. 1977. Manual on mutation breeding. 2nd ed. I.A.E.A., Vienna.

Ismachin, M., and K. Mikaelsen. 1976. Early-maturing mutants for rice breeding and their use in cross-breeding programs. p. 119–121. *In* Induced mutations in cross-breeding. I.A.E.A., Vienna.

Konzak, C. F. 1976. A review of semi-dwarfing gene sources and a description of some new mutants useful for breeding short-stature wheats. p. 79–93. *In* Induced mutations in cross-breeding. I.A.E.A., Vienna.

Lapins, K. O. 1973. Induced mutations in fruit trees. p. 1–19. *In* Induced mutations in vegetatively propagated plants. I.A.E.A., Vienna.

Lundqvist, U., P. von Wettstein-Knowles, and D. von Wettstein. 1968. Induction of eceriferum mutants in barley by ionizing radiations and chemical mutagents II. Hereitas 59:473–504.

Micke, A. 1974. Heterosis bei Kreuzungen von Mutanten derselben Ausgangsform. p. 314–328. *In* Bericht über die Arbeitstagung der Arbeitsgemeinschaft der Saatzuchtleiter in Gumpenstein, Steiermark, Austria.

Micke, A., and B. Donini. 1981. Use of induced mutations in improvement of seed propagated crops. p. 2–9. *In* Induced variability in plant breeding. Centre for Agricultural Publishing and Documentation (Pudoc), Wageningen, The Netherlands.

Muench, S. R., A. J. Lejeune, R. A. Nilan, and A. Kleinhofs. 1976. Evidence for two independent high lysine genes in barley. Crop Sci. 16:283–285.

Muller, H. J. 1927. Artificial transmutations of the gene. Science 66:84–87.

Murray, M. J. 1971. Additional observations on mutation breeding to obtain *Vertillicum* resistant strains of peppermint. p. 171–195. *In* Mutation breeding for disease resistance. I.A.E.A., Vienna.

Nilan, R. A. 1972a. Mutagen specificity in flowering plants: Facts and prospects. p. 141–151. *In* Induced mutations and plant improvement. I.A.E.A., Vienna.

———. 1972b. Induced variation and winter barley improvement. p. 349–351. *In* Induced mutations and plant improvement. I.A.E.A., Vienna.

———. 1974. Barley (*Hordeum vulgare*). p. 93–110. *In* R. C. King (ed.) Handbook of genetics. Vol. 2. Plenum Press, New York.

———, A. Kleinhofs, and C. Sander. 1975. Azide mutagensis in barley. p. 113–122. *In* H. Gaul (ed.) Barley genetics III. Proc. 3rd Int. Barley Genet. Symp. Garching, Germany. 7–12 July. Verlag Karl Thiemig, Munich.

———, C. F. Konzak, R. R. Legault, and J. R. Harle. 1961. The oxygen effect in barley seeds. p. 139–154. *In* Effects of ionizing radiations on seeds. I.A.E.A., Vienna.

Nybom, N. 1970. Mutation breeding of vegetatively propagated plants. p. 141–147. *In* Manual on mutation breeding. I.A.E.A., Vienna.

Persson, G., and A. Hagberg. 1969. Induced variation in a quantitative character in barley. Morphology and cytogenetics of *erectoides* mutants. Hereditas 61:115–178.

Ramirez, E., R. E. Allan, C. F. Konzak, and W. A. Becker. 1969. Combining ability in winter wheat. p. 445–455. *In* Induced mutations in plants. I.A.E.A., Vienna.

Redei, G. P. 1974. Economy in mutation experiments. Z. Pflanzenzuecht. 73:87–96.

Rutger, J. N. 1982. Use of induced and spontaneous mutants in rice genetics and breeding. p. 105–117. *In* Semi-dwarf cereal mutants and their use in cross-breeding. IAEA, Vienna.

Rutger, J. N., M. L. Peterson, C. H. Hu, and W. F. Lehman. 1976. Induction of useful short stature and early maturing mutants in two japonica rice cultivars. Crop Sci. 16:631–635.

Scossiroli, R. E. 1970. Mutations in characters with continuous variation. p. 117–123. *In* Manual on mutation breeding. I.A.E.A., Vienna.

Sears, E. R. 1956. The transfer of leaf rust resistance from *Aegilops umbellulata* to wheat. p. 1–22. *In* Genetics in plant breeding. Symposia in Biology no. 9. Brookhaven National Lab., Upton, New York.

Sigurbjörnsson, B. 1971. Induced mutation in plants. Sci. Am. 224:86–95.

———. 1972. Breeding with natural and induced variability. p. 3–5. *In* Induced mutations and plant improvement. I.A.E.A., Vienna.

———. 1975. The improvement of barley through induced mutation. p. 84–95. *In* H. Gaul (ed.) Barley genetics III. Proc. 3rd Int. Barley Genet. Symp. Garching, Germany. 7–12 July. Verlag Karl Thiemig, Munich.

———, and A. Micke. 1973. List of varieties of vegetatively propagated plants developed by utilizing induced mutations. p. 195–202. *In* Induced mutations in vegetatively propagated plants. I.A.E.A., Vienna.

———, and ———. 1974. Philosophy and accomplishments of mutation breeding. p. 303–343. *In* Polyploidy and induced mutations in plant breeding. I.A.E.A., Vienna.

Singh, J., and R. M. Singh. 1975. Influence of seed moisture content on radiosensitivity of barley, *Hordeum vulgare* L. Var. 'Amber'. p. 222. *In* H. Gaul (ed.) Barley genetics III. Proc. 3rd Int. Barley Genet. Symp. Garching, Germany. 7–12 July. Verlag Karl Thiemig, Munich.

Sparrow, A. H. 1961. Types of ionizing radiation and their cytogenetic effects. p. 55–119. *In* J. D. Luckett (ed.) Mutation and plant breeding. Natl. Acad. Sci.-Natl. Res. Council, Pub. 891.

Stadler, L. J. 1928a. Mutation in barley induced by X-rays and radium. Science 68: 186–187.

———. 1928b. Genetic effects of X-ray in maize. Proc. Nat. Acad. Sci. USA, 14: 69–75.

Stoilov, M., and S. Daskaloff. 1976. Some results on the combined use of induced mutations and heterosis breeding. p. 179–188. *In* Induced mutations in cross-breeding. I.A.E.A., Vienna.

von Sengbusch, R. 1942. Süsslupinen and Öllupinen. Die Entstehungsgeschichte einiger neuen Kulturpflanzen. Landwirtsch. Jahrb. 91:723–880.

PART III.

The Breeding Process

Chapter 9

Domestication and Breeding of New Crop Plants

S. K. JAIN

University of California
Davis, California

Domestication of the major crops dates back several thousand years. A knowledge of a crop's ancestry or geographical origin is of interest in finding clues to its useful genetic resources for modern breeding. The list of cultivated useful plants continues to grow, as noted by Schwanitz (1966), with sugarbeet and strawberry added in the last two centuries, and numerous algae and fungi grown to produce drugs and other industrial products. However, Baker (1970) noted in his book *Plants and Civilization* that ". . .practically no new species have been introduced to cultivation within historical times. Almost all the food and fiber plants that we value today were known to the early agriculturalists in the Old or New World." This is not to deny that most of the "old crops" gained exponentially in both productivity and wide adaptability with the application of modern science and technology in the past few decades. Baker (1970) also recognized numerous "new crops" considered minor in production area or commercial value (but not necessarily both), and emphasized the important role of these crop plants in past human explorations and in political or cultural histories. Simmonds (1979) tabulated crop species as ancient (prior to 5000 B.C.), early (5000 B.C.–0 A.D.), late domesticates, (0 to 1750 A.D.), or recent (after 1750 A.D.), with recent including such crops as sugarbeet, oil palm, pyrethrum, rubber, strawberry, triticale, quinine, kola, jute, and blackberries. Accordingly, the rate of crops developed per 1,000 years (15 to 17) is nearly constant (Table 9.1).

A student of crop plant evolution will certainly be impressed by the role of successful plant introductions accompanying the colonization of new lands. Records of early exchanges of notes and plant materials among writers, politicians, governments, and individual hobbyists provide interesting commentaries on a society's interests in plant diversity and uses. Thus, all our crop plants and their past developments represent a rich heritage. The so-called Green Revolution of the 1960's, more than anything else, drew serious attention to this heritage and to the plant breeders' con-

Table 9.1. Places and times of domestication.†

Type	Age‡	Near East and Europe	Central and East Asia	Africa	America	Totals
Cereals and pulses	A	7	0	0	0	7
	E	4	4	6	9	23
	L	0	1	0	0	1
	R	0	0	0	1	1
	Totals	11	5	6	10	32
Tubers, vegetables and fruits	A	0	1	2	2	5
	E	10	17	2	7	36
	L	6	3	1	3	13
	R	2	0	0	1	3
	Totals	18	21	5	13	57
Other crops	A	1	0	1	0	2
	E	7	10	3	4	24
	L	2	0	2	1	5
	R	2	2	2	1	7
	Totals	12	12	8	6	38
Grand total		41	38	19	29	127

† Simmonds (1979).
‡ A = ancient, 5000–7000 BC; E = early, 5000 BC–0; L = late, 0–1750 AD; R = recent, after 1750 AD. Age totals A, 14; E, 83; L, 19; R, 11.

tributions to world food production. An extremely lucid and timely publication by the U.S. National Academy of Sciences (1975) titled *Underexploited Tropical Plants with Promising Economic Value* drew the attention of researchers and technical advisers interested in new crops. This report was followed by a series of publications, under the notable leadership of Noel Vietmeyer, on new crop prospects such as *Leucaena,* guayule, and winged bean. Likewise, publications of the International Board for Plant Genetic Resources (e.g. *Tropical Vegetables and their Genetic Resources* by G. J. H. Grubben, 1977; *Directory of Germplasm Collections: Food Legumes* by G. Ayad and N. M. Anishetty, 1980) have recently stimulated research interests. Crop scientists all over the world now have a wealth of new information describing the potentials of new genetic resources.

In the decade of the 1970's, interest in research on new plants for agricultural use grew rapidly in the developed nations (particularly the USA) for several reasons. First, the technology of food, feed, fiber, and forest crop production is advanced enough to provide reliable and stable outputs. Secondly, societal problems such as energy shortages, inadequate farm income, pollution of air, water and soil, and changes in diet and nutritional concepts have contributed to the interest in new crops. Examples follow.

1. Interest in biomass production, gasohol, hydrocarbon crops, etc., resulted from uncertainties about the availability of inexpensive fossil energy.

2. Substitutions for products such as sperm whale oil and rubber are being developed to replace shrinking sources of supply. Likewise, declines in production in reliable import sources have stimulated interest in crops like *Plantago* and pyrethrum.

3. Prices of major food crops, such as wheat, maize, sorghum and soybeans, have remained low in comparison to the spiraling costs of input items such as pesticides, fertilizers, fuel, and labor. Cotton hectarage and

production was also reduced when prices were lowered by competition with synthetic fibers.

4. Pollution of soil and water by nondegradable pesticides has created a demand for crops that need fewer applications. Lowered nitrogen inputs are the goal of research on biological nitrogen-fixation sources in terms of new crops as well as new symbiotic systems.

5. Reduction in tobacco consumption in the USA has resulted in the loss of income for farmers dependent on tobacco. Interest in kenaf in the South and Southeast USA has been based on a desire to offset farm income depression owing to the reduced income from cotton and tobacco. A new industrial oil crop, meadowfoam, appears promising in the Williamette Valley in Oregon, where lower income from ryegrass seed production appears critical (Jolliff et al., 1981).

6. Changes in the dietary patterns and concepts of nutrition also generate demands for new foods to substitute for traditional food items. Recent interest in oil from sunflower, rapeseed, and safflower is partly the result of research on cholesterol and its apparent association with heart attacks.

Developing nations also recognize the need for new crops and for them too, new crop development is an attractive idea. The National Botanical Research Institute at Lucknow, India, for example, evolved from a botanical garden to a shift toward new crops research. Awareness of genetic vulnerability of crop plants and of agriculture based on large scale monocultures of a few species has revitalized interest in crop diversification. Crop rotation and mixed cropping often cited as elements of traditional agriculture in developing countries, are considered a part of the needed new agricultural ecology (Cox and Atkins, 1979).

Finally, ethnobotany associated with sociological researches in the cultural diversity of man has served well in pointing toward numerous potential crop plants. Some estimate that there are 3,000 or more potential food plants and well over 20,000 plants for various uses available for domestication. New tools of biochemical and biosystematic research now make it possible to screen thousands of species rapidly for useful products. Duke et al. (1975) initiated a computerized record of ecological and biological features of economically useful plants and updated crop geography beyond a handful of major crops.

Examples of "new crops" which are new to extensive use in commercial agriculture include: 1) species like lupine and triticale developed in this century, (2) species like amaranths and winged bean already domesticated and of considerable importance in certain areas, (3) crops in various stages of domestication (e.g. oil palm, guar, wild rice, and various African and American forage species being developed in Australia), and (4) species in almost native wild form (e.g. buffalo gourd, guayule, meadowfoam, and *Crambe*). Grubben (1976) listed tropical vegetables of Southeast Asia under similar categories. As noted by Simmonds (1979), domestication of pharmaceutical plants and timber trees may continue indefinitely. Some established crop species are grown for new uses and other "new" species have been developed only recently for some particular industrial uses.

CONCEPTS AND APPROACHES

Plant breeders starting from scratch with the new crops face unique challenges. Their work contrasts with that of breeders who work with established crops, where the constraints to production have already been identified, basic understanding of the species behavior exists, special methodologies to solve particular problems are known, and infrastructures exist for seed production, promotion, storage, processing and marketing. Thus to develop successful commercial production, the breeder must assess research needs, identify problems, develop technical solutions, test the efficiency of these solutions, extend the results to the growers, and, equally important, communicate with processing, distribution, and marketing sectors.

The plant breeder who works with new crops will frequently start with basic studies of species behavior to develop strategies for the improvement of traits necessary for agricultural production practices. In fact, as noted by Knowles (1958) and others, the new crops researcher would often spend more time in learning to grow and handle the crop than in actual breeding and cytogenetic work. In many established crops, from 3,000 to 10,000 generations of selection have provided a broad genetic base of cultivated materials for crop improvement in addition to the variation found in the wild and weedy relatives. Steps (and concepts) in the development of new crops have been diagrammed in a flow chart (Fig. 9.1). Breeding work for most of the potential domesticates is currently in phase 1—selection of most promising species for further research. Phase 2 begins with the assembly of plant materials which have been designated as genetic resources (or germplasm) to be used for crop improvement. Seeds, vegetative parts, tissue cultures, and, in the future, DNA clones may be maintained in collections. Living plants in agricultural fields or natural plant communities also represent genetic resources. As the need arises for adaptation to a novel environment or for resistance to natural enemies, hereditary factors in these collections or in natural plant populations could become a critical factor in future crop production. Genetic resources may be discussed under five categories:

1. Primitive cultivars or landraces are cultivars used in agriculture before the era of modern plant breeding (i.e., before the use of systematic selection and controlled breeding systems) and well known in a locality with morphological identity, diversity, and adaptability. A landrace usually evolves in relation to the local people's preferences for taste, color, appearance, etc.

2. Advanced cultivars are lines, hybrids, or populations developed by breeders, using known breeding materials and genetic principles.

3. Genetic stocks include the genetic materials originating from breeding or induced mutation programs and are judged to be of potential value in breeding work.

4. Weedy relatives and crop-weed complexes are often products of evolution in areas where weeds and crops have continued gene exchange in

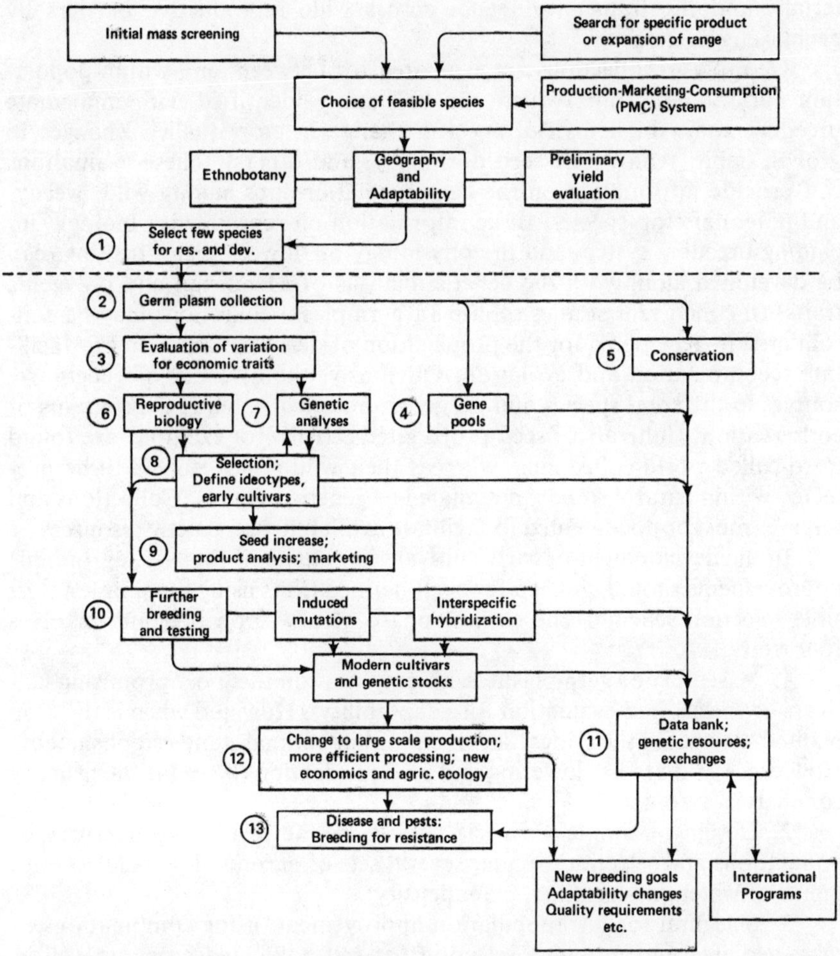

Fig. 9.1. Steps charting new crop development processes.

areas of contact. Weedy plants and crops form dynamic gene pools with evolutionary potential. Agricultural fields and adjoining wastelands allow crop-weed associations to develop over a continuum of habitats with diverse selection regimes.

5. Wild relatives include those related species occurring in the wild which may often have little potential for conventional gene transfer to related crops, but may carry genes for vigor, adaptation to stress, resistance to diseases and insects, and/or unusual biochemical characteristics. Currently, with new genetic engineering developments, interest in wild relatives would most likely expand dramatically.

The use of genetic resources in crop improvement is basic to crop breeding. Even greater strides can be made in the future with efficient screening methods, increased emphasis on heterogeneous breeding materials, and the transfer of genes across wide reproductive barriers by genetic engineering.

Germplasm collections are evaluated for between- and within-population variation of traits which can be readily identified with immediate breeding goals (higher yield, specific change in crop quality, changes in growth habit, removal of seed dormancy, and others). These evaluations can provide information on the genetic relationships among wild, weedy, and potential crop species. Basic information on reproductive biology, including breeding system and the physiology of flowering and fruiting, can be developed along with the genetic analysis of Mendelian and polygenic traits. In concurrent studies (phase 3), germplasm collections may be consolidated in gene pools for the propagation of desirable genes and to facilitate recombination and evolution. Obviously, the diversity of genetic resources in different species and ecogeographical sites requires programs of conservation. Cultivars of seed-propagated cereals, for example, are found in so-called world collections, whereas their wild relatives need to be protected within natural stands, designated as genetic reserves. Collections and reserves must be documented to facilitate availability of genetic resources.

In the development of early cultivars in phase 8 of Fig. 9.1, agronomic improvements should aim at a few well-defined traits using the simplest possible selection scheme. The earliest phase of new crop breeding involves four steps:

1. Assembly of germplasm (with emphasis on the most promising cultivars or species and evaluation for adaptability). How and when is the crop planted? What are the critical shortcomings in optimal stand establishment? How successful are its flowering and seed production (if seed is the primary commercial resource)?

2. Design of simple yield trials to assess the relative productivity of populations and selections, under several sets of agronomic variables (e.g., inputs of water and fertilizer, plant density).

3. Selection for intrapopulation improvement, using combinations of mass-pedigree and pure-line selection for self-pollinated crops as well as mass selection within landraces, and evaluating for high yield and uniformity in plant height, flowering or maturity.

4. Establishment of genetic stocks or source populations for the next cycle of breeding with interpopulation hybrids, new genetic recombinants, selection, and yield trials.

Each stage provides opportunities for ingenuity, keen powers of observation and analysis, and to learn new things.

Throughout this work, an ideal plant form (ideotype) may often be visualized, as the concept will tend to maximize gains under selection, but in reality, the ideotype is used to recognize the alternative character combinations that are related to overall yield or adaptability (Jain and Abuelgasim, 1981). Clearly, most of the flow chart phases (Fig. 9.1) can be repeated in recurrent cycles of crop development. New diseases and pests, or new physiological bases of adaptation, will most likely appear when a crop species expands in use and geographical range. The breeder will be called

Table 9.2. Criteria for selection of some important new crops.†

	National economy benefits					Marketing and processing benefits				
	Import substitution	Productive marginal resource use	Low energy requirements	Market diversification	Potential magnitude of impact	Good market potential	Usable processing system established	Processing system costs established	Consistency of demand	Conservation of natural resources
1. Tepary bean	1	1	1	2	2	1	1	1	2	2
2. Grain amaranth	1	1	1	2	3	2	1	1	2	2
3. Quinoa	2	1	1	2	3	2	2	2	2	2
4. Wild rice	2	1	1	2	2	1	1	1	2	2
5. Buffalo gourd	1	1	1	3	1	2	2	2	1	2
6. Pinyon pine	1	1	1	2	3	1	1	2	2	1
7. Crambe	1	2	2	1	1	1	2	3	1	2
8. Guayule	1	1	2	1	1	1	3	3	1	2
9. Jojoba	1	1	2	1	1	1	1	2	1	2
10. Kenaf	1	3	2	1	1	1	1	3	1	2
11. Limnanthes	1	2	2	1	1	1	3	3	1	2

† Thiesen et al. (1978).
‡ 1 = high rating compared to established alternatives.
2 = neutral rating or denotes lack of sufficient information.
3 = low rating.

upon to use intuition, observation, and scientific tools to make decisions at each step, and as in the past, chance, design and dedication will all be keys to plant breeding accomplishments in new crops.

SELECTION OF MOST PROMISING SPECIES

Several important publications have generated particular interest in new crops; for example, a series of studies published by the U.S. National Academy of Sciences (1975a, 1975b, 1979) evaluated unexploited tropical plants; Duke and Terrell (1974) compiled basic information for selecting crops for adaptive diversification; Martin and Roberte (1975) and Grubben (1976) described new tropical leafy vegetables. Two symposia volumes, edited by Seigler (1977) and Ritchie (1979), are also notable. A soil and land use technology project funded by the U.S. National Science Foundation studied the production-marketing-consumption system and evaluated 54 crops (Thiesen et al., 1978). Table 9.2 lists selected crops from that study along with the criteria used in their evaluation.

White (1977), who reported pioneering studies on new crops, noted that, of the native plants, only maize, sunflower, artichoke, and a few small fruits are of importance in world agriculture. Most major crops in the USA were plant introductions. Guar, safflower, broccoli, kiwi, and pistachio, have gained in importance in recent years.

A systematic search for crops to meet particular needs was discussed by Princen (1977). He identified new crop plants as specific sources of starch, cellulose, edible fats and oils, gums, waxes, hydrocarbons, proteins, and

Table 9.3. Plant species as new sources of industrial oil.†

Species	Component in triglyceride oil	Uses
A. Long-chain fatty acids in seed oil		
Crambe abyssinica Hochst. ex R. E. Fries	60+% C_{22}	A. Polymers, plasticizers, lubricants. Substitutes for petroleum products, sperm whale oil, etc.
Lunaria annua L.	40% C_{22}, 20% C_{24}	
Limnanthes alba Hartw.	95% C_{20} + C_{22}	
Simmondsia chinensis (Link) Schneid.	(liquid wax) long chain fatty acids and alcohols	
B. Hydroxy and keto fatty acids		
Lesquerella gracilis Wats.	(70%) 14–OH–C_{20}	B. Substitute for castor oil.
Cardamine impatiens L.	(23%) dehydroxy C_{22} + C_{24}	
C. Epoxy fatty acids		
Vernonia anthelmintica (L.) Willd.	68 to 75%	C. Plastics and cratings production; supplement soybean oil.
Euphorbia lagascae Sprengel	60 to 70%	
Stokesia laevis (Hill) Greene	75%	
D. Conjugated unsaturation		
Valeriana officinalis L.	(40%) 9, 11, 13 positions	D. Intermediates in industrial processes and substitute for tung oil.
Calendula officinalis L.	(55%) 8, 10, 12 positions	

† Princen (1977).

other medicinal and industrial chemicals. Clearly, there are numerous good candidates for domestication (Table 9.3). For instance, the potential sources of energy from agriculture can be classified as follows: (1) total biomass in kilogram per hectare annually (examples include ramie, sunn hemp, elephant grass), (2) high hydrocarbon or oil, useful in biosynthesis or industrial production of fuels (e.g. gopher plant and natural rubber) and other products (e.g. mustard and rapeseed), (3) high starch or fermentable sugars (e.g. buffalo gourd, Jerusalem artichoke), (4) high cellulose (e.g. kenaf), and (5) dual purpose crops with large biomass residues for energy production (e.g. maize, *Leucaena*).

EXAMPLES OF NEW CROPS

The so-called new crop plants of this century that have drawn interest from the plant breeders are soybean, *Hevea,* lupine, safflower, rapeseed, and triticale. White (1977) noted that soybean should be cited as the "...golden crop of plant introduction in the USA as it rose rapidly from a small acreage in 1900 to about 23 million harvested hectares in 1973." Its success story documents some remarkable parallel developments in genetic improvements, product utilization, and crop production practices. Imle (1977) provided an interesting account of the developments in the rubber industry. An important role was played by chance (or historical quirks) in early dispersals, choice of species, timely developments in processing, and improvement of propagation techniques. Of course, important contributions were made with scientific and technological innovations by design. Current shifts of interest from synthetic to natural rubber that have renewed interest in guayule, suggests that "winners" in new crop development have to be opportunistic or at least ready "ahead of their time."

Lupine was a minor crop prior to four decades of dramatic genetic improvements in Germany. Screening and incorporation of a series of major mutations to achieve a lower alkaloid content, developing non-cracking and non-shattering seed pods, promoting rapid germination, suppressing pod hairiness, placing pods high above the foliage, lightening seed color, producing better flavor, developing early flowering, increasing permeability of the seed coat, and uncovering disease resistance, add up to a fascinating success story (Schwanitz, 1966; Gladstones, 1970, 1975). These traits were discovered in parallel in three different species as predicted by the law of parallel variation first discussed by Vavilov (1926). More recently, Oram (1977) utilized induced genetic variation in his lupine breeding work in Australia. For both forage and feed uses, Gladstones (1970) points out that future developments would have to emphasize higher biomass yields, adaptability, and incorporation of more genetic variation.

Safflower and rapeseed, two ancient crops of the Old World, have become successful new crops in North America. Knowles (1972) reviewed the development of safflower from its primary use as a dye source in the Middle East and as an oil crop in India, Egypt, and Turkey, to a nearly $25 million crop in California. Breeders depended on plant introductions and made systematic searches for desirable genes controlling fatty acid composition of the oil and resistance to diseases and pests. Production-related agronomic research was an important part of this story of crop development (Knowles, 1958, 1960). On the other hand, a chance introduction of turnip rape from Poland into Canada in 1936 became the basis for the development of a new crop. After the blockade kept out European and Asiatic imports of rapeseed oil during early World War II, a dramatic new crop development began. From the first crop of 1,300 ha in 1943, rapeseed became a "Cinderella" crop in 25 to 30 years, with a succession of new cultivars released. Strains low in erucic acid were used for edible oil, whereas low-fiber, high protein cultivars were a promising source of flour with low glucosinolate content (Downey, 1971). Breeding methods for rapeseed improvement included selections from plant introductions, backcrossing, and mass or family selection. Current interests in breeding range from the use of synthetic amphiploids to the use of physiological criteria in defining crop ideotypes.

Triticale, a cereal crop produced by hybridization between wheat and rye, represents a new plant genus created by man through the use of synthetic allopolyploidy that is grown as a feed and food grain in several developed and developing countries. Zillinsky (1973) and Muntzing (1979) reviewed the developments from the first fertile hybrids reported in 1888 to later developments aimed toward improved stability of yield, seed form and fertility, straw strength, disease resistance, and shorter plant stature. First, triticale involved crosses of hexaploid wheat with rye, followed by chromosome doubling, to give $2n = 56$ lines. These types were agronomically inferior and it was not until hexaploid ($2n = 42$) triticales were developed that good agronomic types were found. Such triticales arise from the hybridization of tetraploid wheat ($2n = 28$) with rye ($2n = 14$). Thus, comparative cytogenetic studies of a broad array of triticale materials permitted the use of a wider germplasm, and better selection and testing procedures.

The story of lupines, safflower, and rapeseed provides elegant examples of the role of one or a few genes in new crop development. Several induced mutations in lupines provided new growth forms, and alkaloid-free edible seed types. Low erucic acid, low glucosinolate rapeseed cultivars provide totally new oil and meal sources. A simply inherited pattern of fatty acid composition allowed the selection of high oleic lines in safflower. A crop plant evolutionist learns to appreciate the parallel roles of certain Mendelian traits and of polygenic traits in the processes of domestication and subsequent crop improvement. Therefore, a new crop breeder should have interest and training in the basic tools of gene mapping, chromosome manipulation, biometrical genetics, and developmental genetics.

PROSPECTIVE NEW CROPS

Several prominent new crops—winged bean, buffalo gourd, amaranths, jojoba, meadowfoam, *Crambe, Veronia,* Stoke's aster, guayule, and gopher plant—will now be discussed briefly in terms of their salient biological features, breeding status, and current research needs. For botanical description and agricultural treatment of these crops, the reader is referred to the bibliography provided here (in particular, Thiesen et al., 1978; and U.S. National Academy of Sciences publications).

Some sources of other new crop research reports include: comfrey for feed and biomass energy (Hills, 1976), sunn hemp (White and Haun, 1965), common reed (de la Cruz, 1978) and kenaf (Wilson et al., 1968; Crane, 1947) for paper pulp and fiber, *Dioscorea* yams for sapogenins and contraceptives (Appleweig, 1977; Martin and Elpin, 1965; Martin and Gaskins, 1968), tepary beans (Nathan and Felger, 1978), and chenopods (Singh, 1961; Heiser and Wilson, 1974) for protein in human diet, Jerusalem artichoke (Boswell, 1936; Chubey and Dorrell, 1974), *Leucaena* and other tree crops (Brewbaker and Hutton, 1979), and guar (Whistler and Hymowitz, 1979) for many uses including biomass fuel, and neem for insecticide (Warthen, 1979).

Winged Bean

A common minor vegetable crop, the winged bean, cultivated in Southeast Asia and India for the use of all plant parts in human food has received much attention in recent years. Newell and Hymowitz (1979) concluded ". . .the rediscovery of the winged bean comes at an appropriate time. . .a multi-purpose edible cultigen that grows easily, quickly and yields abundantly." Earlier, Hymowitz and Boyd (1977) commented that ". . .it would be difficult to find another high rainfall-adapted tropical legume crop with as many desirable characteristics as winged bean." Although it is mostly

Table 9.4. Means, ranges and coefficients of variation of seven fruit and seed characters of 85 buffalo gourd accessions.†

Characteristic	Mean	Range	CV‡
Fruit diameter	6.5 cm	5.2 to 7.7	9.1
Seed wt/100 seeds	3.8 g	1.1 to 5.5	20.0
Seed wt/fruit	8.4 g	2.7 to 18.8	36.9
Seed number/fruit	225	87 to 386	28.3
Embryo in seed	67.3%	42.1 to 76.3	9.7
Crude fat in seed	32.9%	21.1 to 43.1	14.3
Crude protein in seed	30.7%	19.5 to 35.4	8.7

† Bemis et al. (1978).
‡ CV = Coefficient of variation, in %.

used as a green pod vegetable, fleshy roots yielding as much as 3 metric tons/ha are sold as a delicacy in Burma and South Pacific Islands; seed yields of 1.4 metric tons/ha were reported in Ghana. High content of protein (37%) and oil (18%) allow the winged bean to compare favorably with soybean which is not adapted to the humid tropics. Many local cultivars of the winged bean are available in Papua, New Guinea and Indonesia, and recently germplasm collections have resulted in nearly 150 lines with considerable morphological and plant growth variation. Their description has included Mendelian traits such as flower color, pod color, and wing shape. A U.S. National Academy of Sciencs study (1975b) compared our present knowledge of winged bean to that of soybean nearly 60 years ago. Basic biosystematic studies in the genus *Psophocarpus* have not been adequate. Very little is known about the breeding structure of winged bean populations or the possibility of quantitative genetic improvements (Karikari, 1972). More studies are needed on nitrogen-fixation, nutritional quality of forage, water requirements, photoperiod response, and relative allocation of energy to vegetative growth versus seed production.

Buffalo Gourd

A feral xerophyte growing abundantly in the Western USA and Mexico, buffalo gourd produces abundant seed that is rich in edible oil and protein, and roots with high starch content. Although described in 1817, the first study of this crop's potential was carried out by Curtis (1946). Important features of the plant are perenniality, prolific seed, asexual reproduction, and occurrence of simply inherited male sterility. The crop has excellent prospects, with the availability of good germplasm (Bemis et al., 1979). Phenotypic variation for several yield traits (Table 9.4) suggests that improvement by selection is possible, but without quantitative genetic analyses such predictions are premature. Bemis et al. (1978, 1979) described schemes for hybrid seed production based on the use of male sterility. Genetic and evolutionary studies of both natural and field-planted materials are still needed. Although buffalo gourd has been mentioned as a high starch crop, Thiesen et al. (1978) ruled out root harvesting and listed its use only for seed oil and protein. Because field experience is limited, no physiological studies on the limits of productivity under field conditions are yet available. Caution is needed to prevent buffalo gourd from escaping cultivation and becoming a weed because of its versatile modes of reproduction.

Amaranths

Several species of the genus *Amaranthus* are widely grown for both grain and vegetable uses in Asia, Africa, the highlands of Central America and Mexico, and the Andean region of South America (Cole, 1979). Grain amaranths probably originated in the New World; early American people recognized their food value. Recent studies have drawn attention to them as a source of protein and to their high productivity in marginal environments. A U.S. National Science Foundation study (Thiesen et al., 1978) named amaranths among the most promising 20 new crops in the USA. The International Board of Plant Genetic Resources placed a high priority on the collection and study of both leafy and grain types because of their ". . .very high actual or potential nutritional importance, and a low present level of research" (Grubben, 1976).

Amaranth grain, used as a popped cereal or flour, provides a valuable source of protein, calories, and minerals. The discovery of lysine content as high as 6% in *Amaranthus edulis* Speg. heightened the nutritional interest in this crop. Compared to an egg protein score of 100, amaranth, maize, and wheat scored 75, 44, and 57, respectively. Grubben (1976) emphasized the value of amaranth, when used as a leafy vegetable in the diet of people in the tropics, and as an excellent source of vitamins A and C.

Although several growers have reported a wide scope for selection based on intrapopulation variation, no systematic collection and evaluation of amaranth germplasm collections have yet been undertaken; in fact, no more than 100 to 300 different accessions were available just a few years ago. Following Sauer's (1950, 1977) painstaking studies on the geography, taxonomy, and agricultural history of amaranth, assembly of germplasm from India and the New World centers of diversity has begun (Jain et al., 1980b). Selected entries from these collections have been studied for biosystematic and agronomic traits, and these entries will soon provide a baseline study for developing computerized data banks (Hauptli and Jain, 1978). Genetic and physiological studies are planned to evaluate both crop and weedy forms and to provide data for the search for evolutionary relationships among the Old and New World accessions (Jain et al., 1980c; Hauptli et al., 1980). The patterns of allozyme variation suggest that

1. Indian grain amaranths originated from one or more Mexican introductions that brought in only a small sample of total genetic variability.
2. The grain and vegetable forms are highly differentiated from each other.
3. Possibly the weedy and vegetable forms share many alleles.

Grain amaranths have a flexible breeding system, with the natural outcrossing rates in the range of 0 to 35%. Selection for higher or lower rates appears effective in terms of correlated changes in male/female sex ratios within the monoecious glomerules. Heterosis and mass selection responses are now being studied by quantitative genetic analysis.

Feine et al. (1979) concluded that amaranth has a long way to go to make a significant contribution to the world food base, but that interest is high and progress is rapid. A promising new crop is at hand, biologically speaking, but the politics and sociology involved in producing and marketing such new crops are still important concerns.

Jojoba

Jojoba, a dioecious perennial shrub, 20 to 25 dm high, was studied in detail by Gentry (1958). Hogan (1979) noted that in the history of agriculture, few plants have received the worldwide interest and publicity in such a brief time as jojoba. If jojoba proves to be a Cinderella crop, the reason lies in the yield (40 to 60%) of a unique liquid wax with many uses, including substitution for the sperm whale oil in polishes, cosmetics, and many other products. In addition, jojoba is adapted to aridity, has economic value to native Americans, and has received good financial support for research programs. Yermanos (1979) stated that it is not difficult to explain its popularity. Jojoba grows in soils of marginal fertility, needs little water, withstands salinity, and is a low labor intensive crop that is easy to grow. Planning, leadership, and communication among jojoba researchers and users are impressive. Yermanos (1974) reported on the seed harvests from 11 natural stands; one gave a commercial yield estimate of 0.23 to 0.50 kg seed per plant with a wax content of 43.2 to 59.8%. Work on germplasm collections and productivity has been reported, and producers are attempting to propagate good selections asexually by stem cuttings, as well as by tissue culture. A recent discovery of a self-pollinating line offers new approaches in jojoba breeding.

Meadowfoam

Meadowfoam has also drawn attention in recent years as another potential crop to substitute for sperm whale oil (Gentry and Millr, 1965; Higgins et al., 1971). At least 95% of its fatty acids are long chains (C_{20} and C_{22}) with a C_{22} component having unique positions of unsaturation (Miller et al., 1964). A research team led by W. Calhoun of Oregon State University, Corvallis, released two cultivars of meadowfoam for commercial production in Oregon, both selected from an accession of *Limnanthes alba* Hartw. that Bentham collected in Northern California. Their breeding and agronomic evaluation emphasized selection for high yield, non-shattering of seed, and large seed size. *Limnanthes alba* Hartw. was noted by Gentry and Miller (1965) as the most promising species in the genus for its lower moisture requirements, adaptation to a wide range of environments, and good seed output per plant. A breeding project, started in 1973 at the University of California, Davis, has uncovered a great diversity of genetic resources in this genus that have been explored by agronomic and population genetic studies (Jain et al., 1977). Numerous traits such as the level of autogamy, growth habit and branching pattern, seed retention, and proportion of C_{22} fatty acid in the oil were highly variable.

Studies of *Limnanthes* have also indicated differences between species for a number of agronomically important characters. Gentry and Miller (1965) described populations of seven *Limnanthes* species including their ecological requirements in terms of soil type, moisture, and pH and compared their growth habit, seed yield, seed retention, date of plant maturity, and oil content. Likewise, Higgins et al. (1971) examined 30 accessions of *Limnanthes*, representing all of the known taxa, for performance at several locations.

Original bulk seed materials of four *Limnanthes* species from 35 collection sites were studied for interpopulational variation in fatty acid composition, oil content, and seed weight (Pierce and Jain, 1977). Significant interpopulational differences in the relative content of C22:1 and C22:2 fatty acids were found for the species, *L. alba* Hartw., *L. floccosa* Howell, and *L. douglasii* C. T. Mason. Differences among species for oil content, fatty acid composition, and seed weight were as large as the estimates of interpopulation variances. Data on several species and cultivars (Miller et al., 1964) and various populations of *L. alba* Hartw. and *L. floccosa* Howell (Jain et al., 1980a) showed a significant correlation between increased seed size and higher oil content. A genetic male-sterility factor has been found in *L. douglasii* var. *nivea* C. T. Mason which would be very useful in hybridization work. In a survey of natural diversity in plant type and productivity, optimal plant types (ideotypes) with desirable combinations of the yield components were defined.

Crambe

This cruciferous genus of nearly 30 species produces a valuable seed oil with 55 to 60% erucic acid. *Crambe abyssinica* Hochst. ex R. E. Fries has good agronomic traits and adaptability, but will require further work on yield improvement, disease resistance, and cultural refinements (Leppik and White, 1975). Three cultivars that varied in yield, seed size, and oil content were developed at the Agricultural Experiment Station of Purdue University. Small commercial plantings have been sustained for several years but uses and processing facilities would grow with more dependable seed supplies (White, 1977). Collection and evaluation of wild and weedy relatives are high priority items, but limited funds have hampered crop development.

Vernonia and Stoke's Aster

As sources of epoxy acids, two prospective new crops have been evaluated for years. Studies of seed yield, seed dormancy, and oil production and quality in *Vernonia anthelmintica* (L.) Willd. provided interesting leads for further improvement, but germplasm availability and funds for further research have been meager (Higgins, 1968; White and Earle, 1971). Stoke's aster is a native of the USA; many natural populations are known for genetic resource studies (Gunn and White, 1974). Since few natural resources are at hand for nearly 57 million kilograms of oil needed in the vinyl plastics industry, Stoke's aster would offer an excellent prospect. Intensive research is recommended to develop improved plant types with high seed retention, hardiness, determinate growth and increased yields of oil, and for a better understanding of its sexual and asexual reproductive systems.

Guayule

Guayule, an erect perennial shrub of the family Compositeae (Asteraceae), produces natural rubber equivalent to *Hevea* rubber in physical and chemical properties (Buchanan et al., 1978). Currently, interest in natural rubber has increased because of energy shortages and the need for a higher quality product in radial tires. Native stands on upland plateaus of southwest Mexico are being considered for commercial harvests. Agronomic experience in the World War II years showed that production of rubber on irrigated plots each year averaged 240 to 280 kg/ha. Large differences occurred in growth form and rubber content (6 to 10% of dry weight), but population research on genetic resources was apparently limited. Genetics of apomictic reproduction was reviewed by Stebbins (1950). Reproduction by sexual as well as apomictic means was related to low vs. high chromosome numbers ($2n=36$ to 100). Grafts and cuttings could be used for plantings, but prolific seed output was an asset. Population genetic studies on germplasm and sexual recombination rates would be rewarding in a program selecting for higher yield of rubber, more rapid growth, adaptation to cold and drought, and longevity under production. Certain interspecific crosses might be useful in breeding for adaptation to cold and other stress factors. Among the cultural practices needing attention are water management and the use of plant growth regulators for optimal rubber production (Siddiqui et al., 1982).

Gopher Plant

Calvin (1978) described the potential of gopher plant for petroleum-like fuel from the latex in its leaves. The molecular weight distributions of polyisoprenes isolated from such plants as gopher plant, rubber tree, and milkbush suggest that such plants may be successful in a "petroleum plantation" (Calvin, 1979). Because most of the enzymes that control the process of polymerization of isoprenes are known, Calvin stated that it is possible to modify the plants genetically to produce molecules that are wanted. He presented from an unselected seed planting a yield estimate of 22 barrels of oil per hectare per year. Data from field plantings and energy considerations of plant processes do not support these projections (Loomis, 1980). Sachs et

Table 9.5. An evaluation matrix for summarizing current research status and needs.

Crop	Product study and processing	Uniqueness of product or some unique factors	Adaptability	Yield	Stage of domestication	Germplasm availability	Variation	Breeding system	Cultural practices	Breeding objectives
Triticale	A	--	A	A	A	A	B	B	A	A
Buffalo gourd	A	✓	A	BO	B	A	A			
Winged bean	A	✓	B	O	B	A	BO	O	B	B
Tepary bean	B	--	A	O	B	B	O	O	B	B
Amaranth	A	✓	BO	O	A	B	O	O	B	B
Quinoa	B	✓	B	O	A	B	O	O	B	B
Wild rice	B	✓	O	B	B	O	O	O	O	--
Mesquite	B	✓	O	B	A	B	O	O	BO	B
Leucaena	B	✓	B	B	O	O	O	O	O	--
Jojoba	A	✓	O	B	B	B	BO	B	A	B
Meadowfoam	B	✓	O	B	B	A	B	A	O	B
Crambe	A	✓	O	B	B	B	O	B	BO	B
Stoke's aster	B	✓	O	B	B	B	O	B	BO	B
Rapeseed	A	✓	B	B	A	B	BO	BO	A	A
Gopher plant	B	✓	O	O	O	O	O	O	O	O
Guayule	A	✓	A	B	B	O	O	B	B	B
Jerusalem artichoke	B	--	O	B	B	O	O	B	B	B
Kenaf	A	--	O	B	B	O	O	O	B	O
Sunn hemp	B	--	O	O	B	O	O	BO	B	O
Dioscorea yam	A	✓	O	O	A	B	BO	BO	A	B

A—well-known and documented; B—partially known; O—important research need. ✓—uniqueness established; ---—no information.

al. (1981) studied numerous agronomic and physiological aspects of gopher plant in field plantings at three sites near Davis, Calif. Rather long growing season, serious weed and disease problems, and current lack of suitable germplasm for improvement in resin yield were cited as limiting factors. If the plant is to be useful, much work is needed on variation and improvement by breeding as well as the development of suitable cultural practices. This new crop is a good example of the need for careful scientific studies prior to raising high hopes and unhelpful publicity.

CONCLUSIONS

The current status and the future potential of the breeding work on prospective new crops are summarized in Table 9.5. Clearly, biological information is needed on reproductive systems, ecology, genetics of economically important traits, and genetic variation in germplasm resources. The plant species reviewed here, and many other potential crop species that are not covered, come from habitats or plant-community types little different from those used in early domestication work of various forms of traditional agriculture.

Genetic systems vary from dioecy of jojoba, a monoecy in buffalo gourd, complex gynomonoecy in grain amaranths, predominant autogamy in winged bean, to various combinations of sexual and apomictic reproduction in guayule and in fruit and forage crops. Population genetics and evolutionary studies will assist plant breeders in planning various controlled mating, selection, and hybridization experiments, and in an accelerated program of genetic conservation.

An understanding of ecology and natural selection will be essential to work with the low-input agriculture of marginal lands. The work of Duke and Terrell (1974), for example, laid the foundation of new crop geography allowing for an efficient computerized search for prospective crops. The biology of adaptations, their origin, genetic make-up, and their interplay with the environmental changes will be a central issue of applied research.

Researchers must study and use the paradigm of production-marketing-consumption systems used by Thiesen et al. (1978) in their evaluation of new crops. Availability of land, suitable price structure, integration with the prevalent agricultural technology, and ethnobotanical variables are all determiners of the success of new crop developments by both public and private sectors.

New products can reduce imports, increase stability of agricultural productivity, safeguard against vulnerability to new diseases and pests, and stimulate wise use of marginal lands. Most predictions of agricultural production consider the likelihood of trade-offs between different plant products and equivalent energy outputs. Promises of higher yields from genetic manipulations by somatic hybridization, or the use of recombinant DNA to achieve breakthroughs in nitrogen fixation or increased photosynthetic efficiency could contribute to solutions of some resource problems. Productivity, however, will still be limited by water, temperature, and growing season.

Our dependence on plants is emphasized by shortages of non-renewable natural resources, increased human population, and renewed interest in nutrition. Domestication of many new crop plants requires an understanding of basic systematics, genetics, ecology, physiology, and plant breeding, along with analyses of production-marketing-consumption systems.

ACKNOWLEDGMENTS

I am grateful to Drs. A. Campbell, C. O. Qualset, D. Wood, and G. White for their helpful comments on the manuscript. In particular, I owe thanks for my interest in new crops to Dr. P. F. Knowles, who first introduced me to *Limnanthes* research and provided many useful discussions.

LITERATURE CITED

Applezweig, N. 1977. *Dioscorea*—the pill crop. p. 149–163. *In* D. S. Seigler (ed.) Crop resources. Academic Press, New York.

Ayad, G., and N. M. Anishetty. 1980. Directory of germplasm collections: food legumes. Int. Board of Plant Genetic Resources. FAO, Rome.

Baker, H. G. 1970. Plants and civilization. 2nd ed. Wadsworth, Belmont, Calif.

Bemis, W. P., J. W. Berry, C. W. Weber, and T. W. Whitaker. 1978. The buffalo gourd: A new potential horticultural crop. HortScience 13:235–240.

———, ———, and C. W. Weber. 1979. The buffalo gourd: A potential arid land crop. p. 65–88. *In* G. A. Ritchie (ed.) New agricultural crops. Westview Press, Boulder, Colo.

Boswell, V. R. 1936. Studies of the culture and certain varieties of the Jerusalem articoke. USDA Tech. Bull. 514.

Brewbaker, J. L., and E. M. Hutton. 1979. *Leucaena*: Versatile tropical tree legume. *In* G. A. Ritchie (ed.) New agricultural crops, Westview Press, Boulder, Colo.

Buchanan, R. A., I. M. Cull, F. H. Otey, and C. R. Russell. 1978. Hydrocarbon and rubber-producing crops: evaluation of U.S. plant species. Econ. Bot. 32:131–145.

Calvin, M. 1978. Green factories. Chem. and Eng. News. 20 March 1978.

———. 1979. Petroleum plantations for fuel and materials. BioScience 29:533–538.

Chubey, B. B., and D. G. Dorrell. 1974. Jerusalem artichoke, a potential fructose crop for the prairies. Can. Inst. Food Sci. and Technol. J. 7:98–100.

Cole, J. N. 1979. Amaranth from the past, for the future. Rodale, Press. Emmaus, Pa.

Cox, G. W., and M. D. Atkins. 1979. Agricultural Ecology. Freeman, San Francisco.

Crane, J. C. 1947. Kenaf—fiber plant rival of jute. Econ. Bot. 1:334–350.

Curtis, L. D. 1946. The possibilities of using species of perennial cucurbits in sources of vegetable fats and proteins. Chemurgic Dig. 5:221–224.

de la Cruz, A. A. 1978. The production of pulp from marsh grass. Econ. Bot. 32: 46–50.

Downey, R. K. 1971. Agricultural and genetic potentials of cruciferous oilseed crops. J. Am. Oil Chem. Soc. 48:718–722.

Duke, J. A., and E. E. Terrell. 1974. Crop diversification matrix: Introduction. Taxon 23:759–799.

———, S. J. Hurst, and E. E. Terrell. 1975. Ecological distribution of 1,000 economic plants. Informacion al dia Alerta. IICA-Tropicos Agronomia No. 1, Turialba, Costa Rica.

Feine, L. B., R. R. Harwood, C. S. Kauffman, and J. P. Senft. 1979. Amaranth: gentle giant of the past and future. p. 41–63. *In* G. A. Ritchie (ed.) New agricultural crops. Westview Press, Boulder, Colo.

Gentry, H. S. 1958. The natural history of jojoba (*Simmondsia chinensis*) and its cultural aspects. Econ. Bot. 12:261–294.

———, and R. W. Miller. 1965. The search for new industrial crops. IV. Prospectus of *Limnanthes.* Econ. Bot. 19:25–32.

Gladstones, J. S. 1970. Lupins as crop plants. Field Crop Abstr. 23:123–148.

———. 1975. Lupin breeding in Western Australia. J. Agric. West. Aust. 16: 44–49.

Grubben, G. J. H. 1976. The cultivation of *Amaranths* as a tropical leaf vegetable with special reference to South Dahomey, Kronenklijs Ind. voor de Tropen.

———. 1977. Tropical vegetables and their genetic resources. Int. Board of Plant Genetic Resources, FAO, Rome. 195 p.

Gunn, C. R., and G. A. White. 1974. *Stokesia laevis*: taxonomy and economic value. Econ. Bot. 28:130–135.

Hauptli, H., and S. K. Jain. 1978. Biosystematics and agronomic potential of some weedy and cultivated amaranths. Theor. Appl. Genet. 52:177–185.

———, R. Lutz, and S. K. Jain. 1980. Collection and ethnobotany of amaranths in South and Central America. p. 117–122. *In* Proc. 2nd Amaranth Conf., Emmaus, Pa. 13–14 Sept. 1979. Rodale Press, Emmaus, Pa.

Heiser, C. B., and H. D. Wilson. 1974. On the origin of the cultivated chenopods (*Chenopodium*). Genetics 78:503–505.

Higgins, J. J. 1968. *Vernonia anthelmintica*: A potential seed oil source of epoxy acid. I. Phenology of seed yield. Agron. J. 60:55–68.

———, W. Calhoun, B. C. Willingham, D. H. Dinkel, W. L. Raisler, and G. A. White. 1971. Agronomic evaluation of prospective new crop species. II. The American *Limnanthes*. Econ. Bot. 25:44–54.

Hills, L. D. 1976. Comfrey: fodder, food and remedy. Universe Books, New York.

Hogan, L. 1979. Jojoba: a new crop for arid regions. p. 177–205. *In* G. A. Ritchie (ed.) New agricultural crops, Westview Press, Boulder, Colo.

Hymowitz, T., and J. Boyd. 1977. Origin, ethnobotany and agricultural potential of the winged bean (*Psophocarpus tetragonolobus*). Econ. Bot. 31:180–188.

Imle, E. P. 1977. *Hevea* rubber: Past and future. p. 119–136. *In* D. S. Seigler (ed.) Crop resources. Academic Press, New York.

Jain, S. K., and E. H. Abuelgasim. 1981. Some yield components and ideotype traits in meadowfoam, a new industrial oil crop. Euphytica 30:437–443.

———, C. R. Brown, and H. Hauptli. 1980a. Variation in *Limnanthes* alba: A biosystematic study of germplasm resources. Econ. Bot. 33:267–274.

———, ———, and B. D. Joshi. 1980b. Collection and evaluation of Indian grain amaranths. p. 123–128. *In* Proc. 2nd Amaranth Conf., Rodale Press, Ammaus, Pa.

———, R. O. Pierce, and H. Hauptli. 1977. Evaluation of meadowfoam as a new oil crop in California. Calif. Agric. 31:18–20.

———, L. Wu, and K. R. Vaidya. 1980c. Levels of morphological and allozyme variation in Indian amaranths: a striking contrast. J. Hered. 17:283–285.

Jolliff, G. D., W. Calhoun, N. Goetze, and J. M. Crane. 1981. Growing meadowfoam in the Williamette Valley. Oregon State Univ. Ext. Circ. no. 1080, 4 p.

Karikari, S. K. 1972. Pollination requirements of winged bean (*Psophocarpus* spp.) in Ghana. Ghana J. Agric. Sci. 5:235–239.

Knowles, P. F. 1958. Safflower. Adv. Agron. 10:289–323.

———. 1960. New crop establishment. Econ. Bot. 14:263–275.

———. 1972. The plant geneticist's contribution toward changing lipid and aminoacid composition of safflower. J. Am. Oil Chem. Soc. 49:27–29.

———, T. E. Kearney, and D. B. Cohen. 1981. Species of rapeseed and mustard as oil crops in California. *In* E. H. Pryde, L. H. Princen, and K. D. Mukherjie (ed.) New sources of fats and oils. Amer. Oil Chemists Soc. 18:255–268.

Leppik, E. E., and G. A. White. 1975. Preliminary assessment of *Crambe* germplasm resources. Euphytica 24:681–689.

Loomis, R. S. 1980. The costs of quality in plant production. p. 1–4. *In* Proc. 1980 Calif. Plant and Soil Conf., Sacramento, Calif. 30 Jan–1 Feb. 1980. Calif. Chapter of the Am. Soc. of Agronomy.

Martin, F. W., and H. Elpin. 1965. Sapogenin production and agronomic potential of *Dioscorea spiculiflora.* Turrialba 15:296–299.

———, and M. H. Gaskins. 1968. Cultivation of the sapogenin-bearing *Dioscorea* species. USDA, Prod. Res. Rep. 103.

———, and R. M. Roberte. 1975. Edible leaves of the tropics. Antillian College Press, Mayaguez.

Miller, R. W., M. E. Doxenbiehler, F. R. Earle, and H. S. Gentry. 1964. Search for new industrial oils. XIII. The genus *Limnanthes.* J. Am. Oil Chem. Soc. 41: 167–169.

Muntzing, A. 1979. Triticale, results and problems. Verlag Paul Parey, Berlin.

Nathan, G. P., and R. S. Felger. 1978. Teparies in southwestern North America—A biogeographical and ethnohistorical study of *Phaseolus acutifolius.* Econ. Bot. 32:2–19.

National Academy of Sciences. 1975a. Underexploited tropical plants with promising economic value. Natl. Acad. Sci. USA, Washington, D.C.

———. 1975b. The winged bean: a high-protein crop for the tropics. Natl. Acad. Sci. USA, Washington, D.C.

———. 1979. Tropical legumes: resources for the future. Natl. Acad. Sci. USA, Washington, D.C.

Newell, C. A., and T. Hymowitz. 1979. The winged bean as an agricultural crop. p. 21–40. *In* G. A. Ritchie (ed.) New agricultural crops. Westview Press, Boulder, Colo.

Oram, R. N. 1977. Breeding lupines for Eastern Australia. Ann. Report Division of Plant Industry, CSIRO, Australia.

Pierce, R. O., and S. K. Jain. 1977. Variation in some plant and seed oil characteristics of meadowfoam. Crop Sci. 17:521–526.

Princen, L. H. 1977. Potential wealth in new crops: Research and development. p. 1–15. *In* D. S. Seigler (ed.) Crop resources. Academic Press, New York.

Ritchie, G. A. (ed.). 1979. New agricultural crops. Westview Press, Boulder, Colo.

Sachs, R. M., C. B. Low, J. D. MacDonald, A. R. Awad, and M. J. Sully. 1981. *Euphorbia lathyris*: a potential source of petroleum-like products. Calif. Agri. 35:29–32.

Sauer, J. D. 1950. The grain amaranths: a survey of their history and classification. Ann. Mo. Bot. Gard. 37:561–632.

———. 1977. The history of grain amaranths and their use and cultivation around the world. p. 9–16. *In* Proc. 1st Amaranth Seminar, Kutztown, Pa. 9 Sept. 1977. Rodale Press, Emmaus, Pa.

Schwanitz, F. 1966. The origin of cultivated plants. Harvard Univ. Press, Cambridge, Mass.

Seigler, D. S. (ed.). 1977. Crop resources. Academic Press, New York.

Siddiqui, I. A., J. L. Connell, and P. Locktov (ed.) 1982. Report on the feasibility of commercial development of guayule in California. Calif. Food and Agric. Dep., Sacramento, Calif. 51 p.

Simmonds, N. W. 1979. Principles of crop improvement. Longman, London.

Singh, H. B. 1961. Grain amaranths, buckwheat and chenopods. Indian Council Agric. Res. Cereal Series, New Delhi.

Stebbins, G. L. 1950. Variation and evolution in plants. Columbia Univ. Press, New York.

Thiesen, A. A., E. G. Knox, and E. L. Mann. 1978. Feasibility of introducing food crops better adapted to environmental stress. Natl. Sci. Foundation, Washington, D.C.

Vavilov, N. I. 1926. Studies on the origin of cultivated plants. Leningrad.

Warthen, J. D. 1979. *Azadirachta indica*: a source of insect feeding inhibitors and growth regulators. USDA, SEA, Agric. Reviews and Manuals, ARM-NE-4.

Whistler, R. L., and T. Hymowitz. 1979. Guar: agronomy, production, industrial use and nutrition. Purdue University Press, West Lafayette, Ind.

White, G. A. 1977. Plant introductions—a source of new crops. p. 17–24. *In* D. S. Seigler (ed.) Crop resources. Academic Press, New York.

———, and F. R. Earle. 1971. *Vernonia anthelmintica*: A potential seed oil source of epoxy acid. IV. Effects of line, harvest data, and seed storage on quantity and quality of oil. Agron. J. 63:441–443.

———, and J. R. Haun. 1965. Growing *Crotolaria juncea,* a multipurpose legume, for paper pulp. Econ. Bot. 19:175–183.

Wilson, F. D., T. E. Summers, and J. F. Joyner. 1968. Everglades 41 and Everglades 71—two new varieties of kenaf (*Hibiscus cannabinus* L.) for fiber and seed. Florida Agric. Exp. Stn. Circular S-168.

Yermanos, D. M. 1974. Agronomic survey of jojoba in California. Econ. Bot. 28: 160–174.

———. 1979. Jojoba—a crop whose time has come. Calif. Agric. 33:4–8.

Zillinsky, F. J. (ed.). 1973. Triticale breeding and research at CIMMYT: a progress report. Int. Maize and Wheat Improv. Center Res. Bull. 24. Mexico, D.F.

SUGGESTED READING

Knox, E. G., and A. A. Thiesen (eds.). 1980. Feasibility of introducing new crops. Production-marketing-consumption (PMC) systems. Soil and Land Use Technology, Inc., Columbia, Md.

Simmonds, N. W. (ed.). 1976. Evolution of crop plants. Longman, London.

Chapter 10

Breeding to Control Pests

A. L. HOOKER

Dekalb-Pfizer Genetics Inc.
Dekalb, Illinois

A wide array of pest problems confronts the crop producer. Each crop plant is subject to characteristic diseases and insects, and any plant part can be affected. Diseases are caused by parasitic fungi, bacteria, viruses, mycoplasmas, spiroplasmas, nematodes, and even other seed plants that grow or replicate in their hosts. These causal agents of plant disease are called pathogens. Injuries also occur to plants through the feeding and reproduction of various insect pests. Weeds compete with crop plants and may injure them in other ways.

Breeding to control pathogens and insect pests is most effective when agronomists, plant breeders, geneticists, plant pathologists, and entomologists work together in the same improvement program. Disease and insect resistance is incorporated into the new cultivar or hybrid, and the pest-resistant cultivar or hybrid integrated into agriculture.

Studies on the genetics and biology of plant-pest interactions and the application of this knowledge to breeding for pest resistance represent some of the most fascinating of today's challenges. Breeding is the most widely used and most effective method of plant disease control. Much of the world's food and fiber supply depends upon the growth of disease-resistant crops. Because of the wide use of insecticides, breeding for insect resistance has lagged behind that for disease control. Nevertheless, such resistance is effective and has been developed to limit pests. Rapid advances are now being made to extend the range of insect species considered in plant breeding programs. Weeds are the third major class of pests confronting the agronomist. Crop tolerance to weeds is not usually considered part of plant breeding programs but is a challenge to future plant breeders.

USE OF RESISTANCE IN PEST CONTROL

Breeding for pest resistance is an integral part of any system of crop pest control. Such efforts interact with systems of crop management and the nature of the crop itself.

Host resistance is an ideal method to control plant pests if yield and other desirable characteristics are maintained. Pest resistance is the least expensive of known alternatives, has no secondary environmental effects, and requires no specific action by farmers to achieve control. Resistant cultivars stand ready to control the pest whenever it appears. There are no added production costs or need for timely decision as to when to apply alternate pest control practices during the growing season (Hooker, 1972b).

When resistance is inadequate, growers must use other means of pest control or suffer losses. In some instances, growers may elect to grow cultivars with partial pest resistance and then apply chemical pesticides or follow some other pest management practice. Each practice helps out the other. For instance, chemical control is frequently more effective on a partially resistant cultivar than on a fully susceptible one. Research has shown, however, that with most resistant cultivars there is no added advantage in the use of pesticides.

Availability of Crop Pest Resistance

Resistance has been found in most kinds of plants to most kinds of pathogens that cause plant disease. It is theoretically possible to produce crop cultivars resistant to all important diseases. The actual attainment of this objective presents many difficulties. In a few instances, it may not be economically feasible to develop a resistant cultivar even though it is technically possible to do so. Several years are required to produce, test, and increase a new cultivar. For crops that are grown over a large acreage, and for diseases that cause at least a moderate amount of damage, the cost of developing a resistant cultivar is usually justified.

It is easier to breed for resistance to some pests than to others. It is relatively easy to achieve resistance to viruses, vascular pathogens, the smuts, certain specialized nematodes, and to the pathogens that infect actively growing leaf tissue such as the rusts, powdery mildews, downy mildews, and *Helminthosporium* fungi. It is more difficult to breed for resistance to ectoparasitic nematodes and the pathogens causing root, crown, culm, and storage tissue rots. Among the insects, resistance to aphids, greenbugs, planthoppers, and certain leaf-feeding larvae is easier to obtain than is resistance to root-chewing or grain-storage insects.

Resistance is most frequently used in the control of pests of field and vegetable crops. Many of these are annual plants grown over large areas. Resistance is less frequently used in fruit, ornamental, and tree crops. Perennial crops would require long breeding programs. High-value specialty crops possess unique features or special quality characteristics. Of course, some flowers have been bred for disease resistance, and breeding programs in apples for scab resistance have been effective. Also, breeding programs are underway for rust resistance in cottonwood and in numerous pines.

History of Disease Resistance

We do not know when breeding for pest resistance began. Undoubtedly, early man selected plants for pest resistance following the domestication of crops over 10,000 years ago. Recorded observations that cultivated crops varied in reaction to disease go back to Theophrastus (371–286 BC). T. A. Knight, a plant breeder in England in the first half of the 19th century, noted rust resistance in wheat. Breeding programs in potato for late-blight resistance were under way after 1851, and these programs had the support of Charles Darwin. In France, breeding for downy mildew resistance of grapes was in progress in 1878. With these exceptions, there were relatively few instances of plant disease control by deliberate breeding.

Before 1900 disease resistance was not considered a stable character or reliable as a control measure. For example, when cultivars resistant to rust in England did not retain their resistance when grown in Australia or in India, it discouraged the plant breeders from launching extensive programs of breeding for disease resistance. We must remember that the concept and significance of physiological races, with specific virulence and avirulence properties to resistant cultivars, was not yet known.

Genetics of Disease Resistance

About 1900 a series of successes occurred in the genetics and breeding of disease resistance. The first evidence that resistance to disease was an inherited character was the work of Biffen, at Cambridge, England. Shortly after the rediscovery of Mendel's work, Biffen (1905) published the results of a cross of the wheat cultivar 'Rivet' that was resistant to yellow rust with the very susceptible cultivar 'Michigan Bronze.' Segregation of a recessive gene for resistance was clearly documented in the F_2 and F_3 generations. In the USA, several workers achieved marked success in the control of soil-borne *Fusarium* wilt diseases. The crops involved were cotton, watermelon, cowpea, flax, tomato, and cabbage. Entire industries, fading because of disease, were saved. Furthermore, these wilt-resistant cultivars were successful wherever grown. Other examples of successful selection were resistance to curly top disease in sugar beet and to southern anthracnose in clover. We must recognize, too, that during this same period other breeding programs were less successful, but the success with the *Fusarium* wilts and a few other diseases provided the needed stimulus for further breeding work.

History of Insect Resistance

Insect resistance in plants has been known and practiced to some extent for over 100 years. An early, often-mentioned example is the control of the grape louse in 1872 in Europe through the introduction of resistant rootstocks from North America (Riley, 1872). The European species of grapes was very susceptible while certain American species were resistant. Prior to that time, about 1831, it had been noted that the 'Winter Majestic' apple cultivar was resistant to the wooly apple aphid. These sources of resistance in grapes and apples have persisted and are still functioning. In the USA, resistance to Hessian fly in wheat was noted as early as 1875. Some of the earliest studies on the inheritance of insect resistance, conducted in 1916 and 1917, involved resistance to black scale and the leaf blister mite of cotton. About this time, genetic studies of the wheat-Hessian fly host/parasite system were initiated. They have become the most carefully studied of any host/insect parasite interaction (Gallun, 1972). Hessian fly-resistant cultivars of winter wheat are available for all major wheat-growing areas in the USA. There are a number of other excellent examples where resistant cultivars are used to reduce the damage caused to crops by insects. Several cultivars of wheat resistant to the wheat stem sawfly are grown in the Northern Plains states of the USA and in Canada. In South Africa, resistance to the leafhopper in cotton has made possible the growth of cotton without using insecticide. Maize with resistance to the European corn borer and, in the South, to the corn earworm is available, and some resistance to rootworm seems possible. Cereal cultivars with resistance to the cereal leaf beetle are also becoming available. Rice cultivars, 'IR28', 'IR29', and 'IR30', named by the International Rice Research Institute (IRRI), have resistance to the green leafhopper, the brown planthopper, and moderate resistance to the striped borer (Pathak, 1970).

Wheat Stem Rust Resistance

Stem rust resistance in wheat is often used as an example of breeding for disease resistance where marked success of the breeder has been followed by bitter disappointments. Some wheat cultivars, selected for stem rust resistance and widely grown, were at first resistant, only to succumb in a few years to new and virulent biotypes of the stem rust fungus. Some important lessons have been learned since breeding programs were begun in the early 1900's. It may have taken us 50 or so years to learn these lessons, but it is no accident that for over 25 years now we have had no wheat stem rust epidemics in such important wheat production areas as Canada, Australia, and the USA.

Stakman and Harrar (1957) in their book *Principles of Plant Pathology,* gave a good account of breeding wheat for resistance to stem rust. Breeding work was started at the Kansas and Minnesota Agricultural Experiment Stations soon after the widespread and destructive epidemic of 1904. Selection for resistance among or within cultivars, although effective for the *Fusarium* wilts, was not effective because, at that time, all bread wheats were susceptible to stem rust. However, workers noted that cultivars of einkorn, emmer, and durum wheats were resistant. A hybridization program was started, crossing the resistant wheats with the susceptible bread wheats. Unfortunately, no segregation of rust-resistant, high-quality bread wheats appeared. Instead, breeders experienced sterility and linkage of rust resistance with undesirable durum characters. This gave rise to the belief that genes for bread-wheat quality and stem rust resistance could not be combined. However, a few breeding lines with quality and resistance were found and from these the cultivar 'Marquillo' was selected. Marquillo was not a popular cultivar with farmers but an experimental sister selection was used in further breeding work and was one of the parents for the cultivar 'Thatcher.' Thatcher became the standard cultivar for quality for many years.

In about 1915, physiologic races were found in the stem rust fungus. This caused little concern at the time because the new race was less virulent than the cultures of rust then used in disease evaluation programs. However, within a short time other pathogenic races were identified. A major stem rust epidemic of wheat raged in 1916.

During the next decade a series of cultivars were released only to succumb to new races of the rust. 'Kanred' was found to be susceptible to race 3. 'Kota' was selected in North Dakota from a Russian introduction and 'Marquis' was an early rust-escaping cultivar developed in Canada. These two cultivars were the parents of 'Ceres' wheat, released about 1926. The stem rust epidemic of 1935 resulted because Ceres was susceptible to race 56 of stem rust. The advantages of the adult plant resistance of 'Hope' wheats were recognized and soon several resistant wheat cultivars were released and grown. The period of 1938–1950 was called the "golden age" of spring wheat in the USA and Canada. Good rust-resistant wheats were in production and the barberry eradication program, in effect for many years, had reduced the frequency of new races of rust. (Barberry is the alternate host of this rust, on which the genetic recombination stage occurs.) Then, in 1950 and 1951, race 15B of stem rust swept over North America. All cultivars being grown except 'McMurchy,' a farmer selection, were susceptible. In 20 of the 50 years following the epidemic of 1904, wheat was protected from rust by genetic resistance. Since 1953, stem rust has not been a problem on wheat in North America because the resistance has worked so well.

Corn Blight Resistance

One of the most noteworthy of recent epiphytotics was the southern corn leaf blight of 1970. This disease caused a larger production loss over a greater area of one crop in one year than any other pest problem facing American farmers throughout all of agricultural history. The disease was infamous for the rapidity of its occurrence and noteworthy for the swiftness of its control.

The southern corn leaf blight was caused by the fungus *Helminthosporium maydis* Nisikado and Miyake. Two races, race T and race O, of *H. maydis* were identified at the Illinois Agricultural Experiment Station through research completed in the early months of 1970 (Hooker, 1972a).

The two races differed in several important respects, the most important of which was their specificity to plant cytoplasm types. Race O showed no specificity to plant cytoplasms and could attack maize having a wide range of cytoplasms and nuclear genotypes. Race T, a new biotype responsible for the epiphytotic, was uniquely adapted to attacking plants containing *cms*-T cytoplasm for male-sterility. This cytoplasm was widely used in the production of maize hybrids in the USA. It had to be replaced. Seed growers increased additional foundation seed having normal cytoplasm during the 1970–1971 winter, so that all of the 1971 seed production involved resistant normal cytoplasm. By 1972 the disease caused by race T was under complete control in the USA.

Early Concepts of Resistance

Some important lessons learned before 1940 about breeding for pest resistance can be summarized. For one, the genetic variability of the pathogen was recognized. The significance of inherited resistance was known, and the difficulties of combining pest resistance and high yield and quality were appreciated. It also became evident that as crop cultivars were changed in production areas to achieve control of one disease or insect, another might become prevalent. It was also recognized that resistance could be of different types, and that the phenomena of disease escape and tolerance existed. During this time the influence of environment on the expression of resistance was also observed.

Modern Concepts of Resistance

Significant developments since 1940 in several areas of resistance are included in the following summary.

A. Pathogen variation
   1. Greater recognition of the importance of pathogen variation.
   2. Genetic systems in pathogens included also heterokaryosis and parasexualiam.
   3. Inheritance of virulence and avirulence in pathogens.
   4. Distinction between virulence and aggressiveness.
   5. Recognition of interspecific and intergeneric hybridization as sources of variability of pathogens.
   6. Development of better systems for race identification and a greater understanding of the significance of races.
   7. Development of race surveys as a means of monitoring pathogen variation.
   8. Development of national and international disease nurseries.

B. Inheritance of resistance
   1. Identification of more genes for resistance.
   2. Knowledge that genes for resistance may reside in allelic series.

3. The significance of various types of gene interactions including modifiers and inhibitors.
4. Recombination between closely linked genes.
5. Greater awareness and appreciation of polygenic resistance.
6. Appreciation that cytoplasm could determine plant reaction to disease.

C. Host/pathogen interaction
1. Studies of resistance in the host simultaneously with studies of virulence in the pathogen.
2. Development of the concept of the gene-for-gene relationship.
3. Use of near-isogenic host stocks for race identification and other pathogen studies.
4. Prediction of host genes for reaction based upon disease-reaction tests with specific biotypes of the pathogen having known genes for virulence and avirulence.
5. Use of computer methods of host gene identification.

D. Environmental influence on resistance
1. The use of growth chambers to advance the understanding of environmental effects on disease.
2. More specific information on the influence of environment on infection and disease development.
3. Use of environmental control in the identification of specific genes for resistance.
4. Awareness that the environment influences survival and prevalence of pathogen biotypes.
5. The relationship of disease resistance to epidemiology.

E. Resistance mechanisms
1. More understanding of various types of resistances.
2. Biochemical nature of resistance including the identification of specific inhibitory compounds produced by the resistant plants.
3. Attempts to link resistance to the action of specific genes.
4. Recognition of the significance of pathotoxins to disease susceptibility and resistance.
5. Studies on tolerance and how to measure it.
6. The recognition of specific and general resistance.

F. Breeding for resistance
1. The identification of a larger number of sources of resistance.
2. An appreciation of centers of origin and of diversity in location of resistance sources.
3. Greater attention to wild relatives of crop plants as sources of resistance.
4. Extensive use of backcross breeding in pest resistance.
5. Exploration of mutation breeding.
6. Importance of genetic uniformity in vulnerability to pests.
7. Exploration of multi-line and other means of gene deployment.
8. Greater attention by breeders to general forms of resistance and to resistances that are complexly inherited.

Pest resistance has been important in international programs of crop improvement. They have also added international testing for pest reaction as a means for identification of broadly functional resistances (Thurston, 1977).

NATURE OF RESISTANCE

In an agricultural sense, "resistant" crop cultivars or hybrids are inherently less damaged or less infested by a pest than other cultivars or hybrids under comparable environments in the field. This resistance can involve several components.

In a completely susceptible plant, nothing impedes pathogen development throughout the entire life cycle of plant infection, establishment in the plant, reproduction on the plant, and survival to the next growing season. In a resistant plant or population of plants, the development of the pathogen is less than optimum.

Components of Resistance

Resistance to establishment of the pathogen is one of the components of resistance to fungi and bacteria. Fewer spores may germinate on the surface of a resistant plant, or the rate of germination may be reduced. Because of stomatal structure, the entrance of some pathogens into the plant may be inhibited. In some plants this first line of defense is the replacement of tissues susceptible to penetration and infection by tissues that are more mature and resistant to infection. In some instances, the pathogen enters resistant plants but the invaded tissues die rapidly, and further development of the pathogen is precluded. This form of reaction is sometimes referred to as a hypersensitive reaction and is most common against pathogens that are obligate parasites.

In addition to a hypersensitive reaction, other components of resistance function after pathogen establishment to prevent further colonization of the invaded tissue. Chemicals inhibitory to the pathogen are sometimes involved here. In a few instances they are preformed and are present in the resistant plant prior to infection, but more commonly they are produced in the resistant plant as the pathogen starts to invade the tissue. These inhibitory compounds are called phytoalexins (phyton = plant; alexin = a warding-off compound). Those isolated and identified are mainly phenolic compounds. The nature of these compounds seems to be determined by the host plant rather than by the pathogen or other agent. In susceptible plants, the compounds are either not produced when invaded by the pathogen or they fail to reach the concentration attained in resistant plants. In roots and stems, another form of resistance can come into play. Meristematic activity may be stimulated by the pathogen in resistant plants so that invaded tissue is "walled off" by a layer of thick-walled or suberized cells. This limits further spread of the pathogen. In some resistant plants, bacteria and other pathogens simply fail to grow well, with no evidence for the presence of mechanical barriers or compounds toxic to the pathogen. The nature of this inhibition remains to be explained.

Several features offer practical resistance to viruses but not all forms of resistance are equally desirable. Immunity to virus is uncommon but does occur. Here the virus is rapidly inactivated after entry into the plant and no symptoms are produced. More commonly, resistance to virus multiplication and spread within the plant occurs. In some plants, a hypersensitive reaction results, and the virus is localized and unable to spread from the necrotic lesion that forms afterward.

Resistance to Nematodes

As with other pathogens, resistance to nematodes can take various forms (Johnson, 1972). Usually, resistance does not become evident until after the initial stages of infection and, in its final form, may be the end result of a complex series of reactions. Some of the general features of nematode resistance are summarized below.

1. The host tissues are not suitable, probably due to some nutritional relationship. Nematodes enter resistant plants but are unable to reproduce or they migrate through the root cortex and then simply leave the root.
2. The host fails to respond to the presence of the nematode. For example, the specialized giant cells needed for the continued development of root-knot nematodes form in susceptible roots but fail to form in resistant plant roots or the cells die quickly and the nematodes cannot live.
3. Resistant plants may react through hypersensitivity, toxin formation, or the formation of cork layers.

Mature Plant Resistance

A plant may not react in the same way throughout its life to individual pathogens and strains of pathogens. Usually, but not always, if plants change they become more resistant as they get older. This is called mature plant resistance, most commonly seen in maize and other cereals in the reaction to rusts and, specifically in cereals, to mildews. Mature plant resistance supersedes seedling susceptibility and is of extreme practical importance. The resistance is usually expressed by a low number of infection points. Individual infection points may be large and sporulate well. Differences among plants is quantitative. In the rusts, plants range from those with most of the leaf surface covered with pustules to those with few if any pustules. This outstanding form of resistance is believed to be the reason that corn rust has been of little consequence in the U.S. Corn Belt (Hooker, 1967).

Stage Resistance

Stage resistance may be considered in relation to growth stage or to plant part infected. Plants may vary as to growth stage when resistance or susceptibility is first expressed. Crop cultivars or hybrids vary in the rapidity with which these changes occur and are more resistant or more susceptible because of this feature. We have mentioned mature plant resistance. Very young seedlings may be susceptible to soil-borne pathogens but acquire resistance within a few days or weeks. Sometimes plants are resistant in a young or a vegetative growth stage but become susceptible during their reproductive or mature stages. This sequence is true for stalk rots of maize and for certain leaf diseases of sorghum. A pathogen may affect several parts of the same plant. *Gibberella zeae* (Schw.) Petch, for example, can cause a seedling blight, root rot, stalk rot, and ear rot of maize. Resistance in maize in one plant part to this pathogen is not well correlated with resistance in other plant parts.

Field Resistance

The term field resistance is used to identify certain forms of resistance expressed by populations of plants in the field. Although the term implies that the resistance can occur only in the field, it can occur and be measured in the greenhouse or in the laboratory. This resistance is a complex of factors comprising various combinations of the following components: (1) longer time period required for pathogen penetration, (2) fewer lesions resulting with a given inoculum dosage, (3) longer time interval between infection and lesion formation; (4) reduction in size of lesions, (5) longer time intervals between lesion formation and sporulation of pathogen, (6) fewer spores produced per unit area of lesion, and (7) reduction in the number of times that the pathogen will produce spores.

Other forms of resistance have been described as slow rusting, slow mildewing, late rusting, etc. These terms describe the field appearance of cultivars having resistance. The fundamental basis is more complicated and may depend upon several components (Parlevliet, 1979).

Disease Tolerance

Disease tolerance is the ability of a cultivar or hybrid, although susceptible to a pathogen, to endure disease attack without suffering a severe loss in yield (Schafer, 1971). To say it another way, tolerance is that condition in which two plant cultivars, exhibiting equal amounts of disease at any given time throughout the infection period, show significantly different quantitative responses to the infection. The non-tolerant cultivar suffers an economic loss while the tolerant one gives satisfactory yields. We might say that tolerance is resistance to the disease, i.e., gives resistance to the products of the pathogen, but not resistance to the pathogen itself. Plants demonstrate tolerance to parasitic nematodes, viruses, rusts, and other pathogens.

Disease Escape

Some plant cultivars, normally susceptible to disease, escape inoculation for some reason. These cultivars may be less damaged by disease than others grown in the same area at the same time and give the appearance of being resistant to disease. Unless the plant breeder is aware of this situation some selections may not perform well when moved to other environments. Disease escape may be a valuable characteristic of a cultivar and a worthy breeding objective.

Environment and Resistance

Sometimes resistance breaks down or fails to be expressed because of some factor in the environment, and the breeder or the farmer may incorrectly guess that a new race of the pest has appeared. Pest resistance can be modified by air and soil temperature, light intensity, light duration, mineral nutrient supply, and other factors. Frequently plants with an intermediate reaction to disease are more sensitive to environmental shifts than are highly susceptible cultivars or hybrids. The degree and duration of reaction change depends upon the host genotype, the pest biotype, and the environmental change. Rust resistance of some genes breaks down at high air temperature. Resistance to some soil-borne *Fusarium* wilts also breaks down at high temperature, as does some nematode resistance.

Categories of Resistance to Insects

The ways in which plants provide resistance to insects can be classified into three broad categories (Painter, 1951, 1958). These are non-preference, antibiosis, and tolerance.

Non-preference

A plant has factors that make it unattractive to insect pests to lay their eggs or to find food or shelter. Shape, size, color, surface texture or chemical constituents are factors of plants which attract or repel insects. The striped borer moths have a strong preference for oviposition on certain rice cultivars. Susceptible rice cultivars generally receive 10 to 15 times as many egg masses as do resistant ones. In a similar manner, the pink bollworm of cotton shows a strong preference for oviposition to susceptible species of *Gossypium*. Morphological features of some plants, such as leaf pubescence in wheat cultivars resistant to the cereal leaf beetle, are factors that deter oviposition.

Non-preference for feeding is another feature. As with oviposition, feeding involves a sequence of steps including host plant recognition, initiation of feeding, and maintenance of feeding. When any of these factors is lacking, the plant is rendered non-preferred. Non-preference to feeding may be so strong that the insect starves to death rather than feed. In rice, when the brown planthopper punctures the tissues of resistant cultivars, it does little feeding. Resistant cultivars contain a much smaller proportion of asparagine, an amino acid strongly attractive to the insect, than do susceptible cultivars.

Antibiosis

The host plants adversely affect the insects feeding on them. Antibiosis is regarded by some as the only true resistance to insects. When the insects attack plants having this resistance, they either die, lay fewer eggs, produce fewer young, or have slower rates of growth. There are many examples of antibiosis; only a few can be mentioned here. Antibiosis is the main mechanism of resistance to the Hessian fly in wheat. A classic example of a biochemical basis of antibiosis is in maize resistant to larvae of the European corn borer. Resistant chemicals isolated and identified as 6-methoxybenzoxazolinone (6-MBOA) and 2,4-dihydroxy-7-methoxy-1,4-benzoxazine-3-one (DIMBOA) are present in the leaves of resistant maize inbreds (Beck, 1965). These plant biochemicals inhibit the growth of young larvae. A number of alfalfa cultivars resistant to the spotted alfalfa aphid have been developed by using the mechanism of antibiosis.

Tolerance

Yield is not significantly reduced even though the plant is supporting an insect population that normally would severely damage susceptible cultivars or hybrids.

Tolerance has been less well studied than the other two forms of resistance. It is probably present most often with insects that have sucking mouthparts. For example, 'Culver' alfalfa has tolerance to the meadow spittlebug, and some of the greenbug resistance in barley is due to tolerance. Part of the rootworm tolerance in maize is attributed to the ability of tolerant inbreds to replace injured roots faster than they are lost.

As with resistance to pathogens that cause plant diseases, resistance to insects involves several types of mechanisms, each of which may be a complex of interacting factors (Beck, 1965). Resistance seen in the field often results from the combination or interaction of several different components.

BREEDING FOR PEST RESISTANCE

Breeding for disease and insect resistance differs fundamentally from breeding for other characters since the objective may involve a change in a relationship with an evolving and variable pest or pathogen population.

Another problem is that genes for resistance cannot be identified unless the plant containing the genes is interacting with the pathogen or insect pest in an environment in which susceptible plants normally would be diseased or injured.

Two categories of problems affect breeding for pest resistance: (1) to produce the genetic variability in the segregating breeding material that is diverse enough to contain the desired combination of characters, and (2) to identify and to select the desired type in such a form that it is suitable for vegetative propagation or genetically stable enough for seed production. These problems are simple to state but, in actual plant breeding, difficult to solve (Sprague and Dahms, 1972).

It is safe to say that a plant breeder never sees a segregating population of plants that contains within it a plant that combines all of the desirable characteristics and is genetically homozygous. The laws of genetics and the large number of desired characters make this so. Assume that the desired plant must possess 21 characteristics and that each characteristic is monogenic in inheritance. Furthermore, the genes all segregate independently of each other. Starting with an F_1 plant heterozygous for all 21 genes, the breeder would have to grow an F_2 population of nearly 4.4×10^{12} plants to expect to find one plant that would be homozygous for each of the 21 genes. In maize, it would involve a crop equal to the total USA maize acreage in a normal year. The logistics of producing this amount of seed are formidable in a crop like maize and impossible in a self-pollinated crop like rice or wheat, notwithstanding the problems in the identification of the desired genotype. Of course the situation is morc complicated because many desired characters are polygenic rather than monogenic in inheritance.

Breeding Techniques

Selection for resistance is the first activity when a new disease or insect problem confronts the plant breeder. Resistance may be found among adapted or exotic cultivars or within crops that are genetically heterogeneous. Selection within adapted cultivars worked well for the *Fusarium* wilts mentioned previously. Selection within homozygous pure-line cultivars of crops is rarely successful but chance mutations for pest resistance may have occurred. In maize, selection for disease resistance within open-pollinated cultivars has been successful. An example is resistance to the rust caused by *Puccinia polysora* Underw. in Africa. This rust fungus is common in the tropical areas of the Western hemisphere. In Africa, native cultivars evolved in the absence of this pathogen. During the 1950's *P. polysora* became established in West Africa, spread across Africa and eventually to the Philippine Islands by way of Madagascar, northern Australia, India, and Southeast Asia. During the first few years, losses in Africa were extensive because most of the native cultivars proved to be susceptible. Enough genetic variability for resistance existed within these cultivars, however, so that by selection of the more resistant plants, maize cultivars were obtained that were little damaged by disease.

Hybridization with a source of resistance and selection among segregating progenies is widely used in breeding for pest resistance in cereals and other self-pollinated crops. The objective is to combine the good properties of two or more parents. These are long-term programs, and breeders may go by steps. The products from one series of crosses are used as parents for the next series. Programs involving large numbers of crosses and evaluation of the segregating material in a wide diversity of environments are the most successful.

Backcross Breeding

The backcross method of breeding is used when it is desirable to transfer a single simply inherited trait to another cultivar or line. It is used in pest resistance where breeding programs are advanced, or where cultivars are available that are quite outstanding but may be susceptible to the single pest. The system works best when resistance to the disease or insect pest is inherited as a single gene. By repeated backcrossing and selection, followed by selfing and selection, a single chromosome segment carrying the desired gene is incorporated into the genotype of the desired parent. The backcross method can also be used with some modifications to transfer disease or insect resistance where several genes are involved. Here several families of backcross progenies and larger populations of plants within each are used rather than when resistance is due to single genes. By intercrossing several partially resistant progenies of different families, much of the original level of resistance in the resistant source can be reconstituted. Backcross breeding is often necessary when the genes for disease or insect resistance are located in wild relatives or other undesirable agronomic types.

Recurrent Selection

Various types of recurrent selection methods have been successfully employed in breeding for resistance where the resistance is inherited polygenically. A notable example is the polygenic resistance to leaf blight in maize. The level of resistance in several populations was increased by intercrossing the most leaf blight resistant plants in each generation for several cycles of reproduction. The greatest progress was obtained during the first few cycles.

Wide Crosses

There are a number of examples of disease resistance genes incorporated into commercial cultivars by means of interspecific and intergeneric crosses. This resistance has been made possible through increased understanding of the cytogenetics and evolution of cultivated crops. In wheat, synthetic species have been created and these have served as bridges to transfer genetic material carrying genes for resistance to a cultivated species. While current experience has shown such resistance may not be superior to that found within the species, the future of synthetic species as sources of resistance to pests seems bright.

Mutation breeding has been used by a number of workers in an attempt to form new genes for disease resistance in agronomic crops. Several successes were reported and this work was first hailed with considerable enthusiasm. Sometimes the resistance genes, thought to be the result of treatment with mutagens, resulted from outcrossing to resistant cultivars and derived from induction of sterility by the mutagens. However, some cultivars have been developed with resistance derived through induced mutation.

TESTING FOR PEST RESISTANCE

In breeding for disease and insect resistance, reliable methods are necessary to determine differences among plants in pest reaction. Plants escaping pest attack would be incorrectly classified, decreasing the rate of progress in the breeding program. It should be possible to select for resistance in each segregating generation, and the results of any greenhouse or laboratory test should be confirmed by field tests.

Certain requirements should be met in the evaluation method for testing pest resistance. The method: (1) must be repeatable, (2) should give different gradations of reaction to the pest, (3) should duplicate the severity of the disease or insect pest infection under natural conditions, (4) should allow evaluation of a large number of plants, (5) should be an easy method to handle, not biologically complicated nor involve the use of expensive equipment, (6) should be adapted to the breeding and selection scheme for the crops, and (7) should use seedling plants if feasible since more plants can be tested to save time and space.

Inoculation Method

There are several ways to inoculate plants with the causal agents of plant disease, including natural infection, direct inoculation, and special laboratory techniques.

Natural infection can always be used. It may not be the most reliable method if the disease is not present in sufficient intensity each year. Natural infection is excellent if a large number of favorable locations are used because it exposes the test material to a wide range of pathogen biotypes in many environments. The method is also good when artificial inoculation methods have not been developed, such as for diseases that are soil-borne or in which an insect is the vector of a virus or virus-like agent that persists from season to season in weed host plants.

Direct inoculation of individual plants is used in the greenhouse and with certain diseases in the field. It tends to be reliable because the investigator has good control of the inoculum, and the greenhouse allows adequate control of the environment to ensure infection and disease development. Considerable time and work is involved if the plant population to be tested is large.

When individual plants are inoculated, the investigator determines the type of inoculum that will induce the disease. Here some care is needed. When it is applied, the inoculum should represent a pure culture of the pathogen as well as be relatively free from other organisms. The cultures should have an optimal degree of virulence or aggressiveness which may not occur with long culture in the absence of host plants. Reduced virulence in cultures may result in reactions of resistance and the breeder may be misled into thinking that the breeding material is resistant to field strains when it is not. It is also important to use the proper biotype or mixture of biotypes of the pathogen. A single biotype may not be adequate to index the presence of resistant genes needed to protect the plant against the diversity of pathogen biotypes that prevail in nature.

Field disease nurseries are commonly used in breeding programs for disease resistance. Individual plants can be inoculated. More commonly, susceptible plants in spreader rows, grown among the test material, are inoculated. The investigator allows and frequently depends upon the secondary spread of the pathogen for disease development. Control cultivars with a known reaction to the disease can be planted at regular intervals throughout the disease nursery to monitor the reliability and uniformity of disease development.

Scoring Disease Reaction

Methods of scoring plant reaction to disease will vary with the way resistance is expressed, and will depend somewhat upon the particular diseases and the host plant species. Usually some type of numerical system or index is used to score plant reaction. Plant pathologists tend to see disease rather than its absence. Hence, the lowest scores commonly represent the lowest levels of disease, and, thus, the highest amount of resistance. To facilitate the calculation of the selection index in breeding programs, the scoring order is sometimes reversed. For diseases in which only two classes are seen, i.e., infected or not infected, as in cereal smuts, usually the number or percentage of infected plants is recorded. The number of diseased plants can be combined with a severity rating for infected plants and a disease index calculated. For foliage diseases, the percentage of uninfected tissue or a rating scale based upon the number and distribution of infected points is used. These data become more meaningful if they reflect actual injury to the plant or loss in yield (Hooker, 1979). For some virus diseases, and in other instances, resistance can be measured by the magnitude of yield loss caused by the disease. Infected and healthy plants of tolerant (resistant) cultivars yield about the same. With rusts, powdery mildews, and other diseases, the infection type on different cultivars will vary. Resistant plants have small chlorotic or necrotic infection points; susceptible plants have larger infection points supporting abundant growth and reproduction of the pathogen. In some instances, resistant plants remain healthy longer than susceptible plants even though eventually both types of plants show an equivalent amount of disease.

A few pathogens produce specific pathotoxins, such as those produced by *Helminthosporium* fungi, including *H. maydis* race T on maize and *H. sacchari* Butl. (Subrain and Jain) on sugarcane. Genotypes may be assessed for resistance by treatment with the specific pathotoxin produced by the parasite. Plant parts exposed to the pathotoxin show the same reaction that they would if exposed to the active pathogen.

Infestation with Insects

Methods of infesting plants with insects and rating the plants for reaction are similar to those methods used with pathogens that cause disease. There are significant differences, of course, in the biology of insects compared to pathogens. Egg masses may be produced, collected, and placed on the plants to be evaluated. Insects can also be reared on susceptible plants and then transferred to the plants to be evaluated. In other instances, several plant cultivars are put in the same cage with the insect pest, and after a suitable feeding or oviposition time, comparisons are made among the cultivars. Cultivars may simply be grown in the field where large populations of a particular insect prevail or special efforts can be made to stimulate the development of an insect population. Various criteria are then used to evaluate resistance (Dahms, 1972; Maxwell and Jennings, 1980).

SOURCES OF PEST RESISTANCE

One of the most important objectives in any program of breeding for disease or insect resistance is to locate the sources of resistant germplasm material for the breeding program. When a new disease or insect problem—or an old problem that continues—is tackled, one of the first questions asked is: Where do I look for a source of resistance? The practice has been to screen material until something of value is found. Discovery is based on both volume and chance; resistance is where you find it. The problem is to find it and to determine the usefulness and effectiveness of resistance.

Several important aspects should be kept in mind when sources of resistance are considered. If the pest problem is new, the best place to look for resistance is within germplasm adapted to the local conditions. Simple selection may reveal a cultivar or hybrid that can be substituted for the susceptible one(s). If breeding is needed, the final objectives are more easily achieved if the source of resistance is in adapted plant material. There is also less danger of introducing susceptibility to another disease or insect pest if a local cultivar is used in the breeding program since it has already been tested under local conditions. Furthermore, growers are familiar with the cultural practices suitable for the local cultivar or hybrid and they may more readily accept a similar one with resistance than they would a new one they know nothing about.

Exotic cultivars from other areas of the world frequently provide excellent sources of disease resistance. Wheats from Kenya and Australia have furnished valuable genes for rust resistance to wheat cultivars in use in the USA and Canada. Oats from the eastern part of the Mediterranean basin have furnished genes for resistance to crown rust for USA and Canadian oats. Barley from Ethiopia has provided resistance to the barley yellow dwarf virus. Sugarcane germplasm from Java and India has provided resistance to mosaic needed for sugarcane in the USA. There are many such examples.

Most crops have primary and secondary centers of diversity—in the area where the crop originated (primary) or where it has been grown for a long time (secondary). Primary areas contain wild relatives of the crop. Genetic diversity is large in both primary and secondary centers. As for crops, plant pathogens also have centers of origin and of diversity. These centers for the crop and the pathogen may coincide or merely overlap in their ranges. Areas where host and pathogen have coexisted for a long period of time are generally good places to find a diversity of resistance to disease (Leppik, 1970). A word of caution, however: these areas may also be sources of virulent strains of pathogens. Unless breeders are aware and cautious, they may introduce new races of pathogens along with resistant germplasm.

Non-cultivated plant species frequently have more resistance to disease and insects than do cultivated species, especially if there has been a long association between host and pest. Through evolution, superior forms of resistance will have survived. Limitations in the use of wild relatives in breeding programs may include sterility, lack of chromosome pairing, and linkage of genes for undesirable characters. Nevertheless, wild relatives of wheat, oat, sugarcane, tobacco, tomato, potato, and other crops have contributed valuable genes for disease resistance to cultivated species.

GENETICS OF HOST/PEST INTERACTIONS

The presence of a gene is determined primarily by observing its effects. Effects of a gene are usually measured in terms of change in phenotypes. Molecular genetics takes us in depth through the route of enzymes, amino acid sequences in polypeptide chains, nucleotides in the DNA of the chromosome, and the translation of DNA through the messenger RNA-ribosome-protein synthesizing system. Classical genetics, on the other hand, is based on segregation and recombination of genes as recognized through their phenotypic expression. Since we cannot hybridize hosts with pathogens (or insects) and observe gene segregation, we must investigate host genetics and pathogen (or insect) genetics. We can analyze the genetic differences between two cultivars of wheat and between two races of stem rust pathogenic to wheat. In all parasitic diseases we find an interaction of host and pathogen genetic systems, in other words, the interaction of the products of the host genes and of the pathogen genes.

In most instances, host genes and pathogen genes are conditional. They remain unexpressed except in the interaction with one another. Furthermore, they are capable of appearing as either of two phenotypes. For example, a gene for resistance in a plant may not be expressed when the pathogen is absent or when it carries the gene for virulence. In the pathogen, a gene for avirulence may not be expressed when the pathogen is growing in a susceptible host. A gene for virulence may not be expressed when the virulence gene is associated with the same genotype with a gene for avirulence.

Resistance in the Host

Resistance in the host has been explained on the basis of numerous genetic models (Hooker and Saxena, 1971). These fit into three basic types: (1) oligogenic, (2) polygenic, and (3) cytoplasmic. The term oligogenic is used to refer to situations in which few genes are involved in the inheritance of a characteristic and the effects of individual genes can be detected. The term polygenic means that many genes are involved, and usually the effects of the individual genes are small and not detectable. Cytoplasmic inheritance involves non-nuclear genes.

Resistance to many pathogens and several insects is oligogenic. The simplest form of inheritance is a single gene that may be either dominant or recessive. Of the two, disease resistance is more often dominant as in resistance to rusts, powdery and downy mildews, viruses, and nematodes. Resistance to the *Fusarium, Septoria, Helminthosporium,* smut and anthracnose fungi, and to bacteria can also be attributed to single genes. In many instances, resistance gene loci are represented by a series of alleles. Sometimes two genes seem to act as alleles but, in fact, are closely linked and can be separated by crossing over.

More complicated resistance involves the interaction of two or more genes. This interaction is considered complementary when all the genes are needed for the ultimate expression of the character. Other forms of gene interaction exist. A gene at one locus may inhibit the effect on a gene at another locus or modify its expression. Some resistance genes are capable of being modified in their action by modifier genes, whereas other resistance genes seem not to be.

Much resistance to disease is polygenic in inheritance (Simons, 1972). Here disease reactions among cultivars or individuals in a segregating population do not fall into discrete classes but range continuously from one extreme to the other. This form of resistance to diseases is highly heritable; much of the variation among individuals is attributable to differences in their genotype and not to environmental effects on disease development. Studies of the number of genes involved in polygenic disease resistance suggest that fewer genes are involved in resistance to disease than are involved in a complex character such as yield. The genes are commonly additive in action. Usually, much progress can be made in breeding programs utilizing polygenic resistance.

In only two diseases have plant cytoplasms been shown conclusively to condition disease reaction (Hooker, 1974). Both diseases are in maize and involve southern corn leaf blight caused by race T of *Helminthosporium maydis* and yellow leaf blight caused by *Phyllosticta maydis* Arny & Nelson. The pathotoxins produced by both these fungi are chemically identical, and both cause malfunction with some of the membranes of the cytoplasm. The membranes involved include those in the mitochondria, with their structure determined, in part, by genetic material in the mitochondria. Mitochondria are organelles that are passed from the seed parent to the offspring through the cytoplasm. Hence, cytoplasmic inheritance of mitochondrial membrane structure may be the reason for disease susceptibility. Mitochondrial membranes found in resistant cytoplasm are relatively unaffected by the pathotoxin. We know of only one maize cytoplasm that conditions susceptibility. This is the *cms*-T (for Texas) cytoplasm for male-sterility. Other cytoplasms, such as normal cytoplasm and male-sterile cytoplasms, designated *cms*-C and *cms*-S, condition resistance.

Variability in the Pathogen or Pest

Breeding for pest resistance differs from breeding for other characteristics in that the character is influenced by the genetic variability of the pest population. This statement is more frequently true for the causal agents of plant disease than for insects. Physiological races are well known in the rusts, powdery mildews, other fungi, and some insects. Their presence means that populations of these pests contain individuals that have the ability to cause disease or injury in resistant plants. Other individuals do not have this ability.

Physiological races are identified by means of differential cultivars. The number of races that can be identified is a function of the number of genetically unlike cultivars. With two cultivars, four races can be identified (Fig. 10.1). Mathematically this can be expressed as 2^n where n is equivalent to the number of differential cultivars. With 3 differential cultivars, 8 races can be identified; 16 with 4 differential cultivars, and 1,024 with 10 differential cultivars.

Modern systems of race identification have become quite important to breeding programs (Green, 1965; Wolfe and Schwarzbach, 1975). Differential cultivars contain known genes for disease reaction; frequently they have been introduced by backcrossing into a common genotype so that unknown genes are not represented or, if represented, are in all differential cultivars. Genes for resistance actually used in cultivars being grown or in the breeding program are represented in the differential cultivars. Furthermore, the number and identity of the differential cultivars is frequently modified to meet current needs.

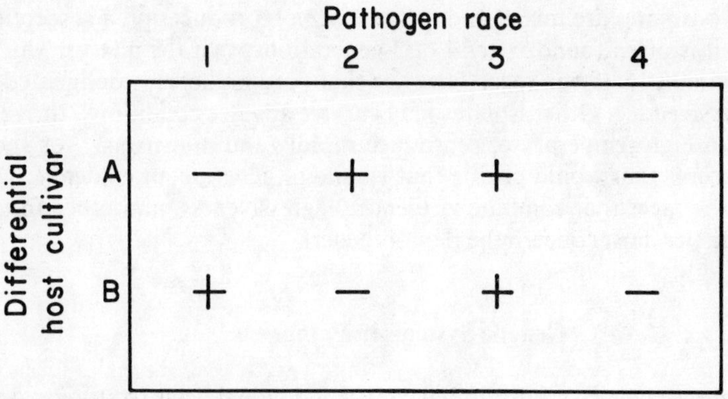

Fig. 10.1. Physiologic specialization shown by four races of the pathogen differentiated by two host cultivars and illustrated by incompatible (−) and compatible (+) interactions.

Physiological Races

The concept of physiological race has helped us understand variability in pathogens in relation to breeding for disease resistance, but one must always keep in mind the genetic nature of a race. Races are abstract entities. They are not pure biotypes of an organism but merely groups of genotypes that express the same reaction when inoculated over a set of differential cultivars established by experiment. All isolates that fall within the limits thus described for a race must be designated as that race. Individuals of the same race, however, can differ from one another. Furthermore, if recombination takes place during sexual reproduction of a race it may yield other races among its descendants. Races may thus exist that cannot be identified by the differential cultivars established (Russell, 1978).

The differential cultivars do give some indication of the virulence characteristics present in a pathogen population in a given area. The virulence characteristics can be determined with reference to known genes or sources of resistance. The plant breeder is interested in knowing the specific genes for resistance that will work for a crop. That is, to what genes for resistance does the pathogen population carry avirulent genes, and what genes for resistance of the host fail because the pathogen has the necessary genes for virulence? As we noted before, with 10 genes for resistance, some 1,024 races could be identified. It is far easier to determine which of the 10 genes work than to distinguish all of the races.

The evolution of races has not occurred in all groups of plant pathogens. In a majority of plant pathogens, races have not been detected. In some the specialization is less conspicuous than it is in the rusts, mildews, and smuts.

Pathogens express other forms of variation in addition to virulence. Those forms of variation, best described as fitness characteristics, are also important in plant breeding. Fitness is a broad term which can be subdivided into a number of parts. For example, two isolates of the same pathogen may appear to be equally virulent on a crop cultivar. When spores of the two isolates are mixed and the pathogen reproduces on a susceptible cultivar, it is often found, after 4 or 5 generations, that the mixture will be composed mostly of one race. The race that predominates is designated as the aggressive race. Other isolates may survive adverse conditions, thrive at a low or a high temperature, reproduce rapidly and numerously, or show other features that would enable that isolate to generate an epidemic. Obviously, the races that combine virulence, aggressiveness, and other fitness characteristics most concern the plant breeder.

Genetic Systems in Pathogens

Plant pathogens as a group have more individual ways to store and release genetic variability than do higher plants. All the familiar oligogenic, polygenic, and cytoplasmic genetic systems of higher plants seem to be operative in pathogens. Gene mutation occurs in all organisms, too. Many pathogens, but not all, have a sexual cycle where gene segregation and recombination occurs (Day, 1974). Except in a few fungi such as *Oomycetes,* the nuclei in the vegetative body (mycelium) are haploid. In the smuts and rusts, each cell has two haploid nuclei that function physiologically as diploid. Other fungi have several nuclei in each cell. These nuclei need not all be genetically alike. When they are unlike, the organism is heterokaryotic (hetero = unlike; karyon = nuclei). Various combinations of whole nuclei can associate in the same cell but result in different phenotypes. They can also disassociate to give these organisms a genetic means for variation. Furthermore, genetically unlike nuclei in heterokaryotic cells sometimes fuse to form diploid nuclei. These nuclei often are unstable and revert by various means, usually chromosome loss, to haploid. In this process, reassortment of chromosomes can occur. Several of these processes can and do occur in the same pathogen.

The inheritance of virulence has been studied in several pathogens. With a few exceptions where cytoplasmic inheritance is implicated, virulence is inherited as nuclear genes. Virulence is often recessive and attributable to one or two genes.

The inheritance of aggressiveness and other fitness characteristics in pathogens has been studied to a limited extent. These attributes tend to be expressed quantitatively; they are called polygenic characters.

Biotypes of Insects

Insects are specialized into races as are pathogens that cause plant disease. In certain groups of insects, there are biotypes that feed on resistant plants and are able to grow larger and more vigorously than do other biotypes on the same plants. In extreme cases, such as the Hessian fly on wheat, certain biotypes are able to feed, while other biotypes are not able to feed at all on resistant plants (Hatchett and Gallun, 1970). Other examples of insect biotypes involved in plant resistance are: spotted alfalfa aphid on

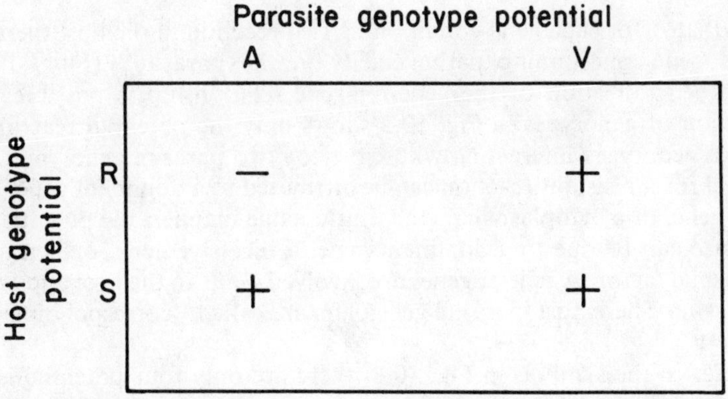

Fig. 10.2. Growth (+) or no growth (−) of a parasite as determined by the interaction of host genes for resistance and of parasite genes for virulence.

alfalfa, corn leaf aphid on sorghum and maize, pea aphid on peas and alfalfa, and the greenbug on wheat. In breeding programs for resistance to these insects, biotypes must be considered (Maxwell et al., 1972; Maxwell and Jennings, 1980).

Host/Parasite Interactions

In previous sections we considered the genetics of hosts and the genetics of parasites. Now let us consider the genetics of host/parasite interactions (Day, 1974; Loegering, 1978). Keep in mind that in host/parasite systems, the genes in the host can be identified only in terms of those present in the parasite and vice versa. Furthermore, the same phenomenon, whether the disease does or does not develop (compatible vs. incompatible), is used to identify genes for resistance or susceptibility in the host as well as genes for virulence or avirulence in the parasite. This unique relationship is illustrated in Fig. 10.2.

Gene-for-Gene Interaction

Genetic studies by H. H. Flor with flax and flax rust at North Dakota identified the gene-for-gene relationship in host/parasite systems. In host and parasite combinations (Fig. 10.2), Flor studied both the genetics of the host and genetics of the parasite. Using rust cultures of the A (avirulent) type to test his plants, Flor found 1-, 2-, and 3-gene segregation ratios for resistance when he crossed resistant (R) and susceptible (S) flax cultivars. Then, using flax plants of the R type as hosts, Flor found segregation ratios for avirulence (A) and virulence (V) in the pathogen that, in terms of gene number, were identical to the number of genes for resistance in the host. Furthermore, when different single genes for resistance in two host cultivars were studied, he was able to distinguish different genes for virulence in the rust fungus. Flor summarized this evidence in the gene-for-gene hypothesis, stating that, "for each gene-conditioning rust reaction in the host there is a specific gene-conditioning pathogenicity in the parasite" (Flor, 1956). Further generalization of the gene-for-gene relationship is possible. The illustration of genotypes in Fig. 10.2 shows only the potential reaction of two host genotypes interacting with either of two parasite genotypes. The potential for a resistant reaction can be attributed to a dominant gene, a recessive gene, or a cytoplasmic factor. In the same manner, the potential for avirulence can be due to a dominant gene, a recessive gene, or perhaps a cytoplasmic factor. A pair of genes are involved—one in the host and one in the parasite. These host/parasite gene pairs are called "corresponding gene pairs."

Refer to the symbols in Fig. 10.2; there are only four potential genotypes and two phenotypes for the host/parasite interaction when only one pair of corresponding genes is considered, i.e., R/A, R/V, S/A, and S/V. Only the R/A combination results in no growth and hence no disease; all other corresponding gene pairs result in growth of the parasite and disease. In some hosts, more than 20 R genes have been identified and many more are probably undetected.

Another type of gene interaction is possible between different sets of corresponding host and parasite genes. Let us say that the host has four genes for resistance designated as R_1, R_2, R_3, and R_4. The pathogen can be either A or V for each of these four genes. When any host R gene is matched with a corresponding A gene in the pathogen, the stop signal is on and the whole system is turned off. This situation is also true when two, three, or four RA combinations occur but only one RA combination is needed to prevent disease. For example $R_1A_1S_2V_2S_3A_3R_4V_4$ will result in no disease because of the R_1A_1 combination. Another feature of the system is that some RA combinations inhibit pathogen growth (or don't favor pathogen growth) more than another RA combination. When both RA combinations occur together in the same system, the greatest inhibition of growth will prevail. In a few instances, two RA combinations will each inhibit pathogen growth to some intermediate, but different, degree when functioning alone in a system. When functioning together in a system, however, they will inhibit pathogen growth to a higher degree. The system becomes more involved, and actually more favorable for the plant breeder, as the number of R genes increases beyond four.

What implications does this gene-for-gene model have for breeding for pest resistance? At least three important areas can be discussed. These include the identification of pathogen biotypes, plant breeding strategy, and host-genetic studies. Because they are interrelated, the areas are treated together in the remaining subsections.

Monitoring Pathogen Biotypes

We have previously mentioned physiological races having different virulence characteristics to a set of differential cultivars. It is evident that the differential cultivars should include the R genes currently used in commercial cultivars within the epidemiology area, as well as those involved in the breeding program or available for use in the breeding program. Green (1965, 1971) working with wheat stem rust in Canada, pioneered in the development of a virulence formula for distinguishing biotypes of the wheat stem rust fungus. He used actual gene designations, but for simplicity let us use the R_1–R_4 symbols. Green's virulence formula for a rust biotype is simply a listing of the effective host genes/the ineffective host genes. The effective host genes provide stop signals to a rust biotype in the host-pathogen system and in the breeder's terms, condition resistance. A rust biotype with the virulence formula R_1R_3/R_2R_4 means that the rust is $A_1V_2A_3V_4$. An inspection of the virulence formula tells the breeder immediately which genes for resistance will be effective and which will be of little value against a specific rust race.

Identifying Appropriate Resistance Genes to Use

The gene-for-gene model also has implications for the strategies used in plant breeding. Obviously, resistant cultivars containing a single R gene have little promise of pest resistance when the correspondent V gene is already present in the area. Furthermore, the use of R_1 alone in a cultivar, when V_1 is believed to be absent, entails a considerable degree of risk that the resistance in the cultivar will be ephemeral because a single mutation in the pest from A_1 to V_1 could occur. The R_1 cultivar would permit the increase of this pathogenic race in the population. However, knowing that the pathogen races having the virulence formula of $R_1R_2R_4/R_3$ (or others having three or four of the R genes functioning) are the only races present in an affected area, then cultivars with $R_1R_2R_4$ genes could be anticipated to remain resistant for quite some time. Why is this resistance to be expected? We said that any one effective R gene serves as a stop signal to prevent disease. With several of these effective stop signals in the host, the only way this resistance can be circumvented by the pathogen would be if it became V at all the loci corresponding to the R genes. It is quite probable that separate mutational events would occur, but less probable that all three mutational events would occur simultaneously in one pathogen-reproductive unit. Recombination of separate mutations is also possible but would require many years. With adequate monitoring of the pathogen population it can be determined when biotypes become prevalent to which only two of the R genes in use are effective. At this time one or two additional R genes can be incorporated into the breeding program.

The gene-for-gene model also has value in host genetics. To identify the genes in a cultivar or resistance source takes considerable time and effort. Several plant generations need to be grown and evaluated for disease or insect reactions. The problem becomes more involved when resistant cultivars have several genes for resistance. Appropriate biotypes of the pathogen must be used to identify them. Some R genes are not expressed if they occur with another R gene that conditions a higher level of resistance. However, a suitable set of pathogen or insect biotypes (each capable of indexing a single corresponding R gene in the host) can be used to test a resistant cultivar having unknown resistance genes. The host genes can be identified without making plant crosses and growing large segregating populations. When the number of cultivars with unknown genes is large, computer models can be used. Of course, they would model only known gene-for-gene interactions. Unknown genes are simply put into a category as being different from the known genes (Loegering et al., 1971).

Use of Specific Resistance

A breeder must be careful when using specific resistance alone (Knott, 1972). Complexity rather than simplicity tends to be the rule to follow. Monogenic specific resistance puts the greatest selection pressure on the pathogen or insect and it is a poor way to use specific resistance. One eventually expects to find a race that is virulent to the resistance. If the pathogen is a "simple-interest" pathogen (the plant is infected only once during the growing season as in the root-infecting wilt fungi) with a slow dispersal rate, the risks are not great. If the resistance is widely used and against a "compound-interest" pathogen (the plant is repeatedly reinfected during the growing season as in the cereal rust fungi) and the pathogen has a rapid dispersal rate, the risks are much greater.

Because some pathogens and insects easily evolve new races to overcome single genes for specific resistance, a number of workers use groups of genes for specific resistance in one cultivar. These multi-gene cultivars seem to be more successful if previously unused genes are involved and if the cultivars receive their inoculum from other cultivars outside of the area. Inheritance studies of cultivars known to have resistance that functions widely throughout the world usually have shown that the cultivars have a combination of several major and minor genes for resistance.

Use of Multiple-Line Specific Resistance

To maximize the effectiveness of genes for specific resistance, some workers have advocated the use of multi-line cultivars (Browning and Frey, 1969; Frey et al., 1973). Multi-lines are composed of a mixture of backcross-derived sublines, each subline containing a different gene for resistance to a pathogen but similar in appearance, maturity, and other respects. The components of the mixture can be changed as new races appear. In addition, since some plants in the mixture are resistant each year, within-field spread of the pathogen is reduced. Thus, disease development on susceptible plants in the mixture is lessened, delayed, or causes little damage. The development of a multi-line is quite expensive, and the final product may be variable in appearance and not improved for yield. The method has been tried for the cereal rusts as a means of disease control, and the results so far are favorable in reducing disease. However, the method has not become popular where agricultural technology is advanced. It seems most suitable to situations in which crop uniformity for quality and other features is less important and the disease problem, such as rust susceptibility, is severe. Multi-lines, as described above, are still uniform, of course, for reaction to other diseases and pests. Some breeders are building in diversity by bulking selections from numerous related crosses that conform in general to a particular cultivar type. When the cultivar type is widely adapted and productive, the system has considerable merit.

Use of General Resistance

Breeders are turning more and more attention to the use of general resistance. General resistance, as its name implies, is resistance against all races of a pathogen or insect and it is usually, but not always, polygenic in inheritance. Each gene exerts a small but additive effect, and high levels of resistance can be developed by selection. Breeders and pathologists have found that a moderate level of general resistance is often adequate to prevent field losses (Nelson, 1973). The most important feature of general resistance is that it does not put the same type and degree of selection pressure on the pathogen as does specific resistance. All biotypes are kept at a low level and little disease develops. The resistance seems to remain effective indefinitely. Whenever possible, breeders should determine whether the resistance they are working with is a specific or a general type. Sometimes the mode of expression is an important indication, whether there are hypersensitive infection points (specific) or a low number of infection points with reduced sporulation (general). Another method of determination is wide area testing, preferably on multi-location basis and over many seasons. A form of resistance that has functioned well for 20 or more years is likely to be a general type. Such resistance has also been referred to as durable resistance (Johnson, 1981) or as horizontal resistance (van der Plank, 1968).

STABILITY OF RESISTANCE IN CROPS

One main difficulty encountered in breeding for resistance to disease and insects is the dynamic nature of the problem (Hooker, 1977). Instabilities of pest resistance can usually be attributed to changes that take place in pests, crop cultivars, or the way crops are grown. Sometimes these changes occur simultaneously, or in a sequence of events, so that the problem becomes acute. Diseases and insects continue to change; they have been described as shifty enemies. Among the changes, those involving the loss of resistance in the crops grown are the most worthy of consideration. Unfortunately for the plant breeder, one disappointment resulting from a severe disease epidemic or pest outbreak on a cultivar or hybrid will rapidly cause that particular genotype to fall from favor, even though it possesses many desirable characteristics.

By far the most common change, resulting in the failure of disease resistance on farms, is in the pathogen, a natural event largely beyond control. Sometimes a new pathogen is introduced. Some pathogens continue to evolve into new races (Watson, 1970). If these races are aggressive and virulent on widely grown cultivars or hybrids, they can increase in prevalence and cause economic losses. The rapid increase and decrease in race prevalence, as with *H. maydis*, is unusual. More often than not, races are present several years before they generate epidemics and they persist for many years later. Races of soil-borne pathogens spread much more slowly than do air-borne pathogens, and breeders have a longer time to cope with new races of soil-borne pathogens.

Where agricultural technology is advanced, plant breeders are continually developing new cultivars and hybrids of crops which are tested for performance before they are released to seed-producing units for multiplication and distribution to farmers. The better cultivars or hybrids soon become widely accepted and grown by farmers. From time to time, breeders and the seed industry may be unaware of the disease susceptibility of a new cultivar. Usually the disease problem is a new one or it has been of only minor importance in the past. An example of this unintentional susceptibility was in oats in the 1940's. Oat breeders in the USA at that time derived good rust and smut resistance from the oat cultivar 'Victoria' and several stiff-straw, high-yielding oat cultivars were released with this cultivar as one of the parents. Unknown to the oat breeders was the fact that extreme susceptibility to another disease, victoria blight, was associated with the resistance to rust. After a few years, the cultivars succumbed to blight and had to be replaced with new cultivars having a different breeding background. Other changes in cultivars are less spectacular in causing fluctuations in disease or insect problems. Sometimes these changes seem to be due to chance when, in fact, new cultivars or hybrids were not fully tested for disease or insect reaction or pest resistance as one of the criteria of the plant breeding program.

Another variable in the stability of resistance in crops being grown can be a change in crop production methods or in climate. This may be less dramatic but does merit attention. Let me give a few illustrations. Typical wheat cultivars respond to high soil-nitrogen fertilization aimed toward greater yield by becoming tall and lodging. Hence, crop production methods could not include large amounts of nitrogen fertilizer. These tall cultivars usually escaped serious damage from mildews and other leaf diseases. Dwarf or short-strawed cultivars, however, respond well to nitrogen fertilizer with high grain yields without lodging. This response, along with short stature, favors leaf diseases. Hence, breeders must now give greater attention to resistance than in the past. Much of the yield increase in maize attributed to genetic improvement is at least partially because modern maize hybrids can be grown with higher populations per unit area of land. This practice also increases the amount of stalk and root rots. Hence, selection of modern hybrids must include both stalk quality and yield ability. Tillage operations that leave much plant refuse on the soil surface are advantageous in certain farming areas. When the same crop is planted, and the previous planting had been infected by leaf-disease pathogens, these diseases can increase unless resistant types are grown. Diseases and some insect pests vary in intensity from season to season because of weather. When the environment is unusually favorable for disease, cultivars or hybrids that normally do not require high resistance to be productive can be injured.

Cultivar Diversification on Farms

When resistance is specific or unknown, cultivar diversification on farms seems desirable. This is genetic diversity in the concept of space; a hedge against hazards known or unknown. The cultivars or hybrids grown on the same farm, or on different farms in the same area, should have different breeding backgrounds for this system to have value.

Zonal Distribution of Resistance Genes or Cytoplasms

For some diseases, the pathogen overwinters in one geographic zone or region and migrates annually along a definite course to another region. In North America, a familiar example is the cereal rust fungi, which overwinter in the South and spread northward each year to the Canadian prairies. Some have dubbed the area from Texas to North Dakota and into Canada as the "*Puccinia* pathway." Specific resistance may be enhanced if different genes for resistance are used at the two ends of the pathway as well as in the middle. By this means, a massive build-up of the rust should not develop. As the rust fungus spreads, it would be confronted with new and different genes for resistance. Virulent biotypes that might increase in one zone are likely to be avirulent in the next. A similar system can be used in different cultivars if the pathogen normally moves from early- to late-maturing cultivars during a single growing season.

Hooker (1972a) suggested that, because *H. maydis*, the cause of southern corn leaf blight, is rarely found in the Corn Belt, a new source of cytoplasm for male-sterility would be less vulnerable to disease in the Corn Belt if a different cytoplasm were used where *H. maydis* is endemic. *Helminthosporium maydis* is a warm-climate pathogen. If a new race virulent to the new cytoplasm did appear, it would not have a suitable susceptible host to allow its multiplication and spread northward into the Corn Belt.

Genetic Uniformity and Vulnerability

The corn blight epidemic of 1970 in the USA prompted a study by a committee of the National Research Council of the National Academy of Sciences on the genetic vulnerability of major crops (National Academy of Sciences, 1972). This report documents the use of germplasm in several major crop groups. The report brings out the fact that relatively few cultivars or parental inbreds of hybrids of any one crop are grown each year. These cultivars and hybrids are genetically and cytoplasmically uniform. Most self-pollinated crop cultivars have their origin from a single seed. Even in cross-pollinated crops such as maize and sorghum, the diversity that was formerly the rule has now been lost through the production of uniform single-cross hybrids. In addition, some parental inbreds of these hybrids are widely used.

Plant breeders, pathologists, and entomologists have long recognized that genetic uniformity of a crop makes it potentially vulnerable to any production hazard whether it be disease, insect, drought, or something else. The danger of cytoplasmic uniformity must now be recognized as well. The cytoplasm of the 1970 maize crop was almost uniformly of the *cms*-T type but, perhaps, no more uniform than the cytoplasms in most other crops. Where cytoplasmic male-sterility is used in hybrid seed production, usually one predominant cytoplasm type is used. This practice holds for sorghum, sugarbeet, and other crops. In soybean and cereal crops, only a few cytoplasms are represented even though the cytoplasms are of the so-called normal type.

Uniformity appears in a high technology agricultural environment (Day, 1973). Economic factors, market and processor demands, farm profits and mechanization, quality standards, seed certification programs and the Plant Variety Protection Act all require or stimulate uniformity. Unless an enormous amount of effort is expended, plant breeders cannot develop cultivars or hybrids that are uniform for a dozen or so important characters and variable in all other respects.

We must remember that crop uniformity has great benefits although risks are involved. As in other decision-making situations, the benefits and risks must be evaluated in relation to each other. Crop uniformity is economically profitable in advanced agricultural technology. Disease epidemics and serious insect infestations are rare. In the USA, profitable yields of good quality are the rule, whereas crop failures are the exceptions. In addition, certain forms of disease and insect resistance benefit from uniformity and there are ways to use specific resistance properly where uniformity is risky. Hence, benefits are gained nearly every year through uniformity.

FUTURE OF PEST RESISTANCE

Future research programs will need to consider ways to: (1) detect new and potentially important disease and insect problems; (2) provide information on the reaction of cultivars, potential cultivars, inbred parents of hybrids, and germplasm resources to the wide range of disease and insect hazards; (3) locate diverse sources of resistance to disease and insect pests; (4) develop easily-applied techniques for measuring pest reaction, including international testing or other means to distinguish specific from general resistance; (5) identify and establish a resource bank for genes and cytoplasms for pest resistance; (6) continue efforts of plant breeding for pest resistance; and (7) continue exploration of optimum ways to deploy genes and cytoplasms in agricultural crops and crop areas.

The plant breeder's first objective in disease and insect control by host resistance is to ensure stability of crop production, that is, prevent the development of destructive epiphytotics and infestations. The second is to reduce, as much as possible, the average disease and insect losses that occur each year. Furthermore, a successful resistant cultivar or hybrid must have resistance to several diseases and insects and must also be satisfactory for yield, quality, and other characters. Unless a disease or insect pest is a limiting factor, pest resistance alone does not improve overall yield and quality. Pest resistance merely keeps a crop healthy, allowing it to achieve its full performance potential (Hooker, 1977).

LITERATURE CITED

Beck, S. D. 1965. Resistance of plants to insects. Annu. Rev. Entomol. 10:207–232.

Biffen, R. H. 1905. Mendel's laws of inheritance and wheat breeding. J. Agric. Sci. 5:4–48.

Browning, J. A., and K. J. Frey. 1969. Multiline cultivars as a means of disease control. Annu. Rev. Phytopathol. 7:355–382.

Dahms, R. G. 1972. Techniques in the evaluation and development of host-plant resistance. J. Environ. Qual. 1:254–259.

Day, P. R. 1973. Genetic variability of crops. Annu. Rev. Phytopathol. 11:293–312.

———. 1974. Genetics of host-parasite interactions. Freeman, San Francisco, Calif.

Flor, H. H. 1956. The complementary genic systems in flax and flax rust. Adv. Genet. 8:29–54.

Frey, K. J., J. A. Browning, and M. D. Simons. 1973. Management of host resistance genes to control diseases. Z. Pflanzenkr. Pflanzenschutz. 80:160–180.

Gallun, R. L. 1972. Genetic interrelationships between host plants and insects. J. Environ. Qual. 1:259–265.

Green, G. J. 1965. Stem rust of wheat, barley and rye in Canada in 1964. Can. Plant Dis. Surv. 45:23–29.

———. 1971. Physiological races of wheat stem rust in Canada from 1919 to 1969. Can. J. Bot. 49:1575–1588.

Hatchett, J. H., and R. L. Gallun. 1970. Genetics of the ability of the Hessian fly, *Mayetiola destructor,* to survive on wheats having different genes for resistance. Annu. Entomol. Soc. Am. 63:1400–1407.

Hooker, A. L. 1967. The genetics and expression of resistance in plants to rusts of the genus *Puccinia.* Annu. Rev. Phytopathol. 5:163–182.

———. 1972a. Southern leaf blight of corn—present status and future prospects. J. Environ. Qual. 1:244–249.

———. 1972b. Breeding and testing for disease reaction. p. 371–372. *In* Encyclopedia of Sci. and Tech. 3rd ed. Vol. 10. McGraw Hill Publications, New York.

———. 1974. Cytoplasmic susceptibility in plant disease. Annu. Rev. Phytopathol. 12:167–179.

———. 1977. A plant pathologist's view of germplasm evaluation and utilization. Crop Sci. 17:689–694.

———. 1979. Estimating disease losses based on the amount of healthy leaf tissue during the plant reproductive period. Genetika 11:181–192.

———, and K. M. S. Saxena. 1971. Genetics of disease resistance in plants. Annu. Rev. Genet. 5:407–424.

Johnson, H. W. 1972. Development of crop resistance to disease and nematodes. J. Environ. Qual. 1:23–27.

Johnson, R. 1981. Durable resistance: definition of, genetic control, and attainment in plant breeding. Phytopathology 71:567–568.

Knott, D. R. 1972. Using race-specific resistance to manage the evolution of plant pathogens. J. Environ. Qual. 1:227–231.

Leppik, E. E. 1970. Gene centers of plants as sources of disease resistance. Annu. Rev. Phytopathol. 8:323–344.

Loegering, W. Q. 1978. Current concepts in interorganismal genetics. Annu. Rev. Phytopathol. 16:309–320.

———, R. A. McIntosh, and C. H. Burton. 1971. Computer analysis of disease data to derive hypothetical genotypes for reaction of host varieties to pathogens. Can. J. Genet. Cytol. 13:742–748.

Maxwell, F. G., J. N. Jenkins, and W. J. Parrott. 1972. Resistance of plants to insects. Adv. Agron. 24:187–265.

———, and P. R. Jennings (ed.). 1980. Breeding plants resistant to insects. John Wiley & Sons, New York.

National Academy of Sciences. 1972. Genetic vulnerability of major crops. Washington, D.C.

Nelson, R. R. (ed.). 1973. Breeding plants for disease resistance, concepts and applications. Penn. State Univ. Press, University Park, Pa.

Painter, R. H. 1951. Insect resistance in crop plants. MacMillan, New York.

———. 1958. Resistance of plants to insects. Annu. Rev. Entomol. 3:267–290.

Parlevliet, J. E. 1979. Components of resistance that reduce the rate of epidemic development. Annu. Rev. Phytopathol. 17:203–222.

Pathak, M. D. 1970. Genetics of plants in pest management. p. 138–157. *In* R. L. Rabb and F. E. Guthrie (ed.) Concepts of pest management. North Carolina State Univ., Raleigh, N.C.

Riley, C. V. 1872. On the cause of the deterioration in some of our native grape vines, etc. Am. Nat. 6:532–544.

Russell, G. E. 1978. Plant breeding for pest and disease resistance. Butterworths, London-Boston.

Schafer, J. F. 1971. Tolerance to plant disease. Annu. Rev. Phytopathol. 9:235–252.

Simons, M. D. 1972. Polygenic resistance to plant disease and its use in breeding resistant cultivars. J. Environ. Qual. 1:232–240.

Sprague, G. F., and R. G. Dahms. 1972. Development of crop resistance to insects. J. Environ. Qual. 1:28–34.

Stakman, E. C., and J. G. Harrar. 1957. Principles of plant pathology. Ronald Press, New York.

Thurston, H. D. 1977. International crop development centers: a pathologist's perspective. Annu. Rev. Phytopathol. 15:223–247.

van der Plank, J. E. 1968. Disease resistance in plants. Academic Press, New York.

Watson, I. A. 1970. Changes in virulence and population shifts in plant pathogens. Annu. Rev. Phytopathol. 8:209–230.

Wolfe, M. S., and E. Schwarzbach. 1975. The use of virulence analysis in cereal mildews. Phytopathol. Z. 82:297–307.

Chapter 11

Breeding for Physiological Traits[1]

D. C. RASMUSSON AND B. G. GENGENBACH

University of Minnesota
St. Paul, Minnesota

As human civilization developed, man began to improve the crops that provided his food. Man's involvement in plant improvement started unknowingly when he grew from seed the plants he depended on for food. As with many ancient practices that have progressed with human civilization, the methods have changed substantially. The rapid changes in plant breeding and crop improvement during the past 50 years closely parallel increased knowledge about genetics and plant development. That yields have not always increased proportionately with increased knowledge about plants underscores the difficulty of translating this knowledge into crop improvement. However, to paraphrase S. W. Johnson's proposal in his 1868 book *How Crops Grow,* one must know how crops grow in order to accomplish effective improvement.

Plant growth is a continuous series of interacting biochemical and physiological processes that are influenced by environmental conditions. The primary value of plant growth is the yield of grain, forage, fruits, vegetables, chemicals, or fiber. Plant breeders historically have emphasized yield and have selected cultivars primarily on the basis of yielding ability in performance trials. Crop improvement also includes the incorporation of resistance to various stresses, such as diseases, insects, and drought into cultivars with high yield potential. This chapter emphasizes those physiological traits of plant growth and development which can be measured or evaluated and which directly or indirectly influence yield. Our definition of physiological traits is broad and includes more plant characteristics than generally are considered to be physiological by plant breeders or plant physiologists.

GENES AND GENE FUNCTIONS

Each plant characteristic and the potential for growth and high yield are determined by the genotype of the plant. Plant improvement involves both gene action and gene inheritance. The action of genes as presented in

[1] This chapter was adapted from another chapter by the same authors, "Genetics and Use of Physiological Variability in Crop Breeding." *In* M. B. Tesar (ed.) *Physiological Basis for Crop Growth and Development.* American Society of Agronomy, Madison, Wisconsin, 1983.

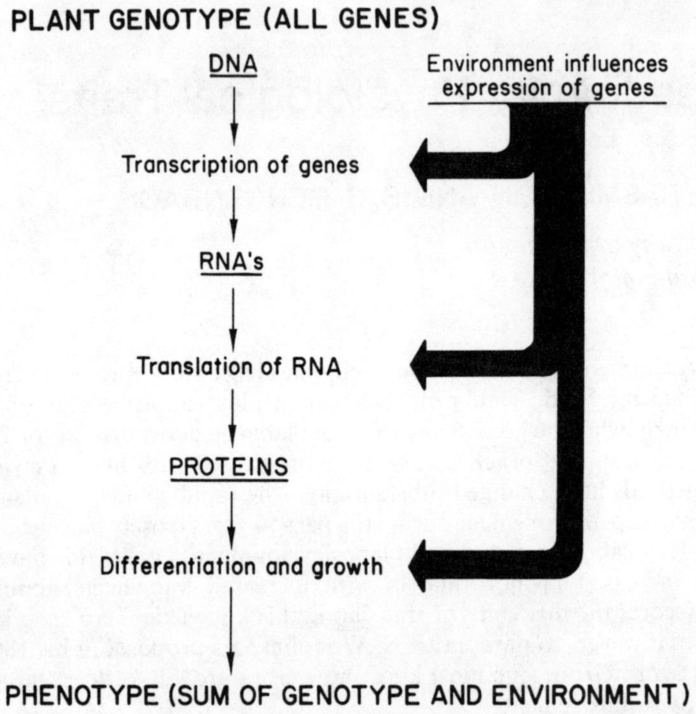

Fig. 11.1. A representation of gene action.

Fig. 11.1 determines every characteristic of a plant, such as its morphology, response to environmental conditions, and yielding ability. Physiological gene action reflects gene differences that provide the basis for selection of desirable genotypes in plant breeding. Efficient recovery and maintenance of desirable genes transmitted from selected plants to their progeny requires knowledge about gene inheritance. Therefore, it is desirable to examine basic aspects of gene function and inheritance and their importance in plant breeding.

Physiological Gene Action

Ability to identify and select plants having genes for desirable traits is important in plant breeding. As shown in Fig. 11.1, physiological processes are under genetic control and are influenced by the environment during the life cycle of a plant. Physiological gene functions determine the manner and extent of the genotypic contribution to the phenotype of the plant.

To understand genetic control of physiological processes we must examine how genes confer specific capabilities to a plant. In general, the deoxyribonucleic acid (DNA) sequence of a gene provides the information for synthesis of an enzyme that enables the plant to carry out a particular metabolic process. Enzymes are proteins that act as biological catalysts to initiate specific changes in one or more cellular metabolites (substrates). An estimated 1,000 to 10,000 enzymes are contained in a typical plant cell with a corresponding number of potential enzyme reactions (Bonner, 1976). Some enzymatic reactions are required periodically while others are required continuously during plant growth and development. All plant processes such as photosynthesis, nutrient metabolism, protein synthesis, and cell division are the result of enzyme actions.

The sequential action of many individual enzymes in biochemical pathways converts raw products, such as CO_2 from the air, into end-products, such as carbohydrates and amino acids, and provides energy for plant growth. This cellular machinery is composed of many small pathways, each synthesizing one or a few specific products as the plant or cell requires them for growth. These products may be used directly to form complex compounds, such as endosperm proteins or starch, or they may become starting materials for synthesis of new products by another biochemical pathway. The interrelationship and interdependency of the pathways in the total plant metabolic system make it difficult, in many cases, to assess the impact on growth caused by changing the activity or the characteristics of one enzyme.

The function of an enzyme is specified by the DNA of the gene that codes for the enzyme. The first step in enzyme (protein) synthesis is the transcription of the DNA of the gene into ribonucleic acid (RNA) molecules (Fig. 11.2). These RNA's are macromolecules containing four ribonucleotide bases: guanine (G), cytosine (C), adenine (A), and uracil (U). In RNA synthesis, the DNA bases G, C, A, and thymine (T) code for the specific complementary RNA bases C, G, U, and A, respectively. Thus the DNA base sequence ACTGAC would code for an RNA segment with the sequence UGACUG.

Several kinds of RNA's are essential for protein synthesis (Fig. 11.3) and all RNA's contain bases copied from the appropriate gene DNA base sequences. Messenger RNA's (mRNA's) contain the specific information necessary for synthesizing proteins; the other two kinds of RNA, ribosomal (rRNA) and transfer (tRNA), assist in this process. Genetic, environmental, physiological, and developmental factors influence RNA synthesis and the subsequent steps leading to protein synthesis. Descriptions of these factors and their interactions are beyond the scope of this chapter, but the reader should know that an understanding of the regulation of physiological processes, an active area of plant research, should help resolve how plant growth responds to genetic and environmental differences.

After synthesis, mRNA moves from the nucleus into the cytoplasm of the cell and is translated into protein (Fig. 11.3). Translation occurs on the ribosomes, which consist of rRNA's and ribosomal proteins. Proteins are large molecules and often contain 75 to 150 amino acids which are linked linearly by covalent peptide bonds. The amino acid sequence of an enzyme is determined by the mRNA base sequence much as the mRNA base sequence is determined by the DNA base sequence. Specific sets of three

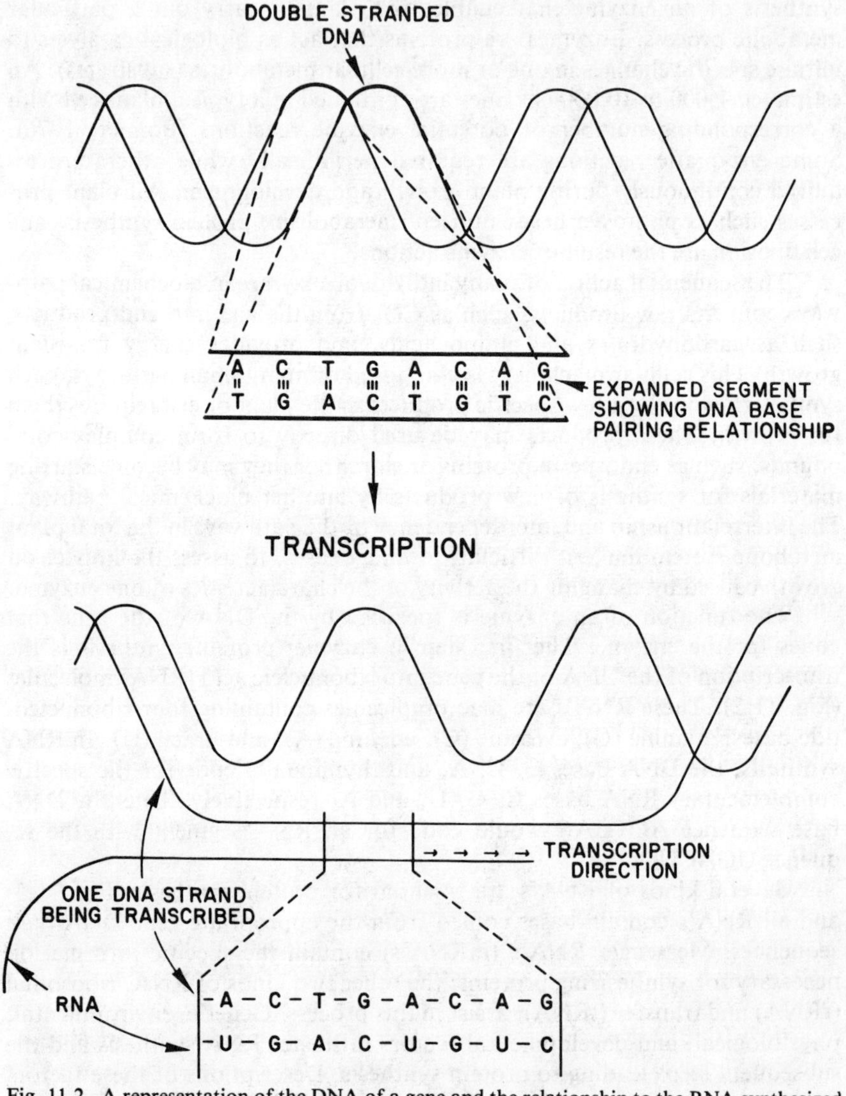

Fig. 11.2. A representation of the DNA of a gene and the relationship to the RNA synthesized during transcription.

bases (codons) in the mRNA are required to specify each amino acid of the enzyme. Each amino acid is specified by a unique codon or set of codons. For instance, the codon UGG specifies the amino acid tryptophan while both GAA and GAG specify glutamic acid.

In cooperation with the mRNA and ribosomes, tRNA's insert the correct amino acid into the protein chain in the order specified by the mRNA codons. There are over 20 different tRNA species, each transferring only one of the 20 different amino acids found in the enzymes. Each tRNA has a specific sequence of three bases (anticodon) that complements one of the codons found in mRNA. The sequential incorporation of amino acids into

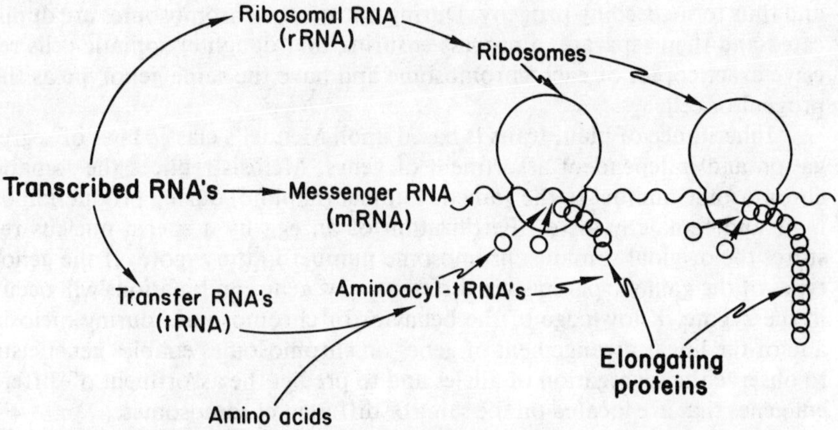

Fig. 11.3. Schematic representation of mRNA translation and protein synthesis.

proteins is determined by mRNA codons pairing with complementary tRNA anticodons. For example, the tryptophan codon is UGG, thus the complementary anticodon is ACC and is found only in the tryptophan tRNA species. The movement of the ribosome along the mRNA coordinates the base pairing of codons and anticodons so that the proper amino acid is attached by a peptide bond to the growing end of the protein. The process is repeated along the entire mRNA molecule, ensuring that each amino acid is inserted in the proper order.

After release of the protein from the ribosomes, interactions between the amino acids determine the final three-dimensional shape of the protein molecule. This shape is determined primarily by the amino acid sequence and specific properties of the different amino acids. The most important part of an enzyme is its active site(s) that reacts with and causes a chemical change in one or more substrates. The shape and charge of the active site(s) determine which substrate is acted upon by an enzyme and how the substrate is changed. Through the reactions they catalyze, enzymes collectively determine total metabolic activity and thereby determine much of the genetic variability in plant traits that the plant breeder needs.

Protein or enzyme synthesis is much more complicated than the above simplified discussion indicates. The purpose is to point out how genes control the synthesis and activity of enzymes. In summary, the information in the DNA base sequence of a gene determines the sequence of amino acids in a protein and consequently its enzymatic activity. The genotype of a plant determines the total enzymatic machinery and thus has a role in determining plant growth.

Gene Inheritance

Gene inheritance is the transmission of genetic information to succeeding generations. This transmission involves replication and distribution of chromosomes both to daughter cells in mitosis and to gametes in meiosis and thus to succeeding progeny. During mitosis the chromosomes are duplicated and then separate, a process ensuring that daughter somatic cells receive exact copies of each chromosome and have the same genotype as the progenitor cell.

Inheritance of plant traits is based upon Mendel's classic laws of segregation and independent assortment of genes. Meiosis reduces the somatic chromosome number to the gametic number (haploid) during production of male and female gametes. Fertilization of an egg by a sperm nucleus restores the original somatic chromosome number in the zygote. If the genotypes of the gametic parents are different, new gene combinations will occur in the zygote. Knowledge of the behavior of chromosomes during meiosis and of the linear arrangement of genes on chromosomes enables geneticists to observe the segregation of alleles and to predict the assortment of different genes that are located on the same or different chromosomes.

The enormous number of plant enzymes is controlled by an equally large number of genes. Since there are many genes and only a few chromosomes in a plant cell, it follows that many genes must be located on the same chromosome. If two genes are located close together on the same chromosome they are said to be linked. Linkage reduces the chance for recombination of the genes during production of gametes. Recombination of linked genes results from exchange of DNA between homologous chromosomes.

An allele is one of two or more alternate forms of a gene occupying the same locus on a particular chromosome. A diploid plant has only two alleles for the same gene, one allele on each chromosome of a pair of chromosomes. In contrast, a population of plants collectively can have many alleles for the same gene. Alleles represent mutations (changes) in the DNA that have been preserved in populations by transmission to succeeding progeny. Many alleles are possible at each locus because a mutation of one base in a gene potentially produces a new allele. Mutations within a gene can cause the substitution of a new amino acid for the original in the protein. The location and type of amino acid substitution determines the effect of the mutation on the protein, such as an altered enzyme activity.

The function of an enzyme in plant growth determines how readily different alleles for that enzyme can be detected. Different alleles can have beneficial or deleterious effects on plant growth and on traits of interest to the plant breeder. Plant breeders not only want to eliminate undesirable alleles from a population, but they also want to increase the frequency of the most desirable alleles. To select desirable alleles effectively, a plant breeder knows or considers the factors discussed in the following sections.

BREEDING FOR PHYSIOLOGICAL TRAITS

The breeder must consider how the particular trait is inherited, that is, whether one, few, or many genes are involved. Plant phenotypes for individual traits, such as plant height, can be described either in qualitative terms, such as tall and short, or in quantitative units, such as centimeters. A trait described in qualitative terms usually is readily separated into two or more distinct and discontinuous classes. Distinct classes arise because of the major effects of one or very few genes. Variation caused by environment is not sufficient to mask the genetic contribution to the trait.

Leaf size and leaf angle can be qualitatively inherited. In soybeans, a gene identified as *na* conditions narrow leaves. A cross between the normal and a narrow-leaf stock gives a normal F_1 and F_2 segregation of 3 normal:1 narrow. A gene called liguleless, symbol *li*, conditions presence or absence of a ligule at the base of the leaf in barley. Another interesting effect of *li* is that it causes leaves to be more erect than those of normal barley. By selecting plants with or without ligules in an F_2 population, plants can be obtained with normal or erect leaves, respectively.

The *B* gene locus determines whether coumarin is formed by sweetclover tissue (Schaeffer et al., 1960). The trait cannot be evaluated by visual observation of the plant, but distinct phenotypes can be identified by chemical procedures. Enzyme assays show that the *B* allele codes for an enzyme that catalyzes the final step in the production of coumarin, whereas the *b* allele codes for a protein that has no detectable enzymatic activity for this reaction. Production of a precursor of coumarin early in the pathway also is controlled by one gene (Goplen et al., 1957). The *Cu* allele at this locus causes high precursor synthesis while the *cu* allele results in much less precursor synthesis. Combining both recessive alleles into one homozygous genotype, *cucubb,* reduces coumarin synthesis at each of the two steps. The resulting low-coumarin type is desired because coumarin reduces the feeding value of sweetclover forage. This example illustrates the important principle that careful study of a character may reveal distinctive phenotypes conditioned by single genes with major effects. Without careful study or appropriate techniques, the character may appear to be quantitative and controlled by many genes.

The majority of physiological characters are quantitatively inherited. The varied expressions of these characters are continuous and cannot be fitted into discrete classes. Yield and primary physiological processes such as photosynthesis, respiration, translocation, and transpiration are quantitative characters. Such characters are often conditioned by many genes with small individual effects and often there is a sizable environmental effect. Because variation is continuous, identifying different genotypes on the basis of their phenotypes often is difficult. For example, most variation in plant height can be attributed to genetic control when the plants are grown under carefully controlled conditions in a phytotron. However, when the same genotypes are grown under variable conditions in the field, a major portion of plant-to-plant variability may be due to environmental causes.

Choice of Physiological Traits

In years past, plant breeders have emphasized breeding for yield per se and for reducing the hazards of production. Breeding for yield has focused on yield trial performance and the most productive entries have become cultivars. Disease and insect resistance, tolerance to weather conditions, and resistance to shattering and lodging reduce the hazards of production and often have been very beneficial in obtaining and maintaining high yields.

The decision to undertake a breeding program to increase yield by breeding for a physiological trait is a very important one for the plant breeder. A breeding effort on a new trait generally will require a commitment of resources for several years; a single breeding cycle will seldom lead to a new cultivar. The current cultivars in many crops are the culmination of 30 to 50 years of breeding effort in which numerous gene combinations already have been tried and utilized if beneficial or have been discarded. Initiation of a breeding program for physiological traits may require a shift of resources from existing successful programs to programs where success is less certain. Genes for desired traits are often found in exotic stocks that are poorly adapted and low yielding in the geographical area served by the breeding program. Thus, in crosses to obtain desirable genes from an exotic source, the breeder may introduce genes for other traits that reduce overall performance of the progeny. This problem is compounded by linkage, which can make it difficult to obtain the desired genes free of the deleterious genes.

Whether selection for a physiological trait such as leaf angle or photosynthesis will increase yield is the major question a breeder wants to answer before proceeding with a breeding program on a physiological trait. A definite affirmative answer is seldom available from existing data, but information on which to base a decision can be obtained from several sources. Such a decision often is best made through the collaboration of plant physiologists and plant breeders. Together they can identify traits of metabolic and developmental importance, methods for their evaluation, and the most appropriate breeding approach. Available evidence must be examined carefully. A published report may have indicated, for example, that an erect leaf trait increased yield under certain conditions. However, varying the conditions such as the genetic background or geographic location may affect the value of erect leaves. Leaf angle may be of limited importance in regions where sunlight is intense or in cultivars with an especially small leaf area. Often the necessary information will have to be obtained for the specific breeding situation.

The contribution of a character to yield can be assessed with specially developed stocks called isolines. Isolines are alike genetically for nearly all genes except for those controlling the character being investigated. If one line of an isoline pair yields more than the other, the character by which they differ will usually be important in determining yield. California researchers (Qualset et al., 1965) determined the effect of awn length on the yield of barley by using four backcross-derived isolines each having a different awn length. Yields, kernel weights, kernel numbers, and spike numbers per plant for the four isolines are given in Table 11.1. Fully-awned and half-awned plants yielded more than quarter-awned and awnless plants. The data showed that longer awns contributed to greater kernel weight and thus to

Table 11.1. Mean performance of awn-length isogenic lines of Atlas barley.†

	Genotype			
Character	Full-awned	Half-awned	Quarter-awned	Awnless
Yield, g/plant	41.4	43.8	38.6	35.6
Kernel weight, mg	39.9	36.5	33.8	32.7
Kernel no./spike	57.8	61.8	60.8	57.8
Spike no./plant	25.8	26.2	26.3	26.9

† Qualset et al. (1965).

higher yields. Therefore, awns are desirable under California growing conditions and barley breeders there select plants with long awns.

When isolines are not available, a useful alternative is to make comparisons among a large number of lines that have contrasting phenotypes for the trait. If sufficient lines that differ for the trait are not available, they can be derived by crossing parents chosen for their differences and then selecting lines with contrasting phenotypes. The researcher hopes that the lines selected for comparison have approximately equal merit in all other genes, except those controlling the character under investigation. A pitfall in this approach is that a set of lines or progenies from a cross may have dozens of genes that distinguish one line from another and these can affect performance. For example, disease reaction, standing ability, maturity, or photoperiod response can be more important in conditioning differences in yield than the trait being investigated. It is especially risky to draw conclusions about the role of a trait in determining yield if the lines in the experiment are not adapted to the area in which they are being tested.

Another way to determine whether a trait influences yield is to observe what happens to yield when environmental factors such as light, temperature, moisture, or CO_2 are varied. For example, if yields are increased by increasing the incident light within a crop canopy, modification of leaf angle might be worthwhile. Another approach is to physically change the plant trait from its normal status and to observe the effect on yield. Altering leaf position, removing leaves at different stages of growth and removing parts of the inflorescence are examples of this type of study. Such studies frequently are done in the laboratory and the treatments often impose conditions on the plants that alter normal development. Hence the results must always be interpreted with care.

ACCOMPLISHMENTS IN BREEDING FOR PHYSIOLOGICAL TRAITS

Plant breeders have devoted considerable attention to breeding for yield. It is also true, however, that plant breeders have for decades sought to improve yield by manipulating what we call physiological components of yield. In this section we give two examples of improvements in crop plants that have been made because of selection of desirable plant types.

Rice Breeding

The development of short-stature rice cultivars in the Philippines began in the 1960's following observations that short-strawed cultivars yielded more than the taller, leafier cultivars when given nitrogen fertilizer (Jennings, 1964). A number of characters were associated with generally higher yields and the nitrogen response. The responsive selections were short and had many tillers and good resistance to lodging. Their leaves were short, thick, relatively narrow, erect, dark-green, and remained functional until shortly before harvest. Subsequent genetic studies revealed that a single gene controlled many of the leaf characters as well as height and tillering. Two additional characters, early maturity and high floret fertility, though not controlled by the above-mentioned gene, were incorporated into the genetic package that contributed to high yield under nitrogen fertilization.

In breeding for new plant types at the International Rice Research Institute, well-adapted tropical cultivars were crossed with introduced cultivars of either japonica or indica types that served as sources of the desirable new traits. Observation of populations resulting from these crosses indicated that the desirable, short, less leafy plants were not competitive and hence were being eliminated from the breeding populations. The importance of this association is demonstrated by the data in Table 11.2. The short, less leafy plants competed very poorly in the mixture with tall plants, but in pure stands the short plants yielded more than the tall. Many of the short plants with the combination of useful characteristics described above have been highly productive. One cultivar, IR-8, gained world-wide recognition because of its very high yields, especially in response to nitrogen fertilization.

Semidwarf Wheat

In 1935 a wheat selection of hybrid origin, 'Norin 10', was released for use in Japan. Norin 10 was one-half to two-thirds as tall as common wheat cultivars and had more heads. In the USA, Norin 10 was not useful as a cul-

Table 11.2. Changes in percent composition in a mixture of five rice cultivars and relative yields of the five cultivars in pure stands.†

Cultivar	Description	Survival (%) 1963	1964	1965	1966	Grain yield tons/ha
TN1	Short-statured, small erect leaves	20	3.7	0.2	0.04	6.46
Ch. 242	Similar to TN1	20	2.1	0.06	0.0	5.53
M6	Tall, leafy, few tillers	20	1.7	0.02	0.0	4.99
MTU	Tall, leafy, many tillers	20	25.7	9.3	5.7	3.53
BJ	Similar to MTU	20	66.9	90.5	94.3	3.11

† Jennings and Jose de Jesus, Jr. (1968).

tivar, but rather as a parent in crosses. O. A. Vogel, a USDA plant breeder at Pullman, Wash., crossed Norin 10 with ordinary wheats such as 'Brevor'. From these crosses came the first highly productive semidwarf wheats in the USA. The first new semidwarf, 'Gaines', and a closely related semidwarf 'Nugaines', soon achieved great popularity and other wheat breeders began to concentrate on semidwarf wheat.

As many as a dozen characters potentially contribute directly to high yield in short stature rice. However, less is known about the physiological basis for high productivity in the semidwarf wheats. Semidwarf wheats are more resistant to lodging than their taller counterparts, but lodging resistance is not the entire reason for the yield difference because semidwarf cultivars are often superior to tall ones even when lodging does not occur. Many of the semidwarf wheats have more heads per unit area than taller wheats and a higher ratio of grain to straw (higher economic to biological yield). The leaves of many productive semidwarf wheats are similar to the droopy, wide, and long leaves of taller wheats. Consequently, the productive semidwarf wheats and short stature rices differ greatly in leaf characteristics. Whatever the basis for the higher yields in the new rice and wheat cultivars, their superiority suggests that crops can be made more productive by modifying their general form.

OPPORTUNITIES IN BREEDING FOR PHYSIOLOGICAL TRAITS

Accumulation of fundamental information about how plants grow, improved techniques and instrumentation, recognition of the opportunities in interdisciplinary research, and continuing interest in higher yields all enhance the potential for crop improvement through breeding for physiological components of yield. In this section we describe physiology-genetics-breeding research that has been done on several physiological components of yield. Emphasis will be on traits that appear to be important in determining yield, on their genetic variability and inheritance, and on evidence indicating their relationship to yield.

Photosynthesis

Photosynthetic Rate

Because of the fundamental relationship between photosynthesis (Ps) and yield, there is considerable interest in enhancing Ps capacity through breeding. Research has concentrated both on direct measurements of Ps rates and on characters, such as leaf area index, leaf angle, leaf orientation, and stomatal frequency, which affect light utilization and CO_2 entry into the plant and Ps.

Cultivar differences in Ps rates have been reported in many crops, including maize, soybeans, wheat, barley, ryegrass, sugarcane, and red

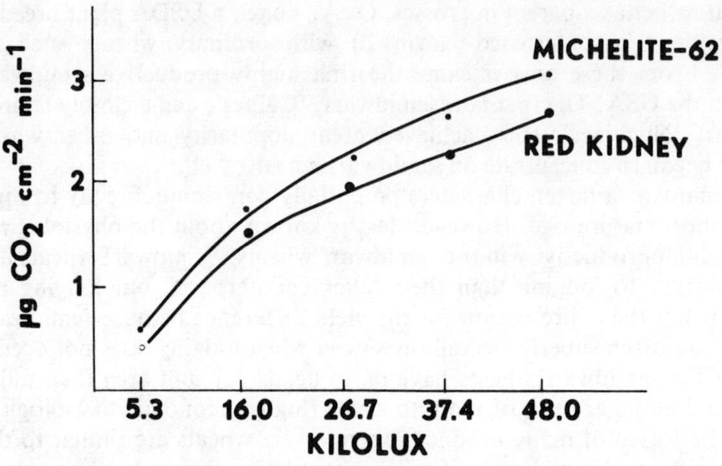

Fig. 11.4. Net CO_2 exchange at five light intensities for Red Kidney and Michelite-62 cultivars of field bean (Izhar and Wallace, 1967).

kidney beans (Wallace et al., 1972). An example of the differences observed is given in Fig. 11.4. The figure shows net CO_2 exchange of two bean cultivars, 'Michelite-62' and 'Red Kidney,' at five light intensities. The consistent difference shows that selection in a breeding program should be possible over a wide range of intensities at least in genetic stocks like Michelite-62 and Red Kidney. Izhar and Wallace (1967) concluded that the genetic mechanism controlling the difference in net CO_2 exchange rate is quantitative, although relatively few genes may be involved.

Heritability estimates for Ps are available. In an early study Wilson and Cooper (1969) crossed six strains of ryegrass and evaluated their progeny. They obtained relatively high estimates of heritability, thus encouraging those interested in breeding for higher Ps rates. Even so, the opportunities for direct improvement of Ps seem limited at present. Measurements of Ps on a single leaf at one stage of growth may not be indicative of Ps rates throughout the life of a plant or of Ps in a field canopy. Replicated Ps measurements on crop canopies in the field usually are impractical for routine breeding purposes. However, a limited number of genotypes might be evaluated in the laboratory or field to identify parents to use in crosses. When such parents are used, genes favorable to high rates of Ps might be fixed by chance in their progenies without selection for Ps.

Leaf Angle

In many plant species, individual leaves become light-saturated at relatively low light intensities. Examples of such species are barley, wheat, oats, soybeans, tobacco, and sugar beet. However, in field canopies, leaves are shaded so they are not often light-saturated. In these situations changes in the canopy architecture could reduce shading and increase canopy Ps. In

Table 11.3. Grain yields of maize plants with normal and upright leaves.†

Comparisons	Yield, kg/ha
Genetic isolines of hybrid C103 × Hy	
Normal leaf	6,202
Upright leaf	8,769
Mechanical manipulation of leaf angle of Pioneer 3306	
Normal (untreated)	10,683
All leaves positioned upright	11,386
Leaves above ear positioned upright	12,202

† Pendleton et al. (1968).

other crops, such as maize and sorghum and most tropical grasses, neither individual leaves nor field canopies are light-saturated at normal light intensities. These circumstances appear to afford a good opportunity to increase Ps by breeding for increased light penetration. Modification of leaf angle and leaf area may potentially increase canopy light penetration.

Genetic variation in leaf angle is common in most crop species. In barley and maize, upright leaf angle can be conditioned by a single gene. In rice, the gene that causes short stature also conditions upright leaf angle. On the other hand, upright leaf angle can be quantitatively inherited and controlled by several genes. Breeding is more difficult when several genes are involved. Some studies indicate, however, that heritability of upright leaves is adequate for breeding progress. In barley, heritability of leaf angle was as high as 40% on the F_2 plant basis and 60% on the F_3 family basis (Barker, 1970).

Since breeding for upright leaf angle is feasible, the crucial question becomes whether it is worthwhile. The success of the short-stature, erect-leaf cultivars of rice discussed above and the results of Pendleton et al. (1968), who studied the relationship of leaf angle and grain yield of maize in the field, may suggest that breeding is worthwhile. These workers used genetic isolines in one experiment. One member of the isoline pair was homozygous for the recessive lg_2 allele and had upright leaf angle; the other line was homozygous for the Lg_2 allele and had normal leaves. In a companion experiment, leaves of a high yielding hybrid were mechanically positioned to obtain three leaf-angle orientations. Data in Table 11.3 show that upright leaf-angle plants had a yield advantage in these two experiments. In the comparison of isogenic lines, erect-leaf plants had a yield advantage of 2,567 kg/ha, which represents an increase of 41% over the normal line. In the mechanical manipulation experiment, yields were increased significantly either when all leaves or when only leaves above the ear were positioned upright. Other investigators working on maize, however, have reported no yield advantage due to upright leaves. It is likely that the value of breeding for upright leaves to influence yield depends on genotype, crop, and environment.

Leaf orientation for some plants is determined by the direction and intensity of the light source. Individual leaflets of soybeans and field beans bend at their point of attachment to the petiole thus changing their orientation in response to direction and intensity of light. On bright days the upper leaves of some soybean cultivars are oriented to reduce the light intercepted to about 50% of the potential, whereas the leaflets assume a more horizontal position on overcast days. Breeding to obtain optimum leaflet orientation might be a worthwhile objective in some crops.

Leaf Area

Genetic variation in leaf area and in its primary components, leaf number and size, has been reported for many crops. In grasses, including cereals, genetic variation is larger for leaf size than for leaf number. In legumes such as alfalfa, soybeans, and field beans, variation is large for both leaf size and number. Genetic studies indicate that inheritance of leaf size may be either qualitative, as discussed previously for soybeans, or quantitative. In studies of several populations of barley, Fowler and Rasmusson (1969) found that heritability of leaf area on an individual plant basis ranged from 18 to 73%, suggesting that selection would be effective in barley populations.

The effect of leaf area on yield in soybeans was studied by Hicks et al. (1969) using isolines having the normal (*Na*) or narrow leaflet (*na*) character. More light penetrated into the narrow leaflet canopies than into the canopies of the normal types, but there was no difference in yield between the two leaflet types. The relationship between yield and leaf area is complex and it is not now possible to predict how changes in leaf area will affect yield.

Stomatal Frequency

Genetic variation exists for number, size, and time of opening and closing of stomates. These factors influence entry into the plant of the CO_2 essential to photosynthesis. The results of Miskin and Rasmusson (1970) are typical of data reported for stomatal frequency and size. Stomatal frequency on the lower surface of the flag leaves of 649 cultivars from the world collection of barley ranged from 36 to 98 stomates/mm^2, with a mean of 64/mm^2 (Fig. 11.5). Among cultivars examined for stomatal size, average guard cell length varied from 41 to 56 μ. Miskin et al. (1972) crossed cultivars that differed in stomatal frequency and obtained inheritance information in the F_2 and F_3 generations. In all populations the estimates of heritability exceeded 25% and one estimate was 74%.

Water Utilization

Many plant breeders are concerned with increasing yields in environments where water is a significant limiting factor. As one would expect, much attention has been given to breeding for drought resistance. Much less breeding effort has been devoted to increasing water use efficiency in crops, that is, attaining higher yields from a given amount of water.

Fig. 11.5. Photomicrographs of the lower surface of leaves of barley cultivars with (top) low stomatal frequencies (Miskin and Rasmusson, 1970) and (bottom) high stomatal frequencies. The low line had 39 stomates per mm^2 and the high line 96 ($\times 150$).

Table 11.4. Dry matter produced, water used, and water efficiency of temperate and tropical grasses (at 25°C).†

Species	Dry matter (DM) (gm dm^{-1} $week^{-1}$)	Water used (WU) (gm dm^{-2} $week^{-1}$)	DM/WU $\times 10^4$
Temperate grasses			
Oats	0.62	380	16
Ryegrass	0.67	485	14
Wheat	1.04	507	21
Avg	0.78	457	17
Tropical grasses			
Proso millet	0.63	171	37
Kikuyu grass	0.87	298	29
Sudangrass	0.60	219	27
Avg	0.70	229	31

† Downes (1969).

Water use efficiency can be improved by using less water to produce a constant yield or by increasing the yield from a constant amount of water. Breeding emphasis could be placed on efficiency, total water requirement, or on total dry matter production. In high rainfall climates and under irrigation, total yield would be more important than efficiency or water requirement, whereas the relative importance might be reversed under dry conditions.

Two approaches can be used in attempting to modify water use efficiency, direct selection for efficiency or selection for traits that affect efficiency. The direct approach seems possible because of the research showing that plant species differ greatly in efficiency. Results from one experiment that are given in Table 11.4 show that temperate and tropical grasses produced about the same amount of dry matter. However, the tropical grasses required much less water per unit of leaf area and hence were almost twice as efficient as the temperate grasses. The superior efficiency of the tropical grasses suggests that the first step in a breeding program should be to select among species, assuming that various species meet requirements for the area. Differences in water use efficiency within a species are smaller than differences between species, but seem to be large enough to justify selection. Water use efficiency of six clones of blue panicgrass ranged from 371 to 575 g of water required to produce 1 g of forage (Dobrenz et al., 1969).

The alternative to direct selection for water use efficiency is to breed for traits that affect efficient use. Rooting traits such as depth and degree of branching might affect efficiency and qualify for a breeding effort. Other plant traits might include stomatal frequency and size, duration of various growth stages, presence or absence of awns, and ratio of economic to biological yield.

In considering ways to alter a plant's ability to control water loss, attention is drawn to the critical control that stomates exert over water movement through plants. The amount of water loss through the cuticle is unknown for many plants, but stomates are more important in controlling

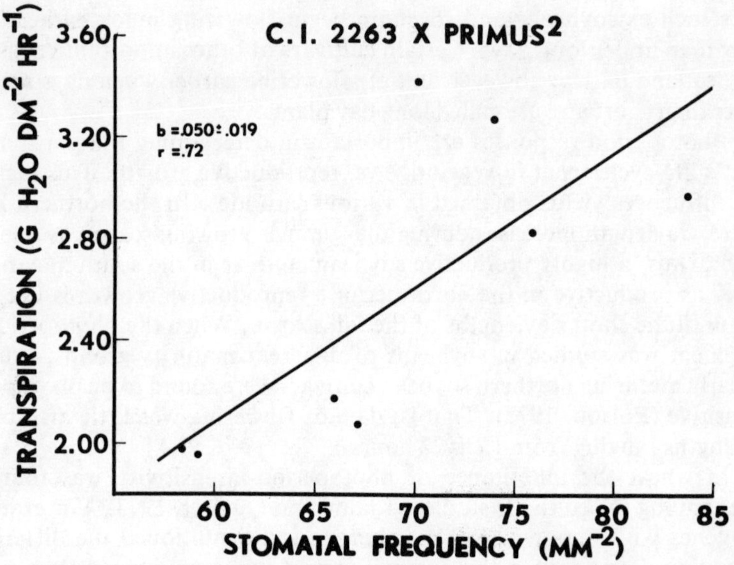

Fig. 11.6. Relationship between stomatal frequency and transpiration in eight barley back-cross lines (Miskin et al., 1972).

water loss. A complication in breeding for optimum stomate frequency, size, or time and degree of opening and closing is that stomates are the major site of CO_2 entry into the plant. If selection is for stomates that are small, infrequent, or that only open partially for short periods of time, not only might water loss be reduced, but water use efficiency also could be reduced if CO_2 entry into the plant were slowed and Ps rates were reduced. Available evidence, however, indicates that stomate frequency exerts greater control over water loss than over CO_2 entry.

Evidence obtained in barley by Miskin et al. (1972) indicates that transpiration can be reduced by lowering stomatal frequency. As illustrated in Fig. 11.6, four lines having few stomates transpired less water than four lines having more stomates. The eight barley lines were obtained by crossing two cultivars differing in stomate frequency and then selecting the extreme high and low stomate lines. Additional research is needed to learn how modifying stomatal frequency through breeding will affect efficiency of water use.

Photoperiod Response

Many crops are affected by the light-dark cycle, or photoperiod regime, in which they are grown. The duration of darkness is the determining factor in photoperiodic responses, although the effect is usually described in terms of daylength. An important photoperiod response is the difference in the time required for flower initiation under various daylengths. Short day crops such as soybeans and sorghum begin flowering much earlier under short than under long days. Certain cultivars of other important crops such as wheat and barley, however, initiate flowering earlier when days are long rather than short and are called long day plants.

Photoperiod responses are important in determining the portion of a plant's life cycle spent in vegetative vs. reproductive growth. This factor in turn influences yields obtained in various latitudes. In the northern hemisphere, daylength increases during the summer growing season as one goes north. Thus, a highly productive soybean cultivar in the south may not be nearly as productive in the north because reproductive growth is not initiated until the short daylengths of the fall season. When the photoperiod requirement was studied in soybeans of different maturity groups, some of the early maturing northern soybean cultivars were found to be photoperiod insensitive (Polson, 1972). That is, date of flowering was little affected by daylengths ranging from 15 to 22 hours.

In wheat the inheritance of photoperiod-insensitivity was found to differ among the cultivars studied (Klaimi and Qualset, 1973). For example, two genes with major effects on photoperiod conditioned the differences in heading date between two insensitive and two sensitive cultivars grown under short daylengths. Insensitivity was conditioned by a dominant gene in one cultivar and by a homozygous recessive gene in the other. Other genes with minor effects also influenced the response. Under short days insensitive cultivars headed within a 7-day interval, whereas sensitive cultivars headed over a 40-day interval. This finding indicates that environmental effects on different genotypes can differ in magnitude—a situation with which breeders often must work.

Four genes that affect maturity and response to long daylengths have been described in sorghum by Quinby (1972, 1973). The recessive allele of each gene conditions a degree of insensitivity to longer daylengths as shown for several genotypes in Table 11.5. Under short days (10 hour), flower primordia were initiated at the same time for the four genotypes, but genotypes homozygous for one or more of the recessive alleles initiated flower primordia much earlier than the dominant genotype under long days (14 hour).

Genetic variability for photoperiod response provides one basis for changing the environmental adaptive range of many crops. Cultivars can be selected for the photoperiod response to match a particular environment and genes for photoperiod insensitivity can be used to increase a cultivar's range of adaptation.

Table 11.5. Responses of sorghum genotypes to photoperiod.†

	Days to floral initiation	
Genotype	10-hour days	14-hour days
ma_1 Ma_2 ma_3 Ma_4	19	35
Ma_1 ma_2 ma_3 Ma_4	19	38
Ma_1 ma_2 Ma_3 Ma_4	19	44
Ma_1 Ma_2 Ma_3 Ma_4	19	70

† Quinby (1972).

Mineral Nutrition

The availability of minerals can greatly affect plant growth. This effect can range from increased yield as the result of increasing the supply of a limiting essential element to reduced growth resulting from a toxic level. Examples of genetic differences for mineral toxicity are found in several crops. For instance, the wheat cultivar 'Atlas 66' is more tolerant to an acidic, Al-toxic soil than 'Thatcher' (Foy et al., 1965, 1967). Tolerance of Atlas 66 might be attributable to its ability to maintain a higher pH in the root zone. Inheritance of tolerance can be controlled by a single gene. For example, a single dominant gene controls Al tolerance in certain barley populations and results in higher grain yields in acid, Al-toxic soil (Reid, 1970). Genetic differences in ability to grow under conditions of mineral deficiency also are known for many crops. Efficient utilization of Fe in soybeans is conditioned by a single gene that affects Fe uptake by the roots (Weiss, 1943). These examples illustrate that crop plants can be bred to overcome a specific soil-regulated mineral toxicity or deficiency rather than having to change the soil environment.

The accumulation of minerals by crops is of interest because of the possible relationship between yield and accumulation. Limited studies suggest that genetic differences can account for differences in mineral accumulation in some crops. Heritability estimates for variation in P, K, Ca, and Mg accumulation in F_3 wheat families were sufficiently high to expect progress from selection (Rasmusson et al., 1971). No relationship was found, however, between ability to accumulate these minerals and yield.

The metabolism of nitrogen, an essential element, is quite important because nitrogen is incorporated into the structure of many cellular constituents, primarily proteins and nucleic acids. In plants that do not fix nitrogen symbiotically, nitrogen is taken into the plant through the roots as NO_3^- which is then reduced to NH_4^+ before incorporation into carbon compounds. The initial step in this conversion process is carried out by the enzyme nitrate reductase (NR), which reduces NO_3^- to NO_2^-. Because nitrate reductase occupies a rate-limiting position in N metabolism it could be important in determining growth and yield potential.

Extensive work on nitrate reductase by Hageman et al. (1967) has centered on determining the genetic control of nitrate reductase activity (NRA) and its relation to grain and protein yields of several crops. Their work with maize indicates a wide range of NRA among different inbred lines. Crosses between inbreds generally resulted in F_1's with NRA values intermediate between the parental values. However, a significantly higher NRA was obtained in the F_1 of a cross between two inbreds with low NRA. The higher NRA in the F_1 was attributed to two genes, one affecting the rate of NR synthesis and the other affecting the rate of NR inactivation. When NRA was measured throughout the growing season and expressed as total seasonal NRA or as input of reduced N, positive relationships with grain protein and yield were obtained. However, to be useful in selection work, a single sampling date is desired. When cultivar rankings obtained from seasonal NRA were compared with the rankings obtained from one sampling at the post-anthesis stage, the same positive correlation was obtained. This indicates that NRA assays from a single sampling date might be useful in selecting genotypes capable of utilizing the higher NRA for increased grain and grain protein yields.

The work with nitrate reductase provides an opportunity to point out two requirements that should be considered when attempting to use enzyme activity as a selection criterion in breeding work. First, a simple assay should be available and a sampling system should be determined which accurately reflects genetic differences and not just environmental or seasonal effects. Second, the characteristics of the enzyme should be known. For instance, NO_3^- within the plant tissue is necessary before nitrate reductase is synthesized by the plant.

Ideotype

The success of semidwarf wheat and short-stature rice, which are new plant types, underscores the opportunities in breeding for optimum ideotypes. Theoretical reasons why modifications in plant type should increase grain yields have been advanced by many researchers, including Donald (1968), who urged that plant breeding be extended to include breeding of model plants. These model plants can be specified in terms of a few important characters or in considerable detail involving numerous characters. Table 11.6 contains a listing of several characters considered in model breeding in barley, oats, rice, and wheat. A rationale has been presented for breeding for optimum expression of characters that influence important processes such as photosynthesis, transpiration, transport, or efficiency of production. In this section we will illustrate how yields might be increased through breeding for optimum expression of plant type characters.

Kernels Per Head

In cereals as well as in some other crops, grain yield is the product of kernels per head, kernel weight, and number of heads per unit area. Breeding for these individual components would not be advantageous if the gains derived from genetically increasing one component were offset by a simultaneous reduction in one or both of the others. Although the evidence is conflicting, it appears that a large sink or storage capacity provided by many kernels per head can be advantageous in obtaining high yields. Kernels per head is an attractive breeding objective because it is relatively easy and inexpensive to measure, and genetic variation is available or easily obtained. Kernels per head is a quantitative character, but heritability is high enough that good progress in increasing kernels per head can be expected in a breeding program.

Table 11.6. Some characters with potential for ideotype research and breeding programs.

Leaf characters	Culm characters	Inflorescence characters
Leaf size	Number of culms	Awn length
Leaf angle	Survival of culms	Awn number
Number of leaves	Diameter of culms	Kernel number
Duration of leaves	Number of ears	Kernel weight
Thickness of leaves	Vascular bundles	
Specific leaf weight		
Type of canopy	Root characters	Other
Height of plants	Volume	Photoperiod response
Harvest index	Depth	Length of growth stages
Angle of ear		

Tiller Number

Tiller number potentially can be used to increase yield because a change in tiller number modifies the leaf area or photosynthetic source, and a change in head number modifies the sink capacity of the plant. Some interesting evidence indicates that a crop community of single-tiller plants should maximize production owing to the absence of internal competition between developing heads and young tillers. This point is supported by new high-yielding oat cultivars in Great Britain that have fewer panicles than other cultivars. In contrast, new high-yielding barley cultivars in Great Britain have more heads than the other cultivars. Survival of tillers is important in determining whether more or fewer heads per plant will maximize yield. Simmons et al. (1982) have described a barley genotype that has high tiller number and low tiller mortality. Breeding for low mortality as well as higher or lower tiller number may be beneficial.

Harvest Index

Harvest index is the ratio of economic yield (grain weight in cereals) to biological yield (total plant dry weight). Part of the genetic improvement in yield of several crops is derived from a higher percentage of the biological yield being partitioned into plant parts comprising economic yield. Economic yields can be increased by increasing biological yields without changing the harvest index or by partitioning more of the dry matter production into economic yield. Wallace et al. (1972) noted that the harvest index had increased from 32% for wheat cultivars grown in the early 1900's to 49% for current high-yielding semidwarf cultivars.

These researchers recommended that a breeding program regularly consider harvest index as well as economic and biological yield. Progress in breeding for high harvest index should be possible based on known heritabilities for both. The challenge is to find the balance between biological yield and harvest index that maximizes economic yield. Biological yield should be considered in a breeding program because it is related to relative growth rate, net carbon exchange, and other factors that should be kept high. Data on both types of yield can be obtained from routine trials.

Leaf Vein Frequency

In 1968 it was reported that accumulation of assimilate in an illuminated leaf may reduce the net photosynthesis of that leaf. We know that assimilate accumulation and photosynthesis frequently are negatively correlated, but the physiological explanation still is not clear. One way to minimize assimilate build-up in the leaf may be to increase the number and size of the vascular bundles (veins) that translocate photosynthate from the leaf. Two-fold differences in leaf vein frequency (2.4 to 4.5 veins/mm) have been observed among cultivars of barley (Hanson and Rasmusson, 1975). Heritability for leaf vein frequency was 20% in the F_2 generation and 45% in the F_3 generation. The essential requirements for genetic alteration—genetic variation and a reasonably high heritability—therefore exist in barley.

SUMMARY

In this chapter we have described the genetic, physiological, and breeding basis for using physiological plant traits to improve crops. We have considered a physiological trait to be any measurable characteristic involved in plant growth and development. Several points should be emphasized.

1. Past selection of improved cultivars has been accomplished mainly by direct selection for yielding ability. Through this approach it is certain that physiological traits with direct effects on yield have been improved concurrently with yield. The emphasis is reversed when selecting for altered physiological traits. For this approach to be worthwhile, however, the change in a physiological trait must result in a yield increase and it must be less costly than direct breeding for yield.

2. The extent to which differences in physiological traits are expressed, and are measurable, depends partly on the genetic potential of the plant. This potential is manifested through the interrelationships among genes, enzymes, and plant growth. A gene contributes the information for biosynthesis of an enzyme that functions in a particular metabolic reaction. The combined effects of many genes, through their control of enzymes, result in physiological traits contributing to plant growth, development, and yield.

3. Variation in a physiological trait has both genetic and environmental origins. A plant breeder works with the genetic component to change the genetic potential of the crop for a particular trait; environmental variation hinders this effort. Adequate genetic variability and heritability both are required to obtain genetic changes by breeding.

4. Through breeding, the expression of many physiological traits can be changed. But, because time and resources are limited in breeding programs, there should be experimental evidence or at least a theoretical basis for expecting a positive effect on yield or performance before a program is started. The expression of a physiological trait that maximizes its contribu-

tion to yield rarely has been determined and no plant trait has a single optimum expression encompassing all possible combinations of environment and genotype. The optimum expression of many traits will differ from crop to crop, cultivar to cultivar within a crop, and from one geographical area to another.

5. Plant scientists—geneticists, breeders, physiologists, and others—should work together to obtain a better understanding of physiological traits and their relationship to yield so that productivity of crop plants can continue to increase.

LITERATURE CITED

Barker, R. E. 1970. Leaf angle inheritance and relationships in barley. M.S. Thesis. Univ. of Minnesota, St. Paul.

Bonner, James. 1976. Cell and subcell. p. 3–13. *In* J. Bonner and J. E. Varner (ed.) Plant biochemistry. Academic Press, Inc., New York.

Dobrenz, A. K., L. N. Wright, A. B. Humphrey, M. A. Massengale, and W. R. Kneebone. 1969. Stomate density and its relationship to water-use efficiency of blue panicgrass (*Panicum antidotale* Retz.). Crop Sci. 9:354–357.

Donald, C. M. 1968. The breeding of crop ideotypes. Euphytica 17:385–403.

Downes, R. W. 1969. Differences in transpiration rates between tropical and temperate grasses under controlled conditions. Planta 88:261–273.

Fowler, C. W., and D. C. Rasmusson. 1969. Leaf area relationships and inheritance in barley. Crop Sci. 9:727–731.

Foy, C. D., W. H. Armiger, L. W. Briggle, and D. A. Reid. 1965. Differential aluminum tolerance of wheat and barley varieties in acid soils. Agron. J. 57: 413–417.

———, A. L. Fleming, G. R. Burns, and W. H. Armiger. 1967. Characterization of differential aluminum tolerance among varieties of wheat and barley. Agron. J. 31:513–521.

Goplen, B. P., J. E. R. Greenshields, and H. Baenziger. 1957. The inheritance of coumarin in sweetclover. Can. J. Bot. 35:583–593.

Hageman, R. H., E. R. Leng, and J. W. Dudley. 1967. A biochemical approach to corn breeding. Adv. Agron. 19:45–84.

Hanson, J. C., and D. C. Rasmusson. 1975. Leaf vein frequency in barley. Crop Sci. 15:248–251.

Hicks, D. R., J. W. Pendleton, R. L. Bernard, and T. J. Johnston. 1969. Response of soybean plant types to planting patterns. Agron. J. 61:290–293.

Izhar, S., and D. H. Wallace. 1967. Studies of the physiological basis for yield differences. III. Genetic variation in photosynthetic efficiency of *Phaseolus vulgaris* L. Crop Sci. 7:457–460.

Jennings, P. R. 1964. Plant type as a rice breeding objective. Crop Sci. 4:13–15.

———, and Jose de Jesus, Jr. 1968. Studies on competition in rice. I. Competition in mixtures of varieties. Evolution 22:119–124.

Johnson, S. W. (ed.) 1868. How crops grow. Orange Judd & Co., New York.

Klaimi, Y. Y., and C. O. Qualset. 1973. Genetics of heading time in wheat (*Triticum aestivum* L.) I. The inheritance of photoperiodic response. Genetics. 74:139–156.

Miskin, K. E., and D. C. Rasmusson. 1970. Frequency and distribution of stomata in barley. Crop Sci. 10:575–578.

———, D. C. Rasmusson, and D. N. Moss. 1972. Inheritance and physiological effects of stomatal frequency in barley. Crop Sci. 12:780–783.

Pendleton, J. W., G. E. Smith, S. W. Winter, and T. J. Johnston. 1968. Field investigations of the relationships of leaf angle in corn (*Zea mays* L.) to grain yield and apparent photosynthesis. Agron. J. 60:422–424.

Polson, D. E. 1972. Day-neutrality in soybeans. Crop Sci. 12:773–776.

Qualset, C. O., C. W. Schaller, and J. C. Williams. 1965. Performance of isogenic lines of barley as influenced by awn length, linkage blocks and environment. Crop Sci. 5:489–494.

Quinby, J. R. 1972. Influence of maturity genes on plant growth in sorghum. Crop Sci. 12:490–492.

———. 1973. The genetic control of flowering and growth in sorghum. Adv. Agron. 25:126–162.

Rasmusson, D. C., A. J. Hester, G. N. Fick, and I. Byrne. 1971. Breeding for mineral content in wheat and barley. Crop Sci. 11:623–626.

Reid, D. A. 1970. Genetic control of reaction to aluminum in winter barley. p. 409–413. *In* Barley genetics II. 2nd Proc. Int. Barley Genetics Symp., Pullman, Wash. 6–11 July 1969. Wash. State Univ. Press, Pullman.

Schaeffer, G. W., F. A. Haskins, and H. J. Gorz. 1960. Genetic control of coumarin biosynthesis and β-glucosidose activity in *Melilotus alba*. Biochem. Biophys. Res. Commun. 3:268–271.

Simmons, S. R., D. C. Rasmusson, and J. V. Wiersma. 1982. Tillering in barley: genotype, row spacing, and seeding rate effects. Crop Sci. 22:801–805.

Wallace, D. H., J. L. Ozbun, and H. M. Munger. 1972. Physiological genetics of crop yield. Adv. Agron. 24:97–146.

Weiss, M. G. 1943. Inheritance and physiology of efficiency in iron utilization in soybeans. Genetics 28:253–268.

Wilson, D., and J. P. Cooper. 1969. Diallel analysis of photosynthetic rate and related leaf characters among genotypes of *Lolium perenne*. Heredity 24:633–649.

Chapter 12

Breeding for Improved Nutritional Quality of Crops

D. D. HARPSTEAD

Michigan State University
East Lansing, Michigan

The ancient Biblical writer was well aware of his environment and his needs when he wrote, ''All flesh is grass'' (Isaiah 40:6). People have always been directly dependent upon the growing plant as food for consumption and feed for animals. Even before people cultivated plants, they gathered plants selectively to meet their needs.

For thousands of years the only tools available to measure the quality of food were trial, error, and observation. Nevertheless, tremendous progress was made in selecting high quality food. Most of the crops cultivated today remain nearly the same, in terms of nutritional quality, as those selected many centuries ago.

In the last two decades increasing world population pressures have focused attention on the nutritional inadequacy of the diets of millions of people. Special concerns exist when nearly all of the daily food intake comes principally from plant sources. When social conditions prevail that restrict diets to one or a few food sources, such diets invariably consist of low cost products, usually a cereal grain or a root crop. While these are not in themselves inferior food sources, when consumed alone they may fail to provide either the quantity or quality of some of the essential dietary components.

Quality means different things to different people. The descriptive terms used to define quality vary from those used at the time of food production until it is finally consumed. Common terms are used to denote appearance, storage quality, processing quality and, in a few cases, nutritional worth.

Nearly all of the investigations of quality in our crops have been upon those factors related to processing quality—milling and baking of wheat, canning of beans, chipping of potatoes, malting of barley, and fermenting of grapes. All of these processes are highly specialized. Special laboratories have been built by federal and state governments and private industry to provide technical support to plant breeders in their quest to improve the quality of crops.

Market quality results from the complex interactions of social, economic, and biological factors that are highly specific to individual crops. A book in this series, *Crop Quality, Storage, and Utilization,* should be consulted by students interested in these quality factors in specific crops.

Using the term malnutrition to describe the complex problems of inadequate food sources is a gross oversimplification of the subject. This has been clearly pointed out by Altschul (1976). Protein-calorie malnutrition best describes the relationship between these two major dietary components. It is not possible to consider one essential component of the diet as being more important than another; however, for many vulnerable groups such as children and pregnant and lactating women, protein may become the most critical food item.

This chapter will be devoted primarily to plant breeding for nutritional components in several of the crops widely grown for human food or animal feed for non-ruminants. Special attention is directed to the protein component. To include nutritional characteristics in our definition of crop quality is to acknowledge the growing concern for the world food supply and the desperate situation of millions of people who do not have adequate diets either in terms of quantity or quality.

A BALANCED DIET FOR PEOPLE

Let me digress for a moment to conceptualize a balanced diet. It is virtually impossible to outline an ideal diet. The amount of food required by a person is determined by age, sex, body size, health, activity, environment, and food source. The principal components of food and also the principal sources of energy are carbohydrates, fat, and protein. Other components of lesser magnitude, none of which supply energy, include water, cellulose (fiber), minerals, and vitamins. Each of these components plays its unique role in balanced nutrition (Fig. 12.1).

In a typical diet in the USA, 50 to 60% of the energy may be provided by carbohydrates, 10 to 15% by protein, and 35 to 40% by fat. To provide a 2500 kilocalories/day diet, slightly more than 500 g of actual food on a moisture-free basis would be required. The components of such a diet are illustrated in Fig. 12.2.

After the energy requirements of the body have been met, protein becomes the next major component in the diet. Proteins are the most expensive component and are subjected to a great variation in actual nutritional quality. In addition, the digestion and utilization of protein is frequently interfered with by chemicals found in some foods, usually referred to as antimetabolites or toxicants.

Proteins are giant molecules consisting of thousands of atoms. These atoms are grouped into amino acids, which are the building blocks or subunits of the protein molecule.

Twenty-three amino acids are generally recognized. These, when linked together in unique patterns and frequencies, give the almost limitless number of proteins found in nature. Eight of the 23 amino acids are recognized as essential amino acids inasmuch as they must be present in adequate

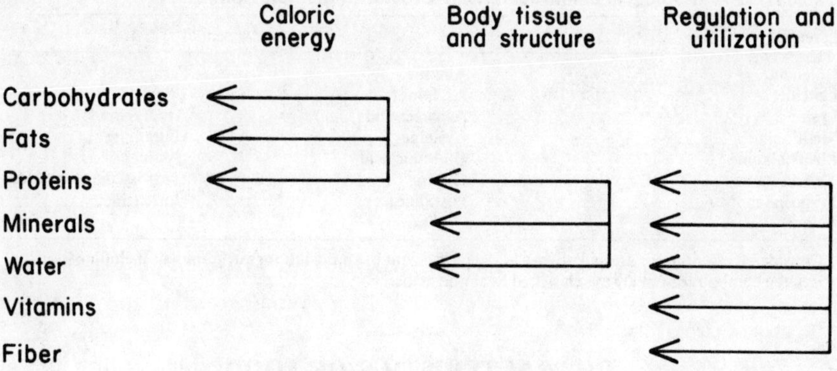

Fig. 12.1. Role of food components in nutrition.

amounts in the proteins of the diet of monogastric (single-stomach) animals, including humans, for normal protein utilization to occur. The ruminant animal also requires these amino acids, but these will be synthesized by bacteria in the rumen during the digestive process (Table 12.1).

A balanced protein is one in which the essential amino acids are present and available for the digestive process in the proportions necessary for proper growth and tissue maintenance. If any one of the essential amino acids is absent or in short supply, the utilization efficiency of the entire protein may be reduced.

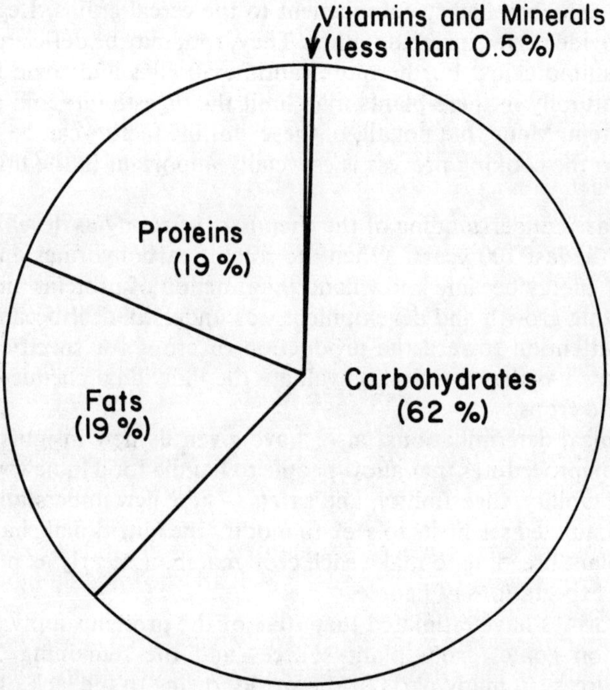

Fig. 12.2. Components of 2500-kilocalorie diet represented on a weight-of-component basis.

Table 12.1. A grouping of amino acids relative to nutritional function.

Essential	Non-essential	
Isoleucine	Alanine	Serine
Leucine	Arginine†	Tryosine
Lysine	Aspartic acid	Asparagine
Methionine	Cysteine	Glutamine
Phenylalanine	Glutamic acid	Cystine‡
Threonine	Glycine	Hydroxyglutamic acid‡
Tryptophan	Histidine†	Norleucine‡
Valine	Proline	

† Considered essential for normal growth and development in young monogastric animals including humans.
‡ Related compounds not always classified as amino acids.

CHARACTERISTICS OF FOOD

Food, as it is normally eaten, consists of a mixture of the necessary dietary components. When all the diet for a monogastric animal comes from plants, the protein fractions become a special concern. The protein content of the edible portions of plants is highly variable both in concentration and nutritional quality. Most cereal grains are relatively low in total protein. Nutrition may be further complicated by deficiencies in several of the essential amino acids required for normal protein nutrition in the nonruminant animal. These two characteristics of cereal proteins become a double-barreled malnutrition threat when a diet is based largely on cereal grains. Legume seeds usually have higher concentrations of protein and serve as a valuable dietary complement to the cereal grains. Legume seeds do not provide perfect proteins either. They, too, may be deficient in certain essential amino acids. Furthermore, antimetabolites and toxic factors occurring naturally in these plants may limit the digestibility and absorption of the protein. Many, but not all, of these limiting factors can be inactivated by heat, so the cooking process is especially important in the utilization of legumes.

Increased understanding of the chemistry of food has developed rapidly during the last 100 years. When the role of carbohydrates and lipids as sources of energy became known and the function of proteins and minerals in supporting growth and development was understood, it became possible to direct attention toward the production of crops for specific purposes. Science now has the capacity to evaluate the individual chemical components of food crops.

Chemical determinations, also, have given us new insights into food preparation procedures that allow people to handle food in new ways to improve palatability, digestibility, and safety. These new understandings have further stimulated scientists to seek to modify the nutritional characteristics through plant breeding to make each crop match, as nearly as possible, the needs and expectations of people.

Authorities have estimated that 70% of the protein supply for human consumption comes from plant sources and the remaining 30% from animal sources. In many parts of the world, owing to the lack of food production and population pressure, animal proteins are not available to most consumers. The plant proteins consumed usually come from few sources, predominantly one or two cereal grains. For this reason it is appropriate to direct our attention first to these crops.

EARLY RESEARCH

A crop selection experiment, destined to become a classical work, was established in 1896 at the Illinois Agricultural Experiment Station to select maize cultivars with high protein and high oil content (Hopkins, 1899). While success was achieved in selecting for both increased and decreased protein and oil content in maize, the work did not produce the anticipated commercially applicable results.

A number of reasons can be cited for the lack of commercial success. Yields of grain began to decline because of inbreeding depression, a genetic effect not understood at the time. Secondly, the poor nutritional characteristics of the increased protein was discouraging.

The early work of Osborne and Mendel (1914) provided an explanation for the poor protein quality and an appreciation for the nutritional characteristics of the proteins in our basic food crops. Osborne (1924) and his colleagues pioneered the fractionation and classification of proteins through solubility characteristics. They were able to show that proteins from various plant sources differed greatly in their chemical makeup. They also established that certain protein fractions lacked sufficient amounts of the essential amino acids to support normal growth and development of laboratory animals.

The cereal grains, maize in particular, became immediate subjects for further investigation. By feeding test animals individual protein fractions, Osborne's work demonstrated that the fraction which made up nearly half of the endosperm protein in maize would not support growth and development in weanling rats. Osborne and his co-workers were further able to demonstrate that this fraction was zein, the prolamin or alcohol-soluble portion. Zein, when supplemented with small amounts of the known essential amino acids, lysine and tryptophan, supported nearly normal growth in rats.

This work pointed an accusing finger at cereal proteins in general, since most of them were known to contain large amounts of the prolamin fraction in their total protein content. Notable exceptions were oats and rice. The superior nutritional quality of these cereal grain proteins could be demonstrated in animal-feeding experiments.

Subsequent investigations revealed that simple selection for increased protein content in maize resulted in a disproportionate increase in zein and did not reflect an increase in nutritive values of the grain when fed to monogastric animals. Showalter and Carr (1922) explained the results of the earlier Illinois selection experiment by showing that the balance of the essential amino acids had not been improved but had, in fact, become less favorable for the monogastric animal.

The work of Osborne also included investigations of the protein fractions of legume and oil seeds. In contrast to the cereals, the predominant proteins were the globulins, which were soluble in neutral saline solutions. In these studies, even legume proteins, which could supply abundant amounts of both lysine and tryptophan, would not support normal growth and development in rats. Work by Waterman and Johns (1921) with a legume seed protein, globulin, demonstrated that this protein, after heat treatment to inactivate antimetabolic factors followed by supplementation with the sulfur-bearing amino acids, could serve as an acceptable protein source in rat diets. We now know that the first limiting essential amino acid in most legume seed proteins is the sulfur-bearing amino acid, methionine.

In the relatively short period of 20 years, Osborne and his colleagues laid the foundation for seed protein investigation, work that is still useful today.

The next three decades saw the development of several new plant breeding efforts designed to increase the nutritional value of food grains through the genetic control of protein synthesis. Frey et al. (1949) and Frey (1951), reported that selection for increased protein content in maize, while successful, resulted principally in an increase in the zein fraction. Work in other cereal crops closely paralleled the results in maize. Total protein content was shown to be a heritable trait, but increased protein did not reflect an increase in essential amino acid content or in greater biological value.

MODIFIED CEREAL PROTEINS

The first significant breakthrough in the modification of nutritional quality of cereals came with the discovery of the effects of the *opaque-2* gene in maize by Mertz et al. (1964). Their work was initiated to investigate the effect of known, recessive, mutant genes which affected the development of the endosperm portion of the maize kernel. The proteins in the endosperm are the lowest in quality of the maize seed proteins and represent more than 80% of the total. The genes Mertz studied had been classified many years earlier by their observable (phenotypic) expressions and were well known by these characteristics.

The discovery of increased lysine content in maize was found initially in kernels that were homozygous for the recessive genes *opaque-2* or *floury-2*. The names of these genes are descriptive terms assigned to their phenotypic expressions. Each of these genes produced a soft, non-vitreous endosperm in the mature kernel when present in the homozygous recessive condition. The normal, dominant counterparts of these genes produced the well known flinty, vitreous endosperm of the mature kernel.

The two genes have slightly different patterns of expression inasmuch as *opaque-2* is expressed as a complete recessive while *floury-2* illustrates a dosage effect. Since the endosperm of maize results from double fertilization, each endosperm cell contains three sets of chromosomes (3n). The various endosperm combinations possible are as follows:

	Normal Protein, Vitreous Endosperm			High Lysine Protein, Soft Endosperm	
Opaque-2 loci	$O_2O_2O_2$	$O_2O_2o_2$	$O_2o_2o_2$	$o_2o_2o_2$	
Floury-2 loci	$Fl_2Fl_2Fl_2$	$Fl_2Fl_2fl_2$		$Fl_2fl_2fl_2$	$fl_2fl_2fl_2$

The protein fractions of kernel endosperms which were homozygous-recessive for these genes were significantly altered as shown in Table 12.2 (Jimenez, 1966).

Lysine determinations made on the endosperm protein of normal kernels show that this amino acid accounts for about 1.6 to 2.0% of the total. In high lysine maize 3.5 to 4.0% of the total endosperm protein is represented by lysine.

The modern amino acid analyzer added a new dimension to this work and made possible the rapid, direct measurement of the quantity of the various amino acids in the protein laid down when the mutant genes were present. Results came rapidly.

The incorporation of the recessive *opaque-2* gene into normal maize increased the lysine content 69% and tryptophan content 100%. The discovery of this genetic effect and later the effect of the *floury-2* gene opened the way for an intensive effort to test the modified maize as a protein source in animal and human nutrition.

Laboratory animal experiments confirmed the superiority of the modified maize as a protein source in diets. Mertz (1966) and Pickett (1966) found that rats and young swine grew three to four times faster on the high-lysine maize than on normal maize. (High lysine is a more accurate descriptive term than *opaque-2* for maize modified to reflect the effects of the *opaque-2* or other mutant genes.) In swine the high-lysine maize diet required only half as much feed per pound of grain as the normal maize diet. Maner (1975) found that even while the high-lysine maize supplied protein at suboptimal levels, swine made healthy growth, although at a reduced rate, on these diets. On the other hand, a parallel diet based on normal maize resulted in seriously malnourished, weakened animals that died after about 100 days.

Harpstead (1971) summarized the initial work leading to the development and testing of commercial high-lysine maize. All of the preliminary

Table 12.2. Soluble protein content for normal, *opaque-2* and *floury-2* maize endosperms.†

Protein fraction	Units	Normal	*Opaque-2*	*Floury-2*
Water soluble	mg	39.6	150.9	160.7
	%	(3.7)	(15.7)	(14.6)
Soluble in a saline solution	mg	17.9	44.8	48.7
(5% NaCl)	%	(1.7)	(4.4)	(4.4)
Alcohol soluble	mg	581.0	260.8	315.5
	%	(54.2)	(25.4)	(28.7)
Soluble in a weak base	mg	432.1	569.8	576.3
(0.2% NaOH)	%	(40.4)	(55.5)	(52.3)

† Percent of soluble protein in endosperm adapted from Jimenez (1966).

biological testing associated with these developments indicated that high-lysine maize was an excellent protein source for the monogastric animal. The economic consequences of these developments are important since proteins are in increasingly short supply in many parts of the world. Protein has always been a high cost commodity in any balanced diet.

The potential impact of an improved maize protein goes far beyond its use as an animal feed. Maize is a principal dietary staple for millions of people in the world. For many, during periods of famine or economic stress, it may be the only food available. The use of high-lysine maize in human diets was summarized by Pradilla et al. (1969, 1972). Not only did the modified maize protein provide a superior diet base for the adult and the normal child, but it could also be used as the sole protein source in the treatment of hospitalized, malnourished children. Diets based on the improved maize protein effected a normal recovery of these critically ill children. Inasmuch as this protein could be used to treat the victims of protein deficiency, its value in food to prevent malnutrition was clearly established.

The actual effect of the *opaque-2* or *floury-2* gene introduced into normal maize has been the subject of extensive investigation. The increased lysine and tryptophan content of the protein is the most readily detected chemical change, while the soft kernel texture is the most common physical manifestation of these recessive genes. There are, however, other known changes which should not escape our attention. A third nutritionally important amino acid, leucine, is decreased when the *opaque-2* gene is present. A favorable ratio between isoleucine and leucine could also contribute to the nutritional enhancement of the protein. The ratios of the various protein fractions are altered. The alcohol-soluble zein fraction is reduced from approximately 50% of the protein in normal maize to 25% in *opaque-2* maize. Since the total protein content in the two types is not changed significantly, the decreases in the zein fractions are compensated for by increases in the saline- and alkali-soluble fractions. Evidence indicates that the alkali-soluble fraction contains significantly more lysine in *opaque-2* maize than the same fraction from normal maize.

Unanswered questions still remain. The association of the higher lysine and tryptophan content with the soft kernel characteristic has caused numerous production problems. This mystery would not be so great if all of the soft kernel types were also high in lysine, but they are not. Recent plant-breeding work has shown that hard kernel types with a nearly normal, flinty endosperm can be selected from within the typical *opaque-2* maize population. These selected populations retain significantly increased levels of the essential amino acids—lysine and tryptophan. The modified maize selections have the advantage of the flint-like kernel type and retain a superior protein in terms of nutritional quality. Extensive research is in progress using other maize endosperm mutants in combination with the *opaque-2* and other protein-modifying genes. It appears technically possible to make further advances in the modification of maize protein by using a combination of these genes (Mertz et al., 1965).

The work with the high-lysine genes has stimulated selection work designed to increase the lysine content in normal-type maize kernels. Zuber and Helm (1975) reported results that should make possible the selection of improved lysine and protein content of normal maize by recurrent selection systems of breeding. The successes achieved in the genetic modification of maize protein sparked renewed interest in the modification of proteins of other cereal grains.

Close on the heels of the discovery of modified protein in maize, Hagberg and Karlsson (1969) reported a strain of barley with significantly higher protein and lysine contents than normal barley. This strain became known as 'Hiproly' and was immediately subjected to intensive investigation. Munck (1972) presented an in-depth review of the effects of this recessive gene in barley. Hiproly had a significant increase in lysine-rich protein fractions of the endosperm with only a small decrease in the lysine-poor protein fractions.

Ingversen et al. (1973) isolated an induced mutant strain of barley, 1508, which almost duplicated the chemical modification found in maize.

Extensive protein quality research in wheat has been reported by Johnson et al. (1969). Historically, protein content among commonly grown wheat cultivars ranged from 7 to 18%. The ability of the plant breeder to select cultivars with increased or decreased total protein was well known. The major barrier to rapid progress in nutritional improvement in wheat was that lysine represented a decreased percentage of the protein in the selected line. Since lysine was the first limiting amino acid in wheat protein (Hegsted et al., 1954), the increased protein content did not necessarily result in an increased biological value of the protein.

Johnson and Mattern (1975) reported results in wheat research specifically designed to select both increased protein and lysine content. Several experimental cultivars were identified and tested. While the increases in lysine were positive, they were small, on the order of 0.5%. The most promising result in the initial research was the isolation of a genetic source which increased the protein content without decreasing the lysine content. This was found in 'Atlas 66' and several closely related cultivars. Selections made from crosses with Atlas 66 and locally adapted cultivars were tested in Nebraska in 1970 (Table 12.3).

A high protein, hard red winter wheat cultivar, 'Lancota', derived from Atlas 66 crosses and developed in Nebraska, was jointly released for

Table 12.3. Yield and protein content of grain of Atlas 66—derived lines grown at three Nebraska locations in 1970.†

Cultivar or selection	Yield $Mg \cdot ha^{-1}$	Percent protein	Protein as percent of the check
Scout 66 (check)	3.4	11.8	100
NB 701132	3.9	14.4	123
NB 701134	3.6	14.3	122
NB 701154	3.5	14.5	124

† Johnson and Mattern, 1975.

commercial production in Nebraska, Kansas, Texas and South Dakota in 1975 (Schmidt et al., 1979). In addition to superior agronomic traits, Lancota had the genetic potential for an increase of 1 to 2 percentage points in high quality protein over the previously adapted commercial cultivars.

Another promising source of germplasm from India was identified. Johnson et al. (1972) reported that 'Nap Hal', in addition to having high protein, consistently had more lysine per unit of protein. The increased lysine content ranged between 0.3 and 0.4 percentage points.

The inheritance patterns of the various nutritional traits in wheat were investigated by making crosses among cultivars which had high protein content, high protein and high lysine content, and high lysine content. When these crosses and their progeny were tested over a range of environments, plant breeders were able to establish the genetic transmission and stability of expression of these qualities. The work was slow and arduous since the selection of the desired genotype could only be identified through chemical analysis. This process was further complicated because varying environments masked or falsely enhanced the phenotypic quality in a particular season. Genotype-by-environment interaction was especially serious when there were small differences between phenotypes. Konzak (1977) provided a comprehensive review of the genetic factors that control the proteins of wheat.

Sorghum has also attracted research efforts to improve its nutritional usefulness. This work is especially significant since the biological value of the protein in normal sorghum is inferior to maize and most other cereal grains. The prolamin or alcohol-soluble fraction represented as much as 60% of the total and, as in maize, this fraction was almost devoid of lysine and tryptophan. The biological value of the protein in sorghum is also known to be influenced by other seed constituents, notably the tannins.

Initial work to isolate sorghum genotypes with an improved protein failed to uncover a genetic effect which would parallel the effects of the *opaque-2* gene in maize. Later reports by Singh and Axtell (1973) identified two sorghum lines from Ethiopia which produced lysine values similar to those found in the opaque-2 maize. The biological value of the selected high lysine sorghum lines was demonstrated by Singh and Axtell (1973) in rat feeding experiments (Fig. 12.3). Axtell et al. (1974) reported the discovery of an induced mutant line in sorghum that produced even more lysine per unit of protein than the lines tested earlier.

The enhancement of the nutritional quality of rice and oats presented unique opportunities to the plant breeder. In rice and oats the prolamin fraction of the protein accounts for only 8 and 12%, respectively, of the total protein, which is very low compared to other cereals (Mosse, 1966). Other protein fractions with more abundant lysine content make up the balance of the proteins. In spite of the higher lysine content, it is still the first limiting amino acid when these crops are the sole protein source in rat diets (Howe et al., 1965).

Rice suffers from a second serious nutritional limitation. The total protein content of normal rice is low when compared to other cereals, con-

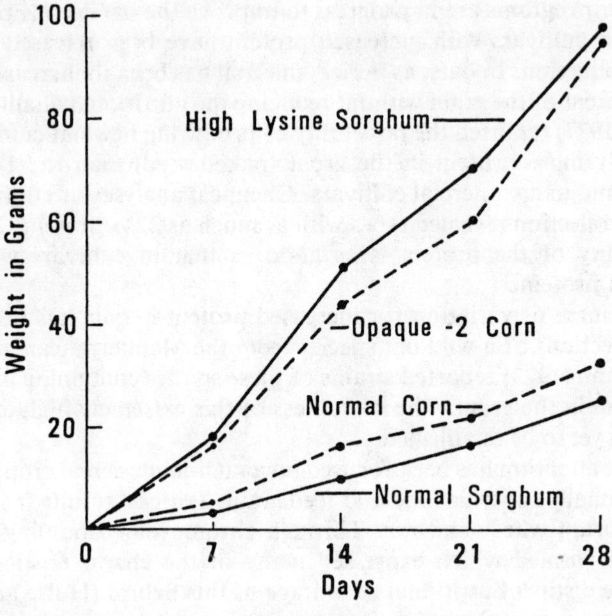

Fig. 12.3. Mean weekly weights of weanling rats fed rations with normal and high lysine corn and sorghum as the protein sources (Singh and Axtell, 1973).

taining only about 8% protein in the brown rice and usually not more than 7% after the milling and polishing processes. This limitation can cause protein-deficiency problems since huge amounts of the grain must be consumed if the minimum protein quantity requirements of the diet are to be met.

In 1966, work was initiated at the International Rice Research Institute to increase the total protein content of rice (Juliano, 1972). A goal of this work has been to increase the total protein content of the milled rice from 7 to 9% without decreasing the desirable biological characteristics of the protein.

Juliano and Beachell (1975) summarized the research that had identified various sources of increased protein of the rice grain. They found considerable variation relative to the potential usefulness of the various sources. Much of this variation was associated with unpredictable genotype-by-environment interactions and the genetic transmission of specific physical characteristics of the kernel—characteristics that adversely affected color or texture of the cooked product.

Rice cultivars that contained 2 to 4% more protein than the standard commercial cultivars showed only a slight decrease in lysine content on a per-unit-of-protein basis. Nutritional trials using the higher protein rice cultivars have been conducted with rats and with human subjects. The nutritional superiority of the high protein rice was confirmed, with no adverse effects found. The use of these cultivars in regions where 40% of the daily protein intake comes from rice could serve to reduce protein deficiencies among people of these areas.

Similar programs are in progress to improve the oat crop (Frey, 1973). Several new cultivars with increased protein have been released for commercial production. In oats, as in rice, the goal has been to increase the total protein content of the grain without reducing the nutritional quality.

Frey (1977) reported the possibility of producing new oat cultivars with significantly more protein in the groat (naked seed) than the 14 to 17% protein found in commercial cultivars. Chemical analyses of cultivars from the world collection revealed types with as much as 22% protein. The nutritional quality of the protein is as good as that in cultivars with lower amounts of protein.

One source of variation for increased protein in oats has been identified in collections of a wild oat species from the Mediterranean area. Ohm and Patterson (1973) reported strains of these species containing as much as 27% protein in the groat. The usefulness of this extremely high amount of protein has yet to be established.

Recent attention has been focused on a man-made cereal crop, triticale, as a nutritionally superior food and feed grain. Triticale results from crosses between durum wheat and rye. Through chromosome doubling, a stable hybrid was formed which expressed many of the characteristics of each parent. The distinct nutritional advantage of this hybrid (Hulse and Laing, 1974), is the fact that the lysine content of the protein is considerably higher than that of wheat, reflecting the influence of its rye parentage. Rye tends to be superior to wheat in protein content of the grain and also provides a protein of superior nutritional quality. Its incorporation into a stable hybrid with wheat holds considerable promise for future plant breeding work with nutritional quality.

Knoblauch et al. (1977) studied the effect of antimetabolites (trypsin inhibitors) upon the nutritional effectiveness of animal diets based on triticale meal. They found that animal growth response to diets from specific triticale genotypes was closely correlated with the level of the trypsin inhibitor present. Animal feeds are normally made from raw, uncooked grain products, so heat inactivation of trypsin inhibitors would not take place. Since the level of trypsin inhibition in the seed appears to be under genetic control, this raises the question: What are appropriate nutritional factors to select for in a plant breeding experiment?

Trypsin inhibitors in seeds are not limited to triticale or to the legumes. They are known to be widespread throughout the plant kingdom. In the future, great care must be taken so that we do not select crop cultivars with an improved protein component but a lower nutritional effectiveness.

MODIFICATION OF OTHER FOOD CROPS

Most of our attention has been directed toward the nutritional quality of the cereals. This is logical when we consider the importance in both human and animal nutrition. However, legume seeds and seeds of other dicotyledonous plants also are essential in world nutrition. Legumes are especially important since they provide a concentrated source of protein that can complement the protein from cereal grains. In general, legume proteins have good nutritional quality per se and provide adequate supplies of the essential amino acids lysine and tryptophan. They are limiting in supplying sulfur-bearing amino acids, especially methionine. Substantial research efforts have been made to improve the amino acid balance of legume seeds, but as yet no new cultivars are available for production. Because agronomic and disease problems in legume breeding have been more pressing, they have been the primary concerns of plant breeders. Growers continue to be paid for increased yield, and not for improved nutritional quality.

The selection of low alkaloid or sweet lupines by von Sengsbusch (1942) was early evidence of success in research toward increasing nutritional quality of legumes. Clark et al. (1970) have shown that at least one of the four known trypsin inhibitors in soybeans is under genetic control, and the presence or absence of this inhibitor can be associated with specific genotypes. This finding could be a first step in the control of soybean nutritional quality through plant breeding procedures.

Without a doubt, most of our commonly used legume seeds have been subjected to many generations of selection by our ancestors in a trial-and-error system. Good evidence indicates that, in the next several decades, much progress can be made in improving the nutritional usefulness of these crops.

Other important food crops have occupied the attention of crop breeders. The potato is one of them. On a dry weight basis, the 2 to 3% protein in the succulent tuber is actually equal to 8 to 10% protein content. The potato as a food source compares favorably with cereal grains in protein content and this protein is known to be superior in nutritive value to that found in most cereals, including rice.

The use of glandless cotton has made available a cottonseed meal free of the alkaloid gossypol, which is destroyed only by precise heat treatment of the cotton seed cake. When the glandless trait is bred into commercial cultivars the seed, a byproduct of lint production, becomes more valuable as a potential foodstuff.

Improving basic food crops to produce more nutritious products still remains one of the greatest challenges facing the plant breeder. To date, limited success has been achieved in the incorporation of the known protein-modifying genes into useful cultivars, largely because of unfavorable agronomic characteristics associated with the modifier genes.

Only recently have plant breeding programs been initiated to select for better nutritive values based on the biological utilization of specific plant products. Even with these programs, plant breeders still depend on chemical analyses to identify the components of nutrition. Systems have not yet been developed to screen the large numbers of genotypes necessary to detect genetic characteristics that exert small effects or occur with low frequency in an observed population.

Successes in the improvement of nutritional quality of our food sources will be the product of research programs that integrate biochemistry, genetics, physiology, agronomy, and nutrition. Each of these disciplines must interact and complement the others in the quest for the production of superior food crops.

The science of plant breeding is on the threshold of significant new improvements in the nutritional quality of the plant parts we eat and feed to our animals. Maize, genetically modified to provide improved nutritional characteristics, is in limited commercial production in Colombia, Brazil, Russia, South Africa, Guatemala, and the USA. These cultivars and hybrids, although far from perfect, have the potential to become even more important factors in protein-deficient diets. Parallel progress appears possible in sorghum, which may make the crop worth more than gold to persons living in sorghum-consuming regions of Africa and Asia. If new cultivars of high-protein rice can be produced with an addition of only two percentage points more protein, their use could mean a 20 to 25% increase in protein consumption for millions of people who live mainly on a rice diet. Barley, wheat, oat, and legume crops can be modified to enhance their usefulness in animal or human diets. All are still new frontiers for the plant breeder.

New insights related to nutritional quality of basic foods are being gained at a rapid rate. New tools are being developed and made available to identify and monitor the component parts of the seed. The challenge is to develop our abilities to recognize opportunities and use the tools available to change our food sources into more useful and healthful products.

LITERATURE CITED

Altschul, A. M. 1976. The protein-calorie trade-off. p. 5–17. *In* Genetic improvement of seed proteins. Natl. Acad. Sci., USA.

Axtell, J. D., D. P. Molian, and D. P. Cummings. 1974. Genetic improvement of biological efficiency and protein quality of sorghum. p. 29–30. *In* Dolores Wilkinson (ed.) Proc. 29th Annu. Corn and Sorghum Conf. Chicago, Ill. 10–12 Dec. 1974. Am. Seed Trade Assoc., Washington, D.C.

Clark, R. W., D. W. Mies, and T. Hymowitz. 1970. Distribution of a trypsin inhibitor variant in seed proteins of soybean varieties. Crop Sci. 10:486–487.

Frey, K. J. 1951. The interrelationship of protein and amino acids in corn. Cereal Chem. 28:123–132.

———. 1973. Improvement of the quality and quantity of cereal grain protein. p. 9–41. *In* Alternative sources of protein for animal production. Natl. Acad. Sci., USA.

———. 1977. Protein of oats. Z. Pflanzenzuecht. 78:185–215.

———, B. Brimhall, and G. F. Sprague. 1949. Effect of selection on protein quality in the corn kernel. Agron. J. 41:399–403.

Hagberg, A., and K. E. Karlsson. 1969. Breeding for high-protein content and quality in barley. p. 17–21. *In* New approaches to breeding for improved plant protein. I.A.E.A., Vienna.

Harpstead, D. D. 1971. High-lysine corn. Sci. Am. 225:34–42.

Hegsted, D. M., M. F. Trulson, and F. J. Stare. 1954. Role of wheat and wheat products in human nutrition. Physiol. Rev. 34:221–258.

Hopkins, C. G. 1899. Improvement of chemical composition of the corn kernel. Illinois Agric. Exp. Stn. Bull. 55. p. 205–240.

Howe, E. E., G. R. Jansen, and E. W. Gilfillan. 1965. Amino acid supplementation of cereal grains as related to the world food supply. Am. J. Clin. Nutr. 16:315–320.

Hulse, J., and E. M. Laing. 1974. Nutritive value of triticale protein. Int. Dev. Res. Center, Ottawa, Canada.

Ingversen, J., B. Koie, and H. Doll. 1973. Induced seed protein mutant of barley. Experimentia 29:1151–1152.

Jimenez, J. R. 1966. Protein fraction studies of high lysine corn. p. 74–79. *In* E. T. Mertz and O. E. Nelson (ed.) Proc. of the High Lysine Corn Conf. Purdue Univ., Lafayette, Ind. 21–22 June 1966. Corn Refiners Assoc., Inc., Washington, D.C.

Johnson, V. A., and P. J. Mattern. 1975. Improvement of the nutritional quality of wheat through increased protein content and improved amino acid balance. p. 1–191. *In* Report of research findings, contract AID/csd-1208 and AID/ta-c-1093. U.S. Agency for Int. Dev., Washington, D.C.

———, ———, and J. W. Schmidt. 1972. Genetic studies of wheat protein. p. 126–135. *In* G. E. Inglett (ed.) Symposium: Seed Proteins Proc., Los Angeles, Calif. 28 Mar.–2 Apr. 1971. Avi Publishing Co., Westport, Conn.

———, ———, D. A. Whited, and J. W. Schmidt. 1969. Breeding for high protein content and quality in wheat. p. 29–40. *In* New approaches of breeding for improved plant protein. I.A.E.A., Vienna.

Juliano, B. O. 1972. Studies on protein quality and quantity of rice. p. 114–125. *In* G. E. Inglett (ed.) Symposium: Seed Proteins, Proc., Los Angeles, Calif. 28 Mar.–2 Apr. 1971. Avi Publishing Co., Westport, Conn.

———, and H. M. Beachell. 1975. Status of rice protein improvement. p. 457–469. *In* High quality protein maize. Dowden, Hutchinson and Ross, Inc., Stroudsburg, Pa.

Knoblauch, C. J., F. C. Elliott, and D. Penner. 1977. Protein of triticale. II. Prediction of protein efficiency indices through chemical analyses. Crop Sci. 17:269–272.

Konzak, C. F. 1977. Genetic control of the content, amino acid composition, and processing properties of proteins in wheat. Adv. Genet. 19:407–582.

Maner, J. H. 1975. Quality protein maize in swine nutrition. p. 58–82. *In* High quality protein maize. Dowden, Hutchinson and Ross, Inc., Stroudsburg, Pa.

Mertz, E. T. 1966. Growth of rats on opaque-2 maize. p. 12–18. *In* E. T. Mertz and O. E. Nelson (ed.) Proc. of the High Lysine Corn Conf. Purdue Univ., Lafayette, Ind. 21–22 June 1966. Corn Refiners Assoc., Inc., Washington, D.C.

———, L. S. Bates, and O. D. Nelson. 1964. Mutant gene that changes protein composition and increases lysine content of maize endosperm. Science 145:279–280.

———, A. O. Vernon, L. S. Bates, and O. D. Nelson. 1965. Growth of rats fed on *opaque-2* maize. Science 148:1741–1742.

Mosse, J. 1966. Alcohol-soluble proteins of cereal grains. Fed. Proc. 25:1663–1669.

Munck, L. 1972. Improvement of nutritional value in cereals. Hereditas 72:1–128.

Ohm, H. W., and F. L. Patterson. 1973. A six-parent diallel cross analysis for protein in *Avena sterilis* L. Crop Sci. 13:27–30.

Osborne, T. B. 1924. The vegetable proteins. Longmans, Green and Co., New York.

Osborne, T. B., and L. B. Mendel. 1914. Nutritive properties of proteins of the maize kernel. J. Biol. Chem. 18:1–16.

Pickett, R. A. 1966. *Opaque-2* corn in swine nutrition. p. 19–22. *In* Proceedings of the High Lysine Corn Conference. Corn Refiners Assn., Inc., Washington, D.C.

Pradilla, A., D. D. Harpstead, F. A. Linares, and C. E. Gonzales-Diest. 1972. Nutritional value of the protein of arepa flour and its improvement by fortification in Colombia. p. 124–130. *In* Nutritional improvement of maize. INCAP, Guatemala City.

———, ———, ———, D. Sarria, and K. Tripathy. 1969. Ensayos analiticas y biologicas de la protein del maize modificado por el gene opaco-2. Antioquia Med. 19(3):201–211.

Schmidt, J. W., V. A. Johnson, P. J. Mattern, A. F. Dreier, and D. V. McVey. 1979. Registration of Lancota wheat. Crop Sci. 19:749.

Showalter, M. F., and R. H. Carr. 1922. The characteristic proteins of high- and low-protein corn. J. Am. Chem. Soc. 44:2019–2033.

Singh, R., and J. D. Axtell. 1973. High lysine mutant gene (hl) that improves protein quality and biological value of grain sorghum. Crop Sci. 13:535–539.

von Sengsbusch, R. 1942. Susslupinen und olupinen. Die entslehum sgeschichte einiger neven kulturpflanzen. Landwirtsch. Jahrb. 91(5):723–880.

Waterman, H. C., and C. O. Johns. 1921. Studies on the digestibility of proteins in vitro. I. The effect of cooking on the digestibility of phaseolin. J. Biol. Chem. 46:9–17.

Zuber, M. S., and J. L. Helm. 1975. Approaches to improved protein quality in maize without the use of specific mutants. p. 241–252. *In* High quality protein maize. Dowden, Hutchinson and Ross, Inc., Stroudsburg, Pa.

SUGGESTED READING

Axtell, J. D. 1981. Breeding for improved nutritional quality. p. 365–432. *In* K. G. Frey (ed.) Plant breeding II. Iowa State Univ. Press, Ames.

Hoveland, C. S. (ed.). 1980. Crop quality, storage, and utilization. Am. Soc. Agron.-Crop Science Soc. of Am., Madison, Wis.

Glossary of Terms

This glossary defines terms used in this book. It was composed by D. R. Wood, editor.

Active Site—A specialized area or portion of an enzyme directly involved in the reaction of substrates.

Additive Genes—Non-allelic genes that have small, equal, and cumulative effects in the control of a quantitative characteristic. Additive genes contribute to the additive genetic variance.

Alkylating Agents—Mutagenic chemicals with one or more reactive alkyl groups which can alkylate DNA. An alkyl group is a univalent radical of the general formula $C_n H_{2n+1}$.

Alien Chromosome—A chromosome from a related species transferred to a crop plant.

Allele—One of the possible forms of a gene.

Amino Acid—An organic acid containing an amino ($-NH_2$) group. Amino acids are components of proteins.

Amphiploid—A polyploid plant that results from doubling the number of chromosomes of a hybrid with two or more non-homologous chromosome sets (amphidiploid).

Antimetabolite—A substance that suppresses or inhibits the utilization of a metabolite and results in the failure of metabolism.

Apomixis—The development of the embryo from maternal cells. The result is vegetative reproduction through the seed.

Auxin—A plant hormone that promotes cell elongation, fruit initiation, callus production, and other plant functions.

Avirulent—A parasite unable to infect and cause disease in a host plant.

Backcross—The cross of a hybrid to one of its parents.

Backcross Breeding—A system of breeding that crosses each successive hybrid recurrently to one parent with concurrent selection for desirable traits from the non-recurrent parent.

Base Analogues—Compounds closely related to purine or pyrimidine bases that can be incorporated into DNA during replication.

Beta Radiation—Electrons emitted by radioisotopes or other sources.

Bulk Breeding—A system of breeding in which a population is grown without regard to pedigree. Seed for the next generation is taken by sampling the harvested bulk seed.

Callus—A mass of unorganized cells produced from a plant explant on a culture medium.

Centromere—The region of the chromosome associated with the attachment of the chromatids to the spindle fibers.

Chimera—A mixture of normal and mutated somatic tissue found in plants after a mutation. An association of tissues having distinct genetic constituents.

Chromatids—The daughter strands of a replicated chromosome held together by a centromere.

Chromosome—A linear nucleoprotein body of the cell nucleus that carries a portion of the genes of an organism.

Clone—A group of organisms propagated asexually from the same individual.

Codon—A sequence of three adjacent nucleotides that code for an amino acid.

Cohesive Ends—The single-stranded extension of DNA molecules that pair with complementary cohesive ends of other DNA molecules.

Cohort—The group of individuals that enter the breeding program in the same season.

Colchicine—An alkaloid drug that arrests spindle fiber formation and disjunction of the daughter chromosomes. In mitotic cells it leads to the doubling of the chromosome number.

Compatible—In plant disease, a congenial interaction between a parasite and its host.

Cotyledon—The first leaf or one of the pair of first leaves developed by the embryo of a seed plant.

Crossing-Over—The interchange of segments between homologous chromosomes.

Cultivar—A category within a species of crop plants. Plants of a cultivar are related by descent and are characterized by morphological, physiological, and adaptation traits.

Cytokinin—A plant hormone that stimulates mitosis.

Cytoplasmic Inheritance—The transmission of traits by non-nuclear DNA through the cytosplasm. Often the result is maternal inheritance.

Degenerate Codons—Two or more codons that code for the same amino acid.

Deoxyribonucleic Acid (DNA)—A polymer of purine and pyrimidine bases linked through deoxyribose by phosphate groups.

Diallel Crosses—All possible hybrids from a set of parents.

Diploid—An organism with two sets of a basic genome.

Diplontic Selection—The competition between different cell types in a chimera.

Dominance—The intraallelic interaction in which the effect of an allele suppresses the effect of another allele in the heterozygote.

Dosage Effect—The effect of the number of times a genetic element is present upon a process or structure of a phenotype.

Double Cross—A hybrid resulting from crossing two single crosses. Each single cross is the result of crossing two inbred lines.

Double Fertilization—The union of one sperm nucleus with the egg nucleus and one with polar nuclei to form the embryo and the endosperm in flowering plants.

Eceriferum—A mutation reducing the waxiness of leaves, stems, and spikes of the plant.

Embryoid—An embryo-like structure seen in cell and tissue cultures.

Environment—The physical and biological factors that interact with an organism in growth.

Enzyme—Proteins produced in living cells that catalyze biochemical reactions.

Epistasis—The interaction of different genes in the expression of a trait.

Erectoides—A mutant characterized by short erect stature and compact appearance of plant and spike.

Ethnobotany—A branch of science dealing with the folklore knowledge of plants.

Eucaryote—A class of life that includes all organisms that have nuclei and membrane-bound cellular organelles.

Explant—A portion of tissue removed from a plant for tissue culture.

Feral—An organism escaped from domestication.

Fractionation—The separation of a mixture into its components by their physical and chemical characteristics.

Gametes—The sex cells of an organism that unite at fertilization.

Gamma Field—A circular area for growing plants centered on a gamma source to provide different radiation doses.

Gamma Source—A radiation source that emits gamma rays, usually containing cobalt-60 (^{60}Co) or cesium-137 (^{137}Cs).

Gene—A sequence of DNA nucleotides that codes for a functional product of RNA or a polypeptide.

Gene Pool—The genes and combination of genes of a population.

Gene for Gene Relationship—A relationship in host-parasite interactions in which every gene conditioning resistance in a host plant is matched by a gene for avirulence or virulence in the pathogen.

General Resistance—Resistance that functions against all biotypes of a pathogen.

Genetic Marker—An allele used to mark or identify a gene or a chromosome.

Genome—The basic set of chromosomes of an organism.

Genotype—The genetic makeup of an individual.

Germplasm—In crop breeding, the totality of genes and genetic materials available for the improvement of a crop.

Globulin—A class of proteins that is soluble in dilute salt solutions.

Groat—Kernel or hulled grain: a) oats, a kernel without the lemma and palea, b) buckwheat, the hulled grain broken into particles of even size.

Haploid—The gametic number of chromosomes of an organism.

Heritability—The portion of the phenotypic variability of a trait that can be assigned to genetic variability.

Heterokaryote—A cell having two or more genetically different nuclei. A common phenomenon in fungi.

Heterosis—The vigor of a hybrid when the performance of the hybrid for a quantitative trait falls outside the range of the parents.

Heterozygous—The condition of having unlike alleles at each of one or more loci of homologous chromosomes.

Hexaploid—A polyploid with six sets of chromosomes.

High-Lysine—A genotype that produces grain with more lysine than usual genotypes of the crop.

Homologous Chromosomes—Chromosomes that pair at meiosis and are alike in structure and function.

Homozygous—The state of having identical alleles at each of one or more loci of homologous chromosomes.

Hormone—A chemical messenger that controls cell functions in growth and development at a site different from the point of synthesis.

Hybrid—The progeny of parents of different genotypes, strains, species, or genera.

Hybrid vigor—The expression of a trait in a hybrid in excess of the mid-point of the amount expressed by the two parents.

Ideotype—An ideal plant form or model formulated to assist in reaching selection goals.

Immunity—A plant condition that does not permit infection by a parasitic organism.

Inbred Line—A breeding line that is considered homozygous after self-pollination over several generations.

Incompatible—In plant disease, a non-congenial interaction between a parasitic organism and its host.

Inoculate—The placement of spores or other infective units of a parasitic organism on or in a host plant so that infection will be achieved.

Inoculum—Spores, cells, particles, or other units of a parasitic organism that initiate the infection process.

Intergeneric Cross—The hybridization of two individuals from species of different genera.

Interspecific Cross—The hybridization of two individuals of different species.

In Vitro—Experiment done in the culture container. Literally, in glass.

Ionizing Radiation—The kind of radiation in which electrons are detached as they pass through tissue. For example, x-rays, gamma rays, neutrons, and alpha or beta particles.

Isochromosome—A chromosome with identical arms.

Isogenic Line—A line of homozygous individuals isolated to compare with a companion isogenic line. The companion lines differ from one another by having different alleles at a single locus.

Isoline—An isogenic line.

Landrace—A cultivar selected and used for cropping before the modern era of crop breeding. A primitive cultivar characteristic of a cropping locality.

Ligate—The bonding of adjacent nucleotides by repairing enzymatically the phosphodiester backbone of a DNA strand.

Linear Energy Transfer (LET)—The energy dissipated per unit length along the track of an ionizing particle.

Linkage—The association of genes in inheritance because they are carried on the same chromosome. Recombination between linked genes is less than recombination between genes on different chromosomes.

Locus—The position of a gene on a chromosome.

Lodging—The settling or collapse of a plant from an upright position.

M1, M2, . . . Mn—The terms used to designate the generations following treatment with a mutagen.

Male-Sterile—The failure of plants to produce functional pollen.

Mass Selection—Selections within a population for a desired attribute. The next generation is produced from the bulked seed of the selections.

Medium—A combination of nutrients, hormones, and other substances, within or upon which cells are grown.

Meiosis—A reduction division of the chromosomes accompanied by a second cell division to produce gametes.

Meristem—A rapidly dividing formative tissue which produces new cells by mitosis.

Messenger RNA (mRNA)—The RNA produced by transcription of DNA that serves as the template for protein synthesis.

Mitosis—The process of nuclear division that forms two new nuclei with the same quantity of chromosomes as the parent nucleus in cell division.

Monogastric—A class of animals having one stomach.

Monoploid—An individual or organism having one basic set of chromosomes.

Multigenic—Inheritance due to many genes each with small, similar, and cumulative effects (polygenic).

Multiline Cultivar—A mixture of lines or cultivars with each differing in genes for resistance to pests.

Mutagen—A physical or chemical agent capable of inducing mutations.

Mutant—An individual with a trait produced by a mutation.

Mutation—The alteration of the genetic material that results in a heritable change in a trait.

Near-Isogenic—Two or more lines that differ for one gene and closely linked adjacent genes.

Necrotic—Dead plant tissue usually brought about by disease, insect activity, or nutrient deficiency.

Nick—A break in the phosphodiester backbone of one strand of a DNA double-stranded molecule.

Oligogenic—Inheritance due to one or a few genes which have discernible effects.

Operon—A genetic unit that consists of a control element—the operator—and associated structural genes.

Overdominance—The effect of the heterozygote of a pair of alleles is greater than either homozygote.

Pathogen—The causal agent of plant disease.

Pathotoxin—A compound produced by a pathogenic organism capable of inducing the disease symptoms induced by the pathogen itself.

Pedigree—The record of the ancestry of an individual or a cultivar.

Phenotype—The observed characteristics of an individual.

Physiological Race—Biotype of a parasitic organism with the same combination of genes for virulence and avirulence.

Phytoalexin—A substance produced by a plant in response to invasion by a pathogen that is toxic to the invading organism.

Plasmid—A small circular DNA molecule, capable of self-replication, that can carry genes which confer resistance to antibiotics. Plasmids are used as vectors in recombinant DNA experiments.

Polycross—A set of open-pollinating lines placed to insure seed production by random mating.

Polygenic—Inheritance due to many genes in which the effect of each gene is assumed to be small, similar, and cumulative (multigenic).

Polyhaploid—Haploid plant derived from a polyploid individual.

Polymer—A large molecule that consists of repeating subunits.

Polymerase—Any of several enzymes that catalyze the formation of DNA or RNA from precursor substances in the presence of DNA or RNA templates.

Polyploid—An organism with more than two sets of chromosomes.

Population—In genetics, a group of individuals characterized by a common gene pool.

Precursor—A substrate molecule whose chemical change is metabolized by an enzyme in a metabolic pathway.

Prolamin—A class of proteins in seeds characterized by solubility in alcohol-water solutions.

Promoter—The region of the DNA at which the RNA polymerase binds to initiate transcription.

Protoclone—A clone derived from a protoplast.

Protoplast—The plant cell exclusive of the cell wall.

Pureline—A strain considered homozygous because of continued inbreeding.

Recessive—A gene allele that is not expressed when heterozygous.

Reciprocal Translocation—The exchange of chromosome pieces between non-homologues.

Recombinant DNA—DNA molecules constructed by joining, outside the cell, natural and synthetic DNA segments to DNA molecules capable of replication in living cells.

Recurrent Selection—A breeding method based upon intercrossing selected individuals followed by continuing cycles of selection and intercrossing to increase the frequency of desired alleles in the population.

Regeneration—The production of plants in vitro from cultured cells.

Regulator Gene—A gene that controls the activity of other genes.

Repressor—A protein coded by a regulator gene that interacts with inducer compounds and the operator to control the transcription of structural genes.

Resistance—Interaction of host plant and pathogen that fails to produce disease.

Restriction Endonuclease—A class of enzymes which break both strands of a DNA molecule at specific points as a result of recognizing base sequences.

Ribosomal RNA (rRNA)—The ribonucleic acid component of the ribosomes.

Selection—The process of choosing some individuals of a population to propagate because they express an inherited trait.

Self-Compatible—The favorable interaction of pollen and the tissues of the pistil which permits sexual reproduction.

Self-Incompatible—The non-congenial interaction between the pollen and the tissues of the pistil that inhibits sexual reproduction.

Semidwarf—A short-stature plant resulting from short internodes.

Species—A category of biological classification below the genus. The individuals within a species are able to intercross.

Specific Resistance—Resistance that functions against certain biotypes of a parasitic organism.

Spindle—The apparatus of dividing cells along which chromosomes are distributed during mitosis and meiosis.

Substrate—A molecule acted on by an enzyme in cellular metabolism.

Susceptible—Interaction between host and pathogen that produces disease.

Synthetic Cultivar—An advanced generation of an open-pollinated population composed of a group of selected inbreds, clones, or hybrids.

Tolerance—The heritable ability of a plant genotype to endure attack by a pathogen or insect pest without suffering severe loss of yield.

Totipotency—The capacity of a cell cultured in vitro to regenerate into a plant.

Transcription—The formation of messenger RNA complementary to the DNA code. The process is catalyzed by RNA polymerase.

Transfer RNA (tRNA)—Molecules of RNA able to combine with a specific amino acid and to base pair with the appropriate codon for that amino acid in the mRNA.

Translation—The process of converting the genetic information of the mRNA into protein structure on the ribosomes.

Translocation—A chromosome carrying a non-homologous chromosome segment.

Trypsin—A proteolytic enzyme from the pancreas.

Variance—The mean squared deviation of a population of variates from their mean.

Variety—The term in common usage synonymous with cultivar.

Vector—A vehicle, often a plasmid, for carrying recombinant DNA into a living cell.

Virulence—The ability of a parasitic organism to infect and cause disease in a resistant plant.

X-Ray—Electromagnetic radiation generated by bombarding tungsten or molybdenum with electrons.

Xerophyte—A plant adapted to life and growth under limited water conditions.

Zein—A seed protein of maize classified as a prolamin. It is deficient in tryptophan and lysine.

Zygote—The cell that results from the union of male and female sex cells.

Glossary of Scientific Names of Crops Plants and Pest Organisms

This glossary gives the common names and scientific names for the various crop plants and pest organisms discussed in this book. The Editor compiled this glossary from common names cited in the individual chapters. He obtained the scientific names from several sources, including *A Checklist of Names for 3000 Vascular Plants of Economic Importance* by Edward E. Terrell (Agric. Handb. No. 505, ARS-USDA, May 1977).

1. Names of Crop Plants

Common name	Scientific name
Alfalfa	*Medicago sativa* L.
Almond	*Prunus amygdalus* Stokes
Alsike clover	*Trifolium hybridum* L.
Amaranth	*Amaranthus* spp.
Annual ryegrass	*Lolium multiflorum* Lam.
Apple	*Malus domestica* Borkh.
Artichoke	*Cynara scolymus* L.
Asparagus	*Asparagus officinalis* L.
Aspen	*Populus tremuloides* Michx.
Autumn crocus	*Colchicum autumnale* L.
Bahiagrass	*Paspalum notatum* Flugge
Banana	*Musa sapientum* L.
Barley	*Hordeum vulgare* L.
Bean	*Phaseolus vulgaris* L.
Belladonna	*Atropa bella-donna* L.
Bermudagrass	*Cynodon dactylon* (L.) Pers.
Blackberry	*Rubus allegheniensis* Porter
Black currant	*Ribes nigrum* L.
Bladderpod	*Lesquerella gracilis* Wats.
Blue panicgrass	*Panicum antidotale* Retz
Broccoli	*Brassica oleracea* L. Italica group
Broadbean	*Vicia faba* L.
Brussels sprouts	*Brassica oleracea* L. Gemmifera group
Buckwheat	*Fagopyrum esculentum* Moench
Buffalo gourd	*Cucurbita foetidissima* H.B.K.
Buffalograss	*Buchloë dactyloides* Nutt.
Buffelgrass	*Cenchrus ciliaris* L.
Cabbage	*Brassica oleracea* L. Capitata group
Cacao	*Theobroma cacao* L. subsp. *cacao*

(continued)

Crop plants continued

Carnation	*Dianthus caryophyllus* L.
Carrot	*Daucus carota* L. subsp. *sativus* (Hoffm.) Arcang.
Cassava	*Manihot esculenta* Crantz
Castor bean	*Ricinus communis* L.
Cauliflower	*Brassica oleracea* L. Botrytis group
Celery	*Apium graveolens* L. var. *dulce* (Mill.) Gaudich.
Chickpea	*Cicer arietinum* L.
Coffee	*Coffea arabica* L.
Comfrey	*Symphytum officinale* L.
Common reed	*Phragmites australis* (Cav.) Steud.
Cotton	*Gossypium hirsutum* L.
Cowpea	*Vigna unguiculata* L.
Crambe	*Crambe abyssinica* R. E. Fries
Crimson clover	*Trifolium incarnatum* L.
Cucumber	*Cucumis sativus* L.
Date palm	*Phoenix dactylifera* L.
Durum wheat	*Triticum turgidum* L. var. *durum*
Eggplant	*Solanum melongena* L.
Elephant grass	*Pennisetum purpureum* Schumach.
Flax	*Linum usitatissimum* L.
Foxtail millet	*Setaria italica* (L.) Beauv.
Gooseberry	*Ribes uva-crispa* L.
Gopher plant	*Euphorbia lathyris* Hook. & Arn.
Grain amaranth	*Amaranthus edulis* Speg.
Grape, American	*Vitis labrusca* L.
Grape, European	*Vitis vinifera* L.
Guar	*Cyamopsis tetragonoloba* (L.) Taub.
Guayule	*Parthenium argentatum* A. Gray
Guineagrass	*Panicum maximum* Jacq.
Honesty	*Lunaria annua* L.
Jerusalem artichoke	*Helianthus tuberosus* L.
Jojoba	*Simmondsia chinensis* (Link) Schnider
Jute	*Corchorus olitoris* L.
Kale	*Brassica oleracea* L. Acephala group
Kenaf	*Hibiscus cannabinus* L.
Kentucky bluegrass	*Poa pratensis* L.
Kikuyu grass	*Pennisetum clandestinum* Chiov.
King Ranch bluestem	*Bothriochloa ischaemum* var. *songaria* (Fisch. & Mey.) Celar. & Harlan

Kiwi	*Actinidia chinensis* Planch.
Kola	*Cola nitida* (Vent.) Schott & Endl.
Ladino clover	*Trifolium repens* L.
Lentil	*Lens culinaris* Medik.
Lettuce	*Lactuca sativa* L.
Leucaena	*Leucaena leucocephala* (Lam.) de Wit.
Lupine	*Lupinus luteus* L.
Maize	*Zea mays* L.
Meadowfoam	*Limnanthes alba* Hartw.
	Limnanthes douglasii C. T. Mason
	Limnanthes floccosa Howell
Milkbush	*Euphorbia tirucalli* L.
Mungbean	*Vigna radiata* L.
Neem	*Azadirachta indica* A. Juss.
Oat, red	*Avena byzantina* (C.) Koch
Oat, white	*Avena sativa* L.
Oil palm	*Elaeis guineensis* Jacq.
Onion	*Allium cepa* L.
Orange	*Citrus sinensis* (L.) Osb.
Pangolagrass	*Digitaria decumbens* Stent
Pea	*Pisum arvense* L.
Peanut	*Arachis hypogaea* L.
Pearl millet	*Pennisetum americanum* (L.) Leeke
Pepper	*Capsicum annuum* L. var. *annuum*
Peppermint	*Mentha piperita* L.
Perennial ryegrass	*Lolium perenne* L.
Petunia	*Petunia hybrida* Vilm.
Pineapple	*Ananas comosus* (L.) Merr.
Pinyon pine	*Pinus edulis* Engelm.
Pistachio nut	*Pistacia vera* L.
Plantago	*Plantago ovata* Forsk.
Potato	*Solanum tuberosum* L.
Pot marigold	*Calendula officinalis* L.
Proso millet	*Panicum miliaceum* L.
Pumpkin	*Cucurbita pepo* L.
Pyrethrum	*Chrysanthemum cinerariifolium* (Trevir.) Vis.
Quinine	*Cinchona* spp.
Quinoa	*Chenopodium quinoa* Willd.

(continued)

Crop plants continued

Radish	*Raphanus sativus* L.
Ramie	*Boehmeria nivea* (L.) Gaudich
Rape	*Brassica napus* L.
Red clover	*Trifolium pratense* L.
Rice	*Oryza sativa* L.
Rubber	*Hevea brasiliensis* (A. Juss.) Muell.-Arg.
Rye	*Secale cereale* L.
Safflower	*Carthamus tinctorius* L.
Sideoats grama	*Bouteloua curtipendula* (Michx.) Torr.
Sorghum	*Sorghum bicolor* (L.) Moench
Soybean	*Glycine max* (L.) Merrill
Stoke's aster	*Stokesia laevis* (Hill) Greene
Strawberry	*Fragaria X ananassa* Duch.
Sudangrass	*Sorghum sudanense* (Piper) Stapf.
Sugar beet	*Beta vulgaris* L.
Sugarcane	*Saccharum officinarum* L.
Sunflower	*Helianthus annuus* L.
Sunn hemp	*Crotalaria juncea* L.
Sweet potato	*Ipomoea batatas* (L.) Lam.
Taro	*Colocasia esculenta* (L.) Schott
Tepary bean	*Phaseolus acutifolius* A. Gray
Tobacco	*Nicotiana tabacum* L.
Tomato	*Lycopersicon esculentum* Mill.
Triticale	*X Triticosecale* Wittmack
Turnip rape	*Brassica campestris* L.
Valerian	*Valeriana officinalis* L.
Vernonia	*Vernonia anthelmintica* (L.) Willd.
Watermelon	*Citrullus vulgaris* Schrad.
Wheat	*Triticum aestivum* L.
White clover	*Trifolium repens* L.
White mustard	*Sinapis alba* L.
White potato	*Solanum tuberosum* L.
White sweet clover	*Melilotus alba* Medik.
Wild rice	*Zizania aquatica* L.
Winged bean	*Psophocarpus tetragonolobus* (L.) DC.
Yam	*Dioscorea cayenensis* Lam.
Yellow sweet clover	*Melilotus officinalis* Lam.

2. Names of Plant Pathogens

Common name	Scientific name
Alfalfa rust	*Uromyces striatus* Schroet.
Bacterial wilt	*Corynebacterium insidiosum* (McCulloch) Jensen
Corn rust	*Puccinia sorghi* Schw.
Crown gall	*Agrobacterium tumefaciens* (E. F. Sm. & Town.) Conn
Crown rust	*Puccinia coronata* Cda. f. sp. *avenae* Fraser & Led.
Diplodia stalk rot	*Diplodia macrospora* Earle
Eyespot	*Helminthosporium sacchari* (Van Breda de Haan) Butler
Flax rust	*Melampsora lini* (Ehrenb.) Lev.
Gibberella stalk rot	*Gibberella zeae* (Schw) Petch
Glume blotch	*Septoria nodorum* Berk.
Leaf blight	*Helminthosporium turcicum* Pass.
Leaf spot	*Pseudopeziza medicaginis* (Lib.) Sacc.
Leaf rust (wheat)	*Puccinia recondita* Desm. f. sp. *tritici*
Powdery mildew	*Erysiphi graminis tritici* E. Marchal
Southern anthracnose	*Colletotrichum trifolii* Bain & Essary
Southern corn leaf blight	*Helminthosporium maydis* Nisikado and Miyake
Southern rust (maize)	*Puccinia polyspora* Underw.
Southern anthracnose	*Colletotrichum destructivum* O'gara
Speckled leaf blotch	*Septoria tritici* Rob.
Stem rust	*Puccinia graminis* Pers. f. sp. *tritici* Eriks. & E. Henn.
Verticillum wilt	*Verticillium albo-atrum* Reinke & Berth.
Victoria blight	*Helminthosporium victoriae* Meehan & Murphy
Yellow leaf blight	*Phyllosticta maydis* Arny & Nelson
Yellow rust	*Puccinia striiformis* West

3. Names of Insects

Common name	Scientific name
Alfalfa weevil	*Hypera postica* Gyllenhal
Black scale	*Saissetia oleae* Bernard
Brown planthopper	*Nilaparvata lugens* Stal.
Cereal leaf beetle	*Oulema melanopus* L.
Corn earworm	*Heliothis zea* Boddie
Corn leaf aphid	*Rhopalosiphum maidis* Fitch
European corn borer	*Ostrinia nubilalis* Hubner
Grape louse	*Phylloxera vitifoliae* Fitch
Green leafhopper	*Nephotettis impicticeps* Ishihara
Greenbug	*Schizaphis graminum* Rondani
Hessian fly	*Mayetiola destructor* Say
Leaf blister mite	*Eriophyes gossypii* Banks
Leafhopper	*Empoasca fascialis* Jack
Meadow spittlebug	*Philaenus spumarius* L.
Pea aphid	*Acyrthosiphon pisum* Harris
Pink bollworm	*Pectinophora gossypiella* Saunders
Rootworm	*Diabrotica longicornis* Say
Spotted alfalfa aphid	*Therioaphis maculata* Buckton
Striped borer	*Chilo suppressalis* Walker
Wheat stem sawfly	*Cephus cinctus* Norton
Wooly apple aphid	*Eriosoma lanigerum* Hausmann

4. Names of Nematodes

Common name	Scientific name
Root-knot nematodes	*Meloidogyne* spp.

Subject Index